Anatomic Study of the Clitoris and the Bulbo-Clitoral Organ

Vincent Di Marino • Hubert Lepidi

Anatomic Study of the Clitoris and the Bulbo-Clitoral Organ

Springer

Vincent Di Marino
UER Médecine
Aix-Marseille Université
France

Hubert Lepidi
UER Médecine
Aix-Marseille Université
France

ISBN 978-3-319-04893-2 ISBN 978-3-319-04894-9 (eBook)
DOI 10.1007/978-3-319-04894-9
Springer Heidelberg Dordrecht London New York

Library of Congress Control Number: 2014939032

Printed on acid-free paper

Springer is part of Springer Science+Business Media (www.springer.com)

We dedicate this book:

To women all over the world, with our highest consideration and admiration

To the generous donors who gave their body to the Body Donation Department of our AMU for advancing science

The authors

Foreword

Vincent Di Marino, helped by **Hubert Lepidi** for histological part, has the great merit of having known how to captivate us by the description of this organ measuring less than 6 mm long, 3,5 in diameter and which is not visible at first sight. Indeed it is not easy to view this organ and in order to see it fully, the cleft where the labia minora and the labia majora meet needs to be spread apart on both sides.

The name clitoris, which was introduced into the French language in the early seventeenth century, most likely comes from the Greek word "kleitoris" which means "key"; some people could think that there is a hidden meaning here or could simply imagine a full programme.

This Lilliputian organ is actually a real treasure trove for the anatomist, histologist, physiologist, sexologist and, more simply, for the sexual partner.

The anatomist in fact describes an organ that consists of a glans covered by a hood, held in place by frenula and which has a similar appearance to a tiny penis. The body of the clitoris is attached to two lateral formations, the "vestibular bulbs", which thus constitutes a bulbo-clitoral organ according to the name proposed by the authors of this book, a term to remember because it conforms to the anatomical and physiological reality. Its vascularisation, its particularly large nerve network and its histology explain its intense physiological activity. In fact, the clitoris is made up of two tissues: the spongy tissue lining the vestibular bulbs so that it can engorge itself with blood and become swollen, whereas the clitoral crura and the body of the clitoris are erectile corpora cavernosa that ensure indeed a discreet but real erection resulting in the externalisation of the glans outside the hood; this erection is the result of a real biochemical cascade triggered during mechanical or manual sexual arousal regardless of whether it is visual or imaginary.

It seems that **Soranus from Ephesus** was the first to have individualised this formation which he described as a wattle with a muscular appearance that sits above the urethra and which he compared to a "nymph" because it is hidden under the labia "like young brides under their veils". In Europe, in the sixteenth century, the successors of the famous Belgian anatomist **Andreas Vesalius**, **Realdo Colombo and Gabriele Falloppio** claimed in turn to have discovered this organ in the sixteenth century and also demonstrated its role in the genesis of female pleasure.

However, female enjoyment, something which women today insist upon with good reason, was not invented by them, and in the West, there have been no problems with the notion of the female orgasm for several decades. In the West, sexual pleasure has been associated with the idea of sin since the early days of Christianity, whereas it was sung of by Persian poets, immortalised in the Kamasutra and revered by Taoists. Curiously, this notion of female pleasure was tolerated by Catholics, Protestants and Jews until the eighteenth century because the religious leaders of these three monotheistic religions were well aware that the search for sexual pleasure, of course, only within the context of marriage, might encourage procreation.

With the age of Enlightenment and the addition of certain scientific knowledge, the notion of sexual pleasure without conception and especially the notion of procreation and renewal of the species without obtaining female sexual pleasure became obvious. The usefulness and possibly the harm of the female orgasm quickly became a key issue that was debated in the most advanced scientific circles. At the end of the nineteenth century, the Belgian biologist **Edouard**

Van Beneden described the mechanism of biological procreation and confirmed that it happens without involving the clitoris and without any orgasms. Over this same period, the media coverage of a mysterious illness, hysteria, became the focus of the attention of the scientific intelligentsia and more particularly the medical intelligentsia. This condition almost exclusively affects women. Its symptoms are highly polymorphic and they have no anatomical substrate, as was clearly demonstrated by **Jean Martin Charcot** using hypnosis sessions; the uterus was considered to be the likely site of the disease, but its removal did not produce any clinical improvement and many scientists, as a result of the work of **Isaac Baker Brown**, thought that the clitoris (this useless and possibly harmful organ) was the cause of the hysteria. **Isaac Brown**, the president of the very serious and learned Society of British Medicine, was convinced of the harmfulness of clitoral masturbation and advocated its removal to cure hysteria as well as to treat and possibly prevent homosexuality and other varieties of "female madness", in the words of **Baker**. All excess leads ipso facto to the opposite extreme and the supporters of clitoral masturbation had an exceptionally creative imagination when they developed the concept of therapeutic vibrating massage first run on steam power, and then gasoline and electricity.

At the same time, **Sigmund Freud's** psychoanalytic concepts delivered the final blow to the clitoris because, for the father of psychoanalysis, female sexuality is marked by the frustration of not having a penis. In his three essays on sexuality (1905), the Master of psychoanalysis wrote that this frustration appears in early childhood when the girl focuses her attention on the clitoris; during adolescence and adulthood, she can experience a "fulfilled" sex life by renouncing the clitoral orgasm and by only obtaining an orgasm via vaginal penetration by the male penis (the vagina acquires a value as it is seen as the place to accommodate the penis). In fact, he distinguished two types of orgasm: a clitoral orgasm which he calls infantile and a "mature" vaginal orgasm. This resulted in two types of women being distinguished (in terms of sexual needs): those who have sexually immature clitoral orgasms and those who have sexually adult and fulfilling vaginal orgasms; this dichotomy persisted for a long time and still exists today.

Marie Bonaparte, a loyal patient, student and friend of Freud, introduced a note of discord to Freud's theory, with whom she founded the Psychoanalytic Society of Paris. Marie Bonaparte, daughter of Prince Roland Bonaparte and wife of Prince George of Greece, suffered from frigidity (probably because of the clitoris); she was convinced of the existence of clitoral frigidity and shared the dichotomous view of sexuality taught by her Master, but as a convinced feminist, she thought that, in some circumstances, frigidity could be due to the peripheral location of the clitoris in relation to the urethral orifice and the anterior wall of the vagina. Based on a study carried out on a cohort of 200 volunteer patients and published under the pseudonym **Narjani**, she defined three types of patients: the "paraclitoridiennes", "mesoclitoridiennes" and "teleclitoridiennes" depending on the distance between the tip of the *glans clitoridis* and urethral orifice. When this distance was larger than 25 mm, it was rare for the persons in this cohort to have an orgasm during intercourse; as she was frigid and part of the disadvantaged group, she decided to surgically correct this handicap by asking Professor **Joseph Halban**, the world renowned vaginal surgeon and Professor of Gynaecology at the gynaecological faculty in Vienna, to perform a surgical procedure called "fixation clitoridienne" (to move the clitoris closer to the vagina), which has since been referred to as the Halban-Narjani operation. Despite four operations, she would always remain frigid.

Since the beginning of the twentieth century, numerous epidemiological and scientific studies have been used to broaden the simplistic idea of female sexuality taught by **Freud** and to define a group that combines the clitoris, vulva, vagina and urethra into a single functional entity, overall helping women to experience sexual pleasure.

This resulted in more and more dissident students of the Master of psychoanalysis who became increasingly critical of their Master.

The American gynaecologist **William H. Masters** and his companion **Virginia E. Johnson**, a midwife, consequently used a cohort of 700 volunteers totalling 10,000 orgasms, to describe a full orgasm on an individual basis starting with the clitoris and finishing with the vagina.

At the same time, the feminist movements that were already present at the end of the nineteenth century continued and intensified their fight in the twentieth century and went on with this fight into the early twenty-first century for the recognition of gender equality. "Their fight also includes the recognition of their own pleasure, namely clitoral pleasure, and the rehabilitation of this organ which only they have" (**M. H. Colson**).

However, the true nature of the vaginal orgasm remains a mystery and "is as silent as the grave" (in French "garde sa tombe inviolée"), in the words of **Jacques Lacan**.

This anatomical and functional unit that regroups the clitoris, urethra and vagina helped us discover and try to resolve another mystery of female sexuality, **the G-spot**.

The G-spot has been described since the beginning of time; we can find traces of it in Tantric texts under the name **Kanda**. Taoist tradition describes it as the "black pearl of eroticism". It is not exactly known where the G-spot is located; it is probably not a point but an area with imprecise limits that are still not clearly defined, located on the front part of the lower third of the vagina near the urethra close to Skene's periurethral glands.

Nowadays, medicine factual cannot yet (?) explain if the G-spot is real or not because the few published studies cannot be used given the very small number of series, their small sample sizes and an insufficient level of evidence. However, there seems to be an anterior vaginal area that is particularly sensitive and responsive as it has a rich nerve and blood supply located near the urethra and clitoris; due to its connections, this area becomes larger during vaginal penetrative sexual intercourse, secondary to the swelling and protrusion of the corpora cavernosa of the clitoris which is associated with the back and forth movements of the penis in the vagina.

Therefore, it seems that the notion of a functional and anatomical entity regrouping the clitoris, vagina and urethra is still valid, which should help to eradicate this restricted dichotomous view of female sexual pleasure.

This anatomical concept should be of interest to surgeons and make them consider the potential negative consequences of surgical procedures in this area: surgical treatment of urogenital prolapse via anterior colporrhaphy, Halban's fascia suture, anterior vaginal resection, putting a prosthetic reinforcement device in place, surgical treatment of urinary incontinence by placing a strip of polypropylene via a suburethral anterior approach as well as reduction nymphoplasties and "cosmetic" vulvo-vaginal procedures when there are, often discrete, abnormalities of the vulvo-vaginal orifice. All of these procedures destroy the vascular nodes and sensory nerve fibres that are a part of a woman's pleasure and may decrease libido, as sometimes observed following surgeries on the urogenital system.

This book fascinated me and I thoroughly enjoyed discovering and browsing through its contents. It is a very complete work that includes the classic chapters that are expected in any anatomy book but which also includes modern inputs such as the place of imaging in the analysis of the organ and the search for a pathology as well as the place of the clitoris in the history of art (painting, sculpture, photography, literature), as well as in the comics where sex plays an important role. The "divine Marquis de Sade" understood this well and did not hesitate to give some of his heroines clitoral hypertrophy.

It is easy to read this book; the text is clear and concise, and it includes remarkable images that fit in very well with the text. On behalf of those people who have dared, like me, to write books, I can testify that **Vincent Di Marino**, with the excellent collaboration of **Hubert Lepidi**, has worked very hard to write this book, something that does not surprise me at all as it reflects his qualities as a passionate teacher: method, concise nature, competence, professional awareness.

I obviously hope that this book gets the success it deserves; it caters to a wide audience: pelvic visceral surgeons, urologists, gynaecologists, plastic surgeons as well as midwives, sexologists, psychiatrists and of course doctors specialising in reproduction.

It should be useful to anyone who advises, monitors or operates on patients with urinary and vulvo-vaginal disorders.

Marseille, France

Bernard Jean Blanc
Professor of Gynecology and Obstetrics at the
UER of Medicine of Marseille, AMU University

Preface

Voir, c'est parcourir les détails, s'arrêter un peu à chacun
et de nouveau saisir l'ensemble d'un coup d'œil[1]

- Alain, Propos sur le bonheur

Our previous anatomical research topics on the levator ani muscles , the striated anal sphincter and the hypogastric plexuses have gradually led us to explore all of the anatomical structures of the pelvis. More recently, the expanded use of new surgical methods to treat urinary incontinence in women has led us to focus our anatomical-surgical research more specifically on the female urethra. Given that this canal is topographically related to the reproductive system, our study is of course extended to the entire female reproductive system. While performing numerous dissections of the female pelvis to help young colleagues who are preparing for a career in gynaecology, our interest in this topic grew due to the ambiguities surrounding a number of anatomical concepts.

Based on the numerous questions asked, primarily about female erectile tissues, we initiated and then developed a scientific and modern approach to these training programmes. This book is the result of these efforts. The more we wrote, the more we realised that not much anatomical research is being carried out about this area, whereas the anatomical substrate is essential to fully understand female sexuality and to develop the necessary sexual medicine that could provide a large number of women with relief.

While working on this book, it became obvious that women are unfamiliar with their own genital anatomy, something which is also regularly expressed by the national newspapers and especially the numerous magazines devoted to women, particularly in articles about the clitoris ("the clitoris, this unfamiliar organ", "the clitoris, underrated organ", "the clitoris, an absolute mystery", "the clitoris is always kept hidden", etc.). The reasons for this lack of knowledge, at least in France, are well known: inadequate sex education that is very poorly handled and selective sexual obscurantism that allows representations of the entire male reproductive system, whereas for women, the only available description of the female genitalia concerns the internal reproductive system. Earth and Life Sciences teachers are afraid to talk to their students about the female reproductive system and female sexuality because they are afraid of the parents' reactions and at last frequent inaccurate drawings and representations found in magazines or on websites in articles relating to the clitoris or any other component of the female external reproductive system![2]

The consequences of this lack of understanding are especially regrettable given that young people are having their first sexual relations at an increasingly early age. A number of young women of childbearing age do not know how to draw a picture of their genitalia and in particular cannot name the different parts! Some girls have sex for the first time when they do not even know that they have a clitoris.[3]

[1]"See means to examine details, stop a bit on each and have again an entire look"Alain (Emile Chartier, dit) French philosopher, 1868–1951.

[2]We recently saw an article in a French women's magazine that placed the Bartholin's glands underneath the glans clitoridis!

[3]According to a survey carried out in 2009 on 376 students, 23 % of students in Year 9/10 (3ème in the French system) did not know they had a clitoris and 65 % were unaware of its function.

Along the way and much to our surprise, we found that men are also profoundly ignorant when they come to "the intimate female anatomy". J.D. Vincent goes so far as to talk about an "unfathomable ignorance" ("ignorance abyssale" in French) of the female genitalia. Again, we wondered about the reasons for this ignorance in an era when taboos are weakening, where pornographic images abound and where young people can, at least indirectly, teach themselves about basic sex education. In fact, due to a lack of genuine education, young men continue to have a frustrating and superficial understanding of the female genitalia such that, in their imagination, many of them still only think of it as a canal for penetration. The clitoris, ignored by most people, is thought by the remaining minority to be a small uninteresting protrusion or something that, according to some sexologists, "scares them" when they learn that it can, all by itself, give sexual pleasure to women and that "it has an unlimited capacity to provide pleasure". For all of these reasons, our book, originally only devoted to providing women with information, has also become a body of work that will enable men to better understand women.

We also prepared this book to teach medical students, as not enough time is devoted to teaching about the pelvic anatomy, but also for various specialists (gynaecologists and surgeons, urologists, obstetricians, sonographers, radiologists, sexologists, doctors practicing sexual medicine, physiotherapists, nurses and caregivers whose daily work involves the genital area of women). In particular, the latter will find presentations and photos that they can use to explain their work to the women in their care.

At last, we hope that this book will be consulted by researchers in the field of sexology who will be able to find an abundance of data for their work in the chapters on histology, embryology and topographic anatomy. There are still many mysteries in the scientific approach to female pleasure and the only way to explain them will be to go back to anatomy with the help of modern exploration techniques, which are increasingly become more efficient. Sexuality plays a very important role. It has earned its place in a holistic approach to a woman's body and spirit, something which scientists, at the time of women's emancipation, have the duty to develop. We are very happy to be able to provide our contribution.

Marseille, France Vincent Di Marino

Contents

1 History of the Clitoris

1.1 General

It would be logical to believe that the history of the clitoris is part of the history of the external female genitalia. However, this is not the case. Knowledge of the clitoris was acquired at a later date and remained inaccurate and incomplete for a long time. More surprising still, after having finally been the subject of remarkable anatomical studies, the clitoris was then ignored or unknown during certain periods! It was only in the twentieth century that the significance of this anatomical formation was acknowledged and integrated into an actual apparatus: the bulbo-clitoral apparatus.

1.1.1 Foundations of Knowledge

Our prehistorical ancestors could already represent the vulva, such as revealed, on the one hand, by the numerous engraved, drawn or painted pubic triangles, discovered on prehistoric sites of the Palaeolithic era and, on the other hand, by the small statues, "Venus impudicae", representing female bodies with generous and disproportionate curves, large breasts and exposed vulva. Anatomical details cannot be observed on these various vulvar figurations as the vulvas, represented by these proto-artists, are systematically summarised as triangles whose incomplete bisecting line represents the "large cleft". Only in rare cases do the edges of the cleft, slightly convex, evoke labia majora.[1] However, the anatomical precision does not go further and therefore knowledge of the clitoris and the labia minora cannot be prejudged. Generally, these are not representations (except in cases where the hips are outlined) but figurations, of which many "confine to abstraction".

Therefore, we may question the meaning of these vulvas, especially when various numbers are represented, at the entrance to a cave or to a room or inside a cave:

- Do they symbolise the reproductive system?
- Do they symbolise the transition from darkness to light?
- Are they an initiation symbol? (Worship glorifying the fertile woman? Worship of the goddess mother?)

As anatomists, we must assist historians in order to provide answers to these questions and must therefore question the general position of vulvar clefts, which have been represented and preferentially placed by the artists of this era, in a pre-pubic ventral position, whereas the men from the Magdalenian era could not be unaware of the fact that the vulvar cleft is almost invisible on an adult woman in a standing position, as it is located in the perineal region.

More close to us, old Egypt remains mysterious as regards knowledge of the labia minora and the clitoris. There is no ostracon specifically representing the external female genitalia. Neither is there any low relief specifically revealing this part of the female anatomy whereas the male apparatus is often represented (representations of ithyphallic god Min,[2] in erection, and scenes of male circumcisions on the lower reliefs of tombs or temples, such as those of Karnak). However, a few small statues dating back from the Neolithic era have been found. They represent nude women[3] but they are very rare, as opposed to the small phallic statues. Moreover, considering the size of these small statues, no conclusion can be drawn in relation to the anatomical knowledge of their authors, only that they had a more harmonious idea of the female body than their ancestors.

Later, Greek physicians made progress in their knowledge of the female genitalia, even if it is known that this knowledge was not direct. None of these famous physicians performed a vaginal examination or dissection on a woman's body. It is well established that their knowledge was only

[1] This is the case of the most famous of these Venuses, the Venus of Willendorf in Austria (24,000 years B.C.).

[2] Min God: one of the oldest divinities in old Egypt (god of fertility and reproduction).

[3] The most beautiful of these statuettes was found in a tomb at Badari. It is made out of ivory and dates back to the Predynastic era (4,500 B.C.). Its sexual symbolism is clear ("pubic" vulva, highlighted by a pattern). This statuette is kept at the British Museum.

V. Di Marino, H. Lepidi, *Anatomic Study of the Clitoris and the Bulbo-Clitoral Organ*,
DOI 10.1007/978-3-319-04894-9_1, © Springer International Publishing Switzerland 2014

based on reports from midwives and women (who became experts in the practice of vaginal self-examination)[4] and perhaps also on what they observed during encounters with their own partners.

It is to be recalled that **Hippocrates**[5] provided a very good description of vaginal transudation, which occurs at the same time as sexual excitation, but was wrongly considered as "female sperm". He used the term "to aidoïon", i.e. "the shameful parts"[6] to refer to the entire genitalia. He had good knowledge of the matrix and its neck. The vagina does not have a specific name denomination and is considered as part of the "aidoïon". The vulvar cleft is known as "nature". According to certain scholars having studied all of his work, he has knowledge of the clitoris but considers it as a simple protrusion, which he refers to as columella or uvula, on the basis of its resemblance with the uvula palatina.

As for **Aristotle**,[7] he denies the existence of female sperm in relation to generation. For him, "when conceiving, the woman provides the matter and this matter consists of the menses". In chapter XIV of the treatise on the generation of animals,[8] he explains that "if some naturalists have believed that women release semen during the act of copulation, it is due to the fact that sometimes women feel as much pleasure as men and that they release a liquid excretion" but "this is an error". "This liquid is not at all spermatic… It is a special fluid….In women who release this liquid, there is not the same amount of liquid as sperm but sometimes there is much more". In this same chapter, Aristotle also mentioned clitoral excitation. He wrote "Which also shows that women do not release sperm, only during copulation, they feel pleasure from being touched in the same place as men, but, in their case, there is no liquid emission".

Even Aristotle has provided an outlined description of what could correspond to the clitoris but with a number of improbabilities. Thus, it is possible to read in the history of animals: "This is how Nature has laid out the pathway followed by sperm in women: They have a canal, such as men have a sex, but it is inside the body. They suction through this canal by means of a small orifice positioned above the location where women urinate. Sperm falls into the uterus from this canal".[9] Overall, Aristotle describes an imaginary canal, but at the same time, he refers to the existence of an anatomical structure, located above the external orifice of the urethra.

As a good observer, Aristotle also describes the modifications of the external female genitalia during sexual excitation. He wrote that "by observing female rats, we can see that the humidified sex of the female swells when she approaches a male".[10]

During the three following centuries, progress was made in the acquisition of gynaecological, anatomical and medical knowledge, and we can only marvel when reading the treatise "Diseases in women", written by **Soranos of Ephesus**,[11] leader of the Methodical School.[12] The description of the female genitalia provided by this author in the first pages of his treatise[13] is astonishingly precise, even if wild ideas remain, in particular concerning "the seminal tract", which exits the matrix, crosses each ovary, skirts the sides of this apparatus up to the bladder and then accesses the neck of the latter.

According to him, "the wings (pterygomata) or labia of the vagina, thick and fleshy, are separated from one another by a cleft. Downwards, they end at the 2 thighs. Upwards, they end at what is called the clitoris (kleitoris) or myrton or nymph". Soranos not only mentions this formation. He also specifies and explains: "The latter, which forms the beginning of the labia, consists of a waffle with a muscular aspect. This small formation is called the nymph because it is hidden underneath the labia such as young brides under their veil". He even provides the following observations: "Slightly below the nymph, another protuberant wattle is hidden, it is the end of the neck of the bladder; referred to as the urethra". Therefore, even if Soranos mixes up the internal and external ostia of the urethra, he positions the clitoris or nymph correctly, above the meatus, which is used for the emission of urine.

Rufus of Ephesus,[14] a contemporary of Soranos, has good anatomical knowledge based on dissections performed on monkeys. He wrote, only in Greek, a book in which the entire nomenclature of anatomical terms of the time is assembled and specified. Thus, he confirms the term "clitoris", which he connects to a verb with an erotic meaning: cleitoriaxo. He confirms that the labia majora or myrtle lips (myrtocheïla) must now be referred to as "pterygomata". As for the term of nymph, it indicates the unit formed by the

[4] "I only know what women have taught me", (Des chairs, XIX) in Hippocrates' complete series of books and written work, Littré, vol.8.

[5] Hippocrates (towards 460 c. BC. in the Kos island–towards 370 c. BC at Larissa): Greek physician, "Father of medicine".

[6] (Hippocrates, De naturâ muliebris, Lib II,).

[7] Aristotle (384 c. BC in Stagire–322 c. BC in Chalcis): Greek philosopher and scientist, one of Plato's students. He considerably influenced Hellenistic medicine through his knowledge gained by performing dissections (animal dissections).

[8] (Aristotle, Treatise on the generation of animals, first book, chapter XIV, 728 b).

[9] (Aristotle, History of animals, Book X, p.637 to 621).

[10] (Aristotle History of animals, p 572 a and b).

[11] Soranos of Ephesus, a Greek physician, from Ephesus, practised in Rome during the second half of the first century and the beginning of the second century, under the reigns of the emperors Trajan then Adrien.

[12] The "methodical" doctrine prohibited, at Soranos, the execution of dissections (whereas dissections will be systematically performed by "dogmatics" such as Galien).

[13] (Soranos of Ephesus, Diseases in women 1, tà gynaikeïa, written in Greek and then translated into Latin).

[14] Rufus of Ephesus, a Greek physician, also lived during the reign of Trajan and wrote "About the name of the parts of the human body".

clitoris and the labia minora. The cleft is called "schisma". All of the internal and external genitalia form the "shameful parts" or aidoïon.

Neither Soranos nor Rufus will achieve the celebrity of their successor, **Galen**,[15] who is also a Greek physician based in Rome, where he will live until the end of the second century (the date of his death is discussed: 204 AD ?).

Galien recalls the bi-spermatic theory made by Hippocrates by adding the essential role (according to him) of blood "matter", whose surplus overflows periodically as menstruations.

For Galen, the female genitalia is the reverse copy of the male genitalia. This copy is interiorised whereas man's genitalia is exteriorised. The male's penis corresponds, according to him, to the uterus. The scrotum occupies the location of the matrix. The female "testicles" are positioned on either side of the matrix. The prepuce of the penis becomes the vagina in women. The latter represents with the vulva, the female shameful part (aidaïon gunaikeïon). No place is made for the clitoris so that the progress made by Aristotle and the physicians of Ephesus as regards the anatomy of the female sex is truly forgotten.

Despite this fact, Galien became the greatest physician of Western civilisation, even more famous than Hippocrates, such that his findings influenced medical knowledge, especially in relation to human anatomy, until the fifteenth century.

One century and half later, **Oribase**,[16] who was also born, such as Galien, in Pergamon, was put in charge, by the Emperor Julian the Apostate (331–363), of gathering all the knowledge of the eminent Greek physicians, since Hippocrates. Oribase completed this task by compiling a magnificent encyclopaedia (70 books), based on the work of his predecessors.

It refers to the female genitalia as pecten. It calls the vulva, "the big cleft". The labia majora are "the wings" of this big cleft. The clitoris is obviously well known. He positions it correctly and describes it as "a muscular waffle, located in the centre". He calls it myrtle or nymph so that the labia minora become "the wings of the myrtle".

Simultaneously to Greek medicine, foreign medicines (Arab and Persian, in particular) have progressed, while profiting from the knowledge contributed by Greek and Latin work. Two physicians from the tenth century, who distinguished themselves among these foreign physicians, knew about the clitoris: a Persian physician, **Avicenna**[17] (980–1037), which calls the clitoris "el bathr" (penis) and an Arabic physician, **Abulcasis**[18] (936?–1013), which calls it "tentigo" (which is placed under tension) but also "softness of love", a term which will be used by Colombo.

The main work completed by Avicenna, a book called *Canon*, includes 5 volumes and the first volume is a panorama of the anatomy and pathology of the various organs.

Abulcasis compiled a medical and surgical encyclopaedia, in 3 parts, and the last part was entirely devoted to surgery with a paragraph concerning clitoridectomy. Abulcasis' reputation (who practised his art at Cordoue) went beyond the Arab world and the borders of Spain, and his work, a reference for teachers for more than one century, was to be republished several times and successively translated into Latin, English, French, Hebrew and the Provençal language.

Few physicians of the following centuries became known for their work and Galen's influence lasted until the Middle Ages. Nevertheless, a few anatomical books were published and included superb prints, which were as beautiful as inaccurate. Among these books, the book written by **Mondino De Liuzzi**[19] entitled *Anathomia* was a real success. Moreover, many readers, on the basis of the author's work, in which he mentions that, in 1316, he performed two dissections on female bodies, consider him as the "protodissector", i.e. the first physician to have restored cadaveric dissection,[20] which had disappeared since antiquity. Actually the first cadaveric anatomical dissection was performed in 1313, in Montpellier by Henri de Mondeville[21] (who was also Mondino's professor). Nevertheless, Mondino's book, based on knowledge from human dissections and not from animal dissections, such as the work completed by Greek

[15] Claudius Galenus (Galen) (131 born in Pergamon in Asia Minor, died in 204?, in Rome) was a well-known physician and philosopher. He was the physician of the emperors Marc Aurele and Commode. He was also a great animal anatomist (as the dissection of human cadavers was prohibited in Rome). He is considered as the founding father of pharmacy (oath of Galen....galenics!). His work, which has been translated into Arabic, has largely contributed to the development of Arab medicine and then, in turn, to medieval medicine.

[16] Oribase (born around 325 at Pergamon, deceased around 395, in Constantinople) was the last of the famous Greek physicians.

[17] Avicenna (980–1037), a Persian physician, scientist and philosopher, wrote a book which has remained well known: *Kitab Al Quanûm fi Al-Tibb* (Book of the law concerning medicine), known under the abbreviation *Quanûm* (Canon). This book was written in Persian and was translated, one century and half later, into Latin by Gerard de Crémone, under the title *Canon médicinae*. It was only printed in Arabic in 1593.

[18] Abulcasis (Aboul Kasim Al Zahravi) (936?–1013), a physician and surgeon, rapidly became well known and his reputation reached all of Europe after his death. His major work, *Kitab Al-Tasrif* (Book of the medical method), is an encyclopaedia: 30 volumes! It was translated into Latin, such as Avicenna's work, by Gerard de Crémone, during the twelfth century.

[19] Mondino de' Liuzzi, an Italian physician (born in Bologna in 1270 and deceased in the same city in 1326).

[20] Until then, cadaveric dissection for gaining anatomical knowledge had been prohibited by the Church. Only a few autopsies were authorised for legal purposes. Thereafter, the progressive transgression of prohibitions was going to finally allow the development of human anatomy.

[21] Henri de Mondeville (born in Mondeville in 1260 and deceased in 1320) was a French anatomist, who was the surgeon of kings of France, Philippe le Bel and Louis le Hutin. In 1304, he was named professor in Montpellier (where his students included, among others, the famous Guy de Chauliac). He was also a professor in Bologna.

authors and Galen's in particular, was going to initiate the future upheaval of anatomical knowledge of the human body.

Two other books using engraving were to be published a few years later and initiate the expansion of anatomical culture in all Europe of the early Renaissance. This was an indication of the development of this fundamental subject of medicine in the near future:

Firstly, a book written by **Jacopo Berengario da Carpi**,[22] with a promising Latin title *Isagogae breves perlucidae ac uberrimae in anatomiam humani corporis a communi medicorum academia usitatam*, was published in Bologna in 1527 then a book written by **Charles Estienne**,[23] a contemporary of Vesalius, who published the *Dissection des parties du corps humain* (Dissection of the parts of the human body), successively in 1545, in Latin, then in 1546, in French. This book, which included even more anatomical charts than Berengario's book, used such as the latter, the technique of staging human bodies.

Sadly, this staged representation, although pleasant and elegant to look at, outweighed the anatomical reality and made the body structures very difficult to observe as they were too small, inaccurate and generally incorrect. The actual descriptions are inaccurate (e.g. that of woman's "shameful member", which seems to more or less refer to a description of the clitoris).

The fact remains that the three above-mentioned authors had established the foundations of modern anatomy, based on the direct observation of the human body and the reproduction of this reality in anatomical charts, which Vésalius rapidly enhanced and developed.

But what happened to the anatomy of the clitoris during the Middle Ages?

In spite of the relative small amount of anatomical discoveries made during this period, the assumed role of the clitoris in fecundation spread in the thatched cottage homes during the Middle Ages. The majority of physicians of the time adopted the theory of the existence of female sperm and considered that this secretion, called cyprine, was produced by the clitoris. The Catholic Church, however, which expressed reservations on the topic of carnal pleasure, had to accept the medical advice favouring fecundation: "before the sexual intercourse, the husband shall delicately rub the button of love with a finger moistened with perfumed oil, in a circular motion".

[22] Jacopo Berengario da Carpi, a physician and famous Italian anatomist (born in 1460 in Carpi and deceased in 1530), is one of the first to have included anatomical illustrations with his written work.

[23] Charles Estienne, a French physician, anatomist and printer (1504–1564), wrote many books in several fields and [in] especially *La Dissection des Parties du corps humain* ("the Dissection of the Parts of the human body" (divided into 3 books), published by Simon de Colines (who was his father-in-law), in 1546. This book was the translation of "Dissectione partium corporis humanis", whose engravings on wood, executed by Estienne de la Rivière, a surgeon and engraver, had been ready since 1539 but whose publication had been interrupted due to a lawsuit for plagiarism. Charles Estienne was a great anatomist who especially discovered the nutrient foramina of the bones and their role.

1.1.2 Rebirth and "Discovery" of the Clitoris

The Renaissance, a period of renewal and all types of discoveries, was going to the era during which anatomical knowledge, including that of the clitoris, was developed. Little by little, prohibitions concerning cadaveric dissection were going to be finally reduced thus allowing a direct approach of the human body and thus of its knowledge.

It is in this new environment that the Belgian, **Andreas Vesalius**,[24] who was going to become the most famous anatomist, was going to rapidly emerge and define the foundations of modern anatomy, in his masterly exercise at the faculty of Padua:

- By transforming dissections on human bodies from exceptional "public place anatomies", such as was the case, in Antwerp or Venice[25] (Fig. 1.1) into a current practice, carried out in dedicated locations: initially, anatomical theatres, dismountable installations and then rapidly permanent structures[26]
- By only transmitting to his students' concepts verified through direct anatomical observations

[24] Andreas Vesalius, a physician, humanist and especially a famous anatomist (born in 1514 in Brussels and deceased in 1564 on the Island of Zakynthos, in Greece). He was a Professor of Anatomy and Surgery at Padua and also taught at Pisa and Bologna. It is in 1543 that he published his major work *De humani corporis fabrica*, in which he dared expose nearly 200 errors made by Galien, which generated considerable enmities. One year later, he left his position at Padua to become the surgeon of the emperor Charles Quint then of Philippe II of Spain. He died on the Island of Zakynthos from typhus when returning from a pilgrimage to the Holy Land.

[25] In Venice, a free city where nobody feared the anger of the Inquisition, we can still see the "Place de l'anatomie" ("Anatomy square"), where "public anatomies" took place (Fig. 1.1). This square is adjacent to a small canal were a gondola would come at night time to recover the leftovers from the dissection and evacuate them further at sea.

[26] The first "permanent anatomical amphitheatres" with a central dissection table, concentric observation circles for the students and a pulpit from which the Master anatomist gave his lesson, were only built in 1550 (Salamanca). The very beautiful and very famous amphitheatre of Padua, was only built in 1584, on the initiative of the anatomist Girolamo Fabrizi d'Acquapendente.

Fig. 1.1 "Piazza dell'Anatomia" in Venice (**a**) and his accesses "sottoportego dell' anatomia" (**b**): the stairs for going down from the bridge of anatomia (ponte dell'anatomia), and the porch that allows the access to the little river (rio). The first and exceptional public anatomies were carried in that place, hidden from sight! The cadavers and the anatomical remains were transported by gondola across the rio

- By performing the dissection himself during his anatomy lessons, such as shown by the front page of his book
- By being the first to write an exceptional book, *De humani corporis fabrica*, published in Basle in 1543, which provided a significant anatomical accuracy, enhanced by the presence of an exceptional iconography (charts prepared by students of the school of Titien, which had attended the cadaveric dissections) and, furthermore, which was printed (following the recent discovery of the printing technique by Gutenberg)

Nevertheless, in the *De humani corporis fabrica* book, libri septem, there is no detailed description of the external female genitalia and even less precisions on the clitoris, which Vesalius seems to ignore (a superb image of the entire genitalia shows a vulva, which is simply represented as 2 greater lips, with a nice representation of combed body hair) and even considers as an anomaly, which is only visible in hermaphrodites.

It is true that Vesalius was only able to study this anatomical region on a small number of specimens, which were only available for a very short time (autopsies).

It was thus necessary to wait for Vesalius' successors as Professor of Anatomy at the University of Padua, Colombo then Fallopio, who conducted accurate studies on the clitoris, which clarified its function and structure.

Realdo Colombo[27] was successively Vesalius' assistant (1541), temporary substitute in 1543 and then permanent substitute as from 1544 (when Vesalius left to take up his position at the court of Charles Quint). He became the Professor of Anatomy at the University of Pisa in 1545 and wrote a remarkable book on anatomy *De re anatomica*, which was only published after his death in 1559.

Sadly, this book included no iconography, except for a superb front page (Fig. 1.2) by Veronese.[28] The book (which was extremely documented) was a great success and its fame went beyond the borders of the Republic of Venice, such as represented by a frontispice painting "A Lesson of Anatomy", commissioned by the famous English anatomist, John Banister, lecturer in anatomy of the Barber Surgeons' Company (licensed by the king Henry VIII in 1540). On this painting, visible at the Glasgow's University, Banister is performing an anatomical demonstration in front of his students while reading the book written by Colombo, which is open on his desk. Numerous discoveries are owed to Colombo, an exceptional anatomist, who performed many dissections (apparently, he has personally performed 1,000 dissections). He was also a talented physiologist. One of his most famous discoveries is that of pulmonary circulation[29] (which was used as a basis, 30 years later, for the

Fig. 1.2 Frontispiece of the book De re anatomica (engraving woodcut of Nicolai Bevilacquae, 1559, probably from a picture of Paolo Veronese). Realdo Colombo is dissecting a human cadaver in front of both his students and the famous physiologist, Andrea Cesalpino

[27] Matteo Realdo Colombo (1515/1516–1559) was a student of Vesalius, who he replaced to teach anatomy at Padua. He then rowed with Vesalius, who learned that in his absence, his pupil had criticised one of his publications! Colombo then became a permanent Professor of Anatomy at the University of Pisa (in the service of Cosimo de Medici) then a Professor of Anatomy in Rome, at the papal University of Sapienza (in the service of the Pope Jules III). His discoveries (clitoris, stapes bone, pulmonary circulation, position of the kidneys) were all disputed by critics, on the basis that they had already been made by other anatomists. He also completed important research work on the pleura and the peritoneum. He recorded all of his "discoveries" in a remarkable book entitled *De re anatomica*, which sadly, was only published in 1559, after his death, and which did not contain any diagrams (which were to be completed by his friend, Michelangelo, with whom he shared a passion for cadaveric dissection in Rome). In 1550, he was appointed surgeon of the pope Jules III and remained in his service until his death. Among other things, he performed a dissection on a saint (such as proclaimed by the Church): Ignace de Loyola, deceased in Rome in 1556.

[28] This front page represents Colombo performing a dissection in the presence of the famous naturalist-physiologist, Andrea Cesalpino.

[29] Similarly to Colombo, Michael Servetus (1511–1553), a Spanish physician and theologist, also described pulmonary circulation, before being burnt to death for being a heretic, in Geneva, in 1553. However, it seems that the original discovery of this small circulation took place beforehand, such as mentioned in the work completed by Ibn al-Nafis (1210–1288), a Syrian physician and anatomist, which had been translated by a friend of Colombo's.

descriptions of the large and small circulation by William Harvey). We also owe him remarkable work on the clitoris, which he claimed to have discovered, whereas, such as we have noted, Greek (but so Arab and Persian) physicians and surgeons already had knowledge thereof, such as proved by their texts.

In spite of these obvious facts, Colombo unequivocally stated that he was the first "discoverer" of the clitoris, such was also the case of his homonym and fellow citizen, the famous Italian navigator, Cristoforo Colombo (Christopher Columbus), who discovered America, in 1492! ....

What remains certain is not only that he accurately described the external female genitalia but, in particular, that he highlighted the clitoris in this description and celebrated, for the first time, its physiological properties, thus demonstrating, in terms of quasi-poetic exaltation, the exceptional role played by this formation in the genesis of female pleasure (De Visceribus, Liber XI.243,[30] in Fig. 1.3). However, he assigned a (female) sperm secretion function to the clitoris (which he referred to as "Amor Veneris"), which is an idea inherited from former anatomists and which is totally false.

Gabriele Falloppio,[31] also one of Vesalius' students, rapidly claimed the paternity of the discovery of the clitoris, as he wrote in his book *Observationes anatomicae*, published in 1561 (but which he is supposed to have written in 1550), "this part is hidden and ignored by anatomists,… it is so hidden that I was the first to discover it and if others speak of it, please know that they learnt about it from me and my students!" In this statement, he is nearly accusing Colombo of plagiarism. Nevertheless, there was an enmity and competition between the 2 students of the Master, which is easily explained by the fact that Fallopio was always the eternal second, as he succeeded Colombo, firstly as Professor at the University of Pisa in 1548 and then as Professor at the University of Padua, in 1551.

During the next century, another anatomist of talent, **Thomas Bartholin**,[32] was going to restore the truth in relation to Vesalius' 2 successors by writing in chapter XXXIV

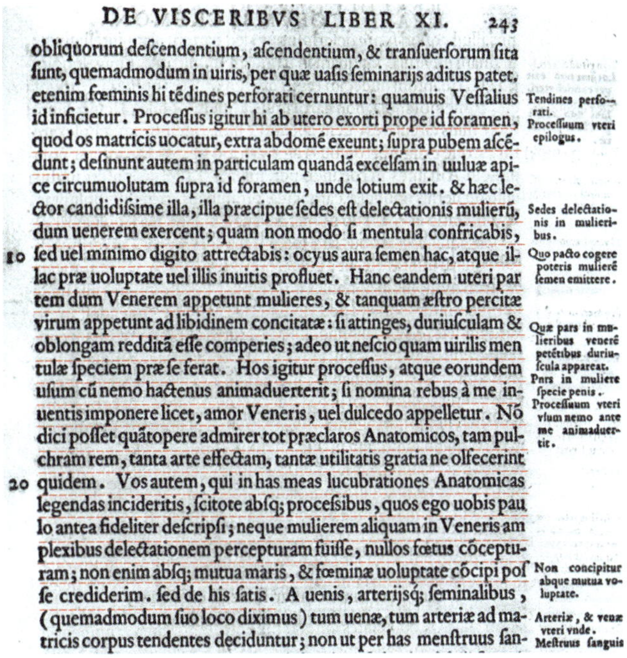
DE VISCERIBVS LIBER XI. 243

obliquorum deſcendentium, aſcendentium, & tranſuerſorum ſita ſunt, quemadmodum in uiris, per quæ uaſis ſeminarijs aditus patet. etenim fœminis hi tēdines perforati cernuntur: quamuis Veſſalius id inficietur. Proceſſus igitur hi ab utero exorti prope id foramen, quod os matricis uocatur, extra abdomē exeunt; ſupra pubem aſcē-dunt; deſinunt autem in particulam quandā excelſam in uuluæ apice circumuolutam ſupra id foramen, unde lotium exit. & hæc lector candidiſsime illa, illa præcipue ſedes eſt delectationis mulierū, dum uenerem exercent; quam non modo ſi mentula confricabis, ſed uel minimo digito attrectabis: ocyus aura ſemen hac, atque illac præ uoluptate uel illis inuitis profluet. Hanc eandem uteri partem dum Venerem appetunt mulieres, & tanquam æſtro percitæ virum appetunt ad libidinem concitatæ: ſi attinges, duriuſculam & oblongam redditā eſſe comperies; adeo ut neſcio quam uirilis mentulæ ſpeciem præ ſe ferat. Hos igitur proceſſus, atque eorundem uſum cū nemo hactenus animaduerterit; ſi nomina rebus à me inuentis imponere licet, amor Veneris, uel dulcedo appelletur. Nō dici poſſet quātopere admirer tot præclaros Anatomicos, tam pulchram rem, tanta arte effectam, tantæ utilitatis gratia ne olfecerint quidem. Vos autem, qui in has meas lucubrationes Anatomicas legendas incideritis, ſcitote abſq; proceſsibus, quos ego uobis paulo antea fideliter deſcripſi; neque mulierem aliquam in Veneris amplexibus delectationem percepturam fuiſſe, nullos fœtus cōceptuгam; non enim abſq; mutua maris, & fœminæ uoluptate cōcipi poſſe crediderim. ſed de his ſatis. A uenis, arterijsq; ſeminalibus, (quemadmodum ſuo loco diximus) tum uenæ, tum arteriæ ad matricis corpus tendentes deciduntur; non ut per has menſtruus ſan-

Tendines perforati.
Proceſſuum vteri epilogus.
Sedes delectationis in mulieribus.
Quo pacto cogere poteris mulierē ſemen emittere.
Quæ pars in mulieribus venerē petētibus duriuſcula appareat.
Pars in muliere ſpecie penis.
Proceſſuum vteri vſum nemo ante me animaduertit.
Non concipitur abque mutua voluptate.
Arteriæ, & venæ vteri vnde.
Meſtruus ſanguis

Fig. 1.3 Superior part of the magnificent page of De re anatomica in which Realdo Colombo describes, for the first time, the clitoris as "amor Veneris", the sublime organ of female pleasure

of book I, "De clitoride", from his "Anatomia reformata",[33] written in Latin but easily understandable: "Hujus inventionem sibi arrogat Fallopius: Columbus, vero gloriose,… sibi tribuit. cum tamen ejus fecerint mentionem Avicenna, Albucasis, Ruffus, Pollux § alii" thus recalling all of the scientists who had referred to this topic before our dear Venetian anatomists. In addition, in this chapter, a very interesting outline of the description of the clitoris can be found with Schemes.

1.1.3 The Pinnacle of Knowledge

During the seventeenth century, other famous anatomists (among whom we can count **Jean Riolan**[34] from France) have significantly contributed to the study of the anatomy of the clitoris. Undeniably, the biggest contribution was made

[30]This caused two serious problems for him. He had to appear before the court of the Inquisition and only owed his salvation to his vehement defence, which he ensured himself, supported by his anatomical discoveries. At the end of his inquisitorial lawsuit, he was declared innocent but the publication of his book *De re anatomica* was prohibited (it was only to be published after his death, thanks to the tenacity of his 2 sons).

[31]Gabrielle Falloppio (Fallopius), known in France as Gabriel Fallope (1523–1562), was not only a great anatomist (professor at both Pisa and Padua) but was also a talented botanist. We owe him numerous discoveries, some of which refer to his name: the facial canal (aquaductus Fallopi), the uterine tubes (fallopian tubes) and also the "vagina" (a designation he introduced). He also made several important discoveries in the internal ear region.

[32]Thomas Bartholin (1619–1680), a Danish doctor and anatomist, was one of a long line of scientists in his family and is best known for his work on the lymphatic system. It was his son Caspar (the Younger), who first described the greater vestibular gland, which is now referred to as: "Bartholin's gland".

[33]*Anatomia Reformata*, which was first published in 1655, was in fact a revised and annotated edition of the work of his father, Caspar (the Elder), *Anatomiae Istitutiones Corporis Humani*, published in 1611.

[34]Jean Riolan the Younger (1577–1657) was doctor to Marie de Medicis and Louis XIII. In 1618, as a professor at the Faculty of Medicine in Paris, he published *Anthropographia and Osteologia*. The book, which was written in Latin, covered the anatomical description of man. A second complete edition was published in 1626. The second edition was translated into French in 1629 by G. Constant, who published it under the title *Les œuvres anatomiques de Monsieur Jean Riolan* (The Anatomical Works of Mr Jean Riolan).

by **Regnier de Graaf**,[35] who was most successful at compiling all the research work describing this formation, as well as the entire external female genitalia, in his book *Histoire anatomique des parties génitales de l'homme et de la femme* (Anatomic history of the genitalia of man and woman) published in 1649 in Latin, but which was soon to be translated into French. It is especially the anatomy of the female genitalia which owes a lot to de Graaf, who discovered the spongious bulbs he called the reticular plexus or retiform plexus. He was the first to describe and include these bulbs in a book, which was published in 1672: *De Mulierum Organis Generationi Inservientibus Tractatus Novus Cum Figuris*, (even if Swammerdam wrongly disputed the priority given to this discovery in his opus, *Miraculum Naturae Seu Uteri Muliebris Fabrica*, which was published the same year).

Following the example of the anatomists, obstetricians began to describe the female genitalia that the anatomists had self-appropriated. One of these obstetricians, **Pierre Dionis**,[36] a surgeon appointed to the King's Garden by Louis XIV to teach "anatomy according to the circulatory system and new discoveries", left a remarkable description of the clitoris in his "Traité Général des Accouchements qui Instruit de Tout ce Qu'il Faut Faire Pour Etre un Habile Accoucheur" (General Treatise on Childbirth for the Training of an Accomplished Obstetrician). For the same period, we can also mention two books including anatomical charts with legends including excellent representations of the external female genitalia, which specify the accurate location of the clitoris: the works of Giulio Casserio and Govard Bidloo.

- A book written by **Giulio Casserio**[37] *Tabulae Anatomicae LXXIIX Omnes Novae Nec Ante Hac Visae*, published in 1627, 11 years after the author's death, is a work of great precision. It includes charts which are true paintings and stylised representations of the human body, in a similar manner to Vesalius. The six figures of table XVIII and of book VIII allow the author to represent what can be seen of the clitoris when observing the vulva but also the entire make-up of the clitoris from the two crura (roots) which the author called "nerve bodies".
- A book written by **Govard Bidloo**,[38] *Anatomia de Humani Corporis Demonstrata et Illustrata*, was published in 1685 and includes "Centum et quinque tabulis, per artificiossis **G. de Lairesse** ad vivum delineates", as well as superb illustrations of dissections sketched in real time. In de Lairesse's illustrations representing the vulva (Fig. 1.4), the clitoris is perfectly identified and drawn. If one compares, for example, the illustration referred to as "Quinquagesima Primae" from Bidloo's book to illustration 60 of De Humani Corporis Fabrica of Vesalius, we realise the extent of progress achieved by anatomists in the precise description of the clitoris within less than a century and a half. Other well-known French authors also decided to provide complete descriptions of the clitoris. Similar to de Graaf, the reticular plexus was included in their anatomical descriptions. However, French work had a difficult time crossing the channel and thus we are not to be surprised to read that, during the eighteenth century, **John Bell**,[39] a great English anatomist and author of *Anatomy of the Human Body* (Volume 3), wrote that the clitoris does not include spongious tissue!

The nineteenth century should have been the "clitoris's century", thanks to **Georg Ludwig Kobelt**,[40] who in a highly remarkable book, *Die männlichen und weiblichen Wollustorgane des Menschen und einiger Säugetiere* about

[35]Regnier de Graaf, a Dutch doctor and anatomist (1641–1673) was taught by Sylvius in Paris and then worked at Delft. The year of his death, he and his colleague, Stenon, discovered the ovarian follicle, which adopted his name, "De Graaf follicle". He also wrote the *Histoire anatomique des parties génitales de l'homme et de la femme qui servent à la génération* (Anatomical History of the Male and Female Genitalia employed for Generation) published in 1649, at the same time as a sucpancreatic treatise.

[36]Pierre Dionis (1643–1718), obstetrician, was named by Louis XIV as surgeon responsible for teaching anatomy at the King's Garden in 1672. In 1680 he became surgeon in ordinary to Queen Marie Thérèse, before being named first surgeon to the children of France in 1712. In 1713, he became the Duchess of Berry's obstetrician. In 1715, he was called upon to assist Louis XIV in his last moments.

[37]Casserio Giulio (Julius Casserius Placentinus), whose date of birth is unknown (1552 ?) and who died in 1616, was a remarkable anatomist. He was a Professor of Anatomy, Physics and Surgery at the University of Padua, and like his great predecessors, he left numerous works on anatomy enriched with superb illustrations. His works on the sensory organs and the larynx are still relevant today.

[38]Govard Bidloo (1649–1713), Dutch doctor and anatomist, was a professor at Leiden and published an anatomical treatise, "Ontleding des menschelyken lichaams" in 1685, which contained 105 plates of anatomical drawings engraved from G. de Lairesse's remarkable original drawings, which had themselves been published by their author by 1680 (Fig. 1.4). These drawings were reused by English anatomist William Cowper in his *Anatomy of the human bodies* of 1698, which earned this author an accusation of plagiarism and a violent quarrel with his colleague Bidloo.

[39]John Bell (1763–1820) was not only an excellent anatomist and a great vascular surgeon (he contributed to this specialisation's beginnings) but also a talented artist. He was responsible for many of the illustrations to his own works. He wrote "Anatomy of the human body" in collaboration with his younger brother, Charles Bell, who became a celebrated neuroanatomist.

[40]Georg Ludwig Kobelt (1804–1857), a German anatomist, only joined the Faculty of Medicine after abandoning his legal studies that held no interest for him. He became a Professor of Anatomy at the University of Freiburg and also taught at Heidelberg. A keen dissector, he left some remarkable anatomical specimens, preserved in the anatomical collections of the two cities. He died of tuberculosis.

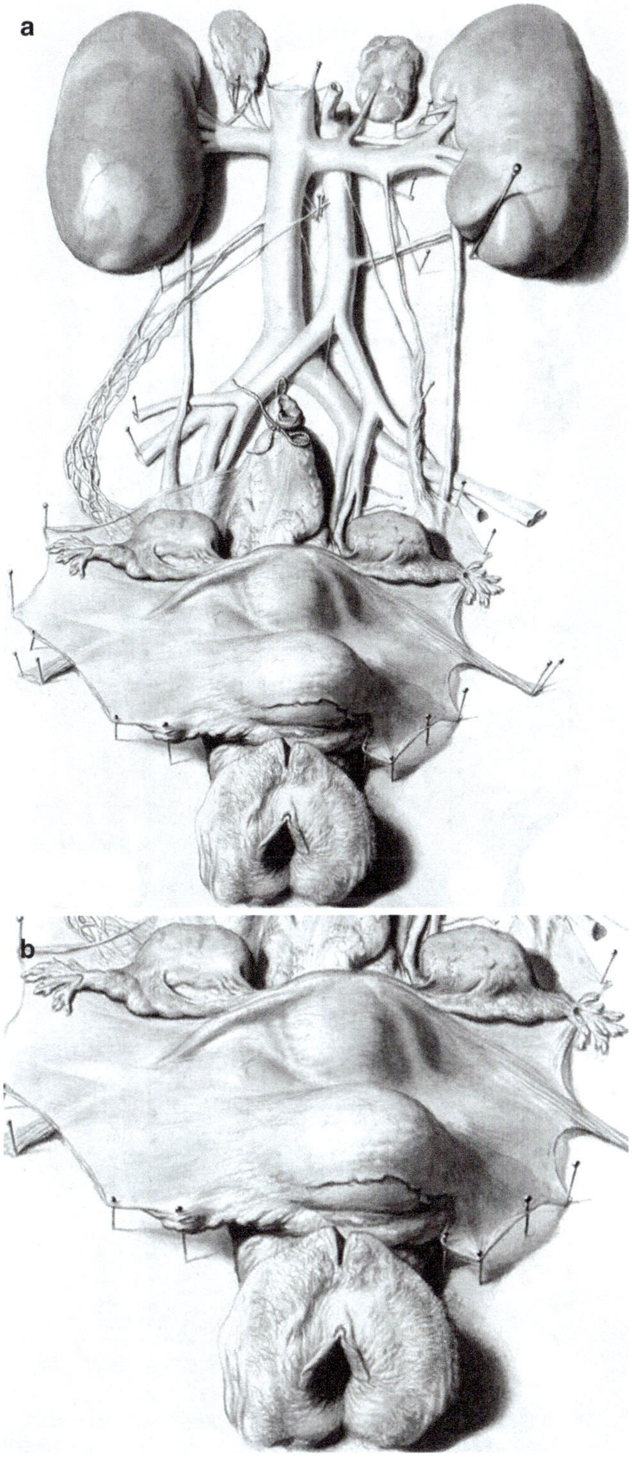

Fig. 1.4 An example of the magnificent drawings of G. de Lairesse (De Lairesse's drawings illustrated two books of Anatomy, Govard Bidloo's, in 1685, and William Cowper's, in 1698) (**a**) Connections between the urinary apparatus and the female genital apparatus. (**b**) Magnification of the previous, showing the precision and accuracy of the drawings. Notice the perfect representation of the external genitalia especially the clitoris

the male and female organs of sexual arousal in man and certain other mammals), superbly illustrated with drawings based on his own dissections, described the complete and precise anatomies of the organs of sexual arousal in both men and women, with a few references to their compared anatomy. If the study of the male genitalia (the first part of the book) does not reveal major medical progress in comparison to previous work, the second part, dedicated to the study of the female genitalia, undoubtedly demonstrates for the first time (with the support of vascular injections) that this region is comprised of two areas, the clitoris and the spongy bulbs, and that the two areas are connected by a vascular network called the "pars intermedia". After these anatomical chapters, the author, in some remarkable sections that retain their relevance, begins a functional study of the "connection" between the male and female organs of sexual arousal, thus giving an impetus to future studies on the physiology of coitus. Kobelt's work, some of whose drawings we show (Fig. 1.5), appeared in Freiburg in 1844. It was not an especially successful publication until its French edition in Strasburg in 1851, in a translation by **Von H. Kaula**.

From this point on, the anatomy of the clitoris was perfectly known, and one might have expected that this knowledge would spread on a worldwide scale. There were a few physicians who, having read Kobelt's work, attempted to bear witness to the clitoris's supremacy. Thus, in France, Dr. Guyot, who, as soon as 1852, had written the surprisingly titled *Breviary of Experimental Love*, which after "secretly making the rounds" was finally published in 1882 (10 years after the author's death). In this recently republished work,[41] Dr. Guyot insists on the importance of the preludes that must precede all penetration and strongly criticises men who misunderstand the clitoris and its role in women's sexual excitement: "There exists an immense number of ignorant, egoistic, brutal men who do not bother to study the instrument that God has entrusted to them". Alas, there were also books, which discredited the clitoris. For instance, as sexologist Sylvie Chaperon reports, Dr. Pierre Garnier, a prolific and successful author, became more and more critical, in each of his books, until in 1891 he wrote: "to claim that this miniscule apparatus, which is most often an insensitive little button so long as it has not been touched or artificially manipulated, is the most active erogenous centre, is to

[41] The *Breviary of Experimental Love* (Payot) was written "on the occasion of the marriage of Prince Napoleon and Princess Clothilde of Savoy. It was discretely read by numerous physicians and a privileged few (Prince Napoleon, Claude Bernard, the actress Virginie Dejazet, Georges Masson, Georges Sand, Sainte-Beuve and the archbishop of Rheims.)

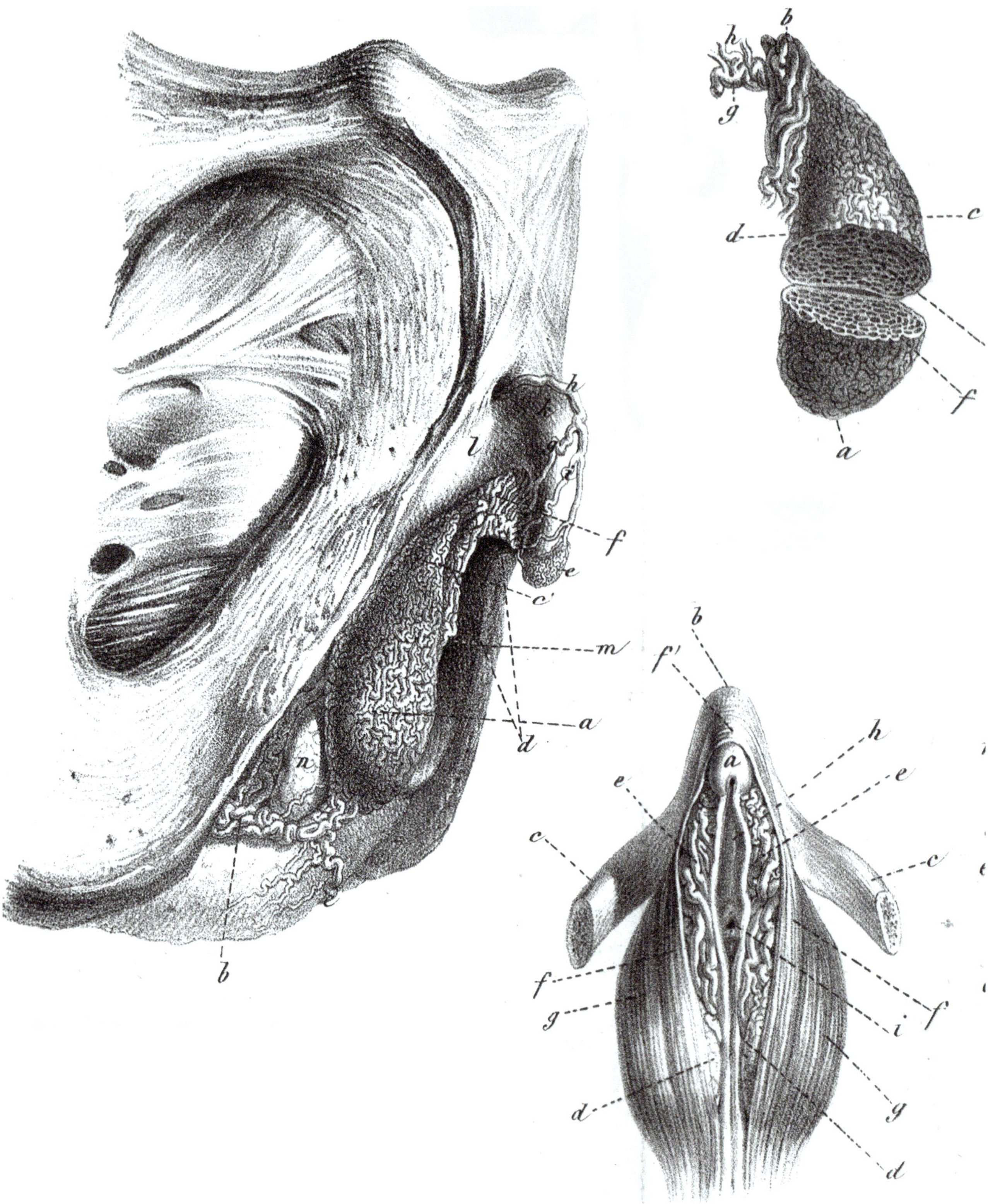

Fig. 1.5 An example of the remarkable illustrations (Notice the accuracy and the precision of the drawings (Taken from the own dissections of the author)) of Georg Ludwig Kobelt's book: Die männlichen und weiblichen Wollustorgane des Menschen und einiger Säugetiere (part of the third page of illustrations, concerning human female)

implicitly accuse all girls of once having resorted to masturbation or of one day becoming debauched".

What's more, an excessive prudishness took over in scientific work and the clitoris was forgotten in certain anatomical treatises. Kobelt's discoveries, for their part, were most often considered pointless and occluded.[42]

The turbulent years afflicting the clitoris did not only affect France!

1.1.4 Decline and Disappearance of Clitoris

It has to be believed that the "magnification" of the clitoris resulting from Kobelt's work was also ignored or misunderstood by English physicians (including those boasting of being specialists in obstetrics or gynaecology), given that within a few years, the unfounded and monstrous idea that the clitoris was useless spread among them. From uselessness to harmfulness, there was only a small step, and it was soon taken. Henceforth, many English physicians would claim that the clitoris was an absolutely useless apparatus and that in certain cases it should be excised! It was in this context that a renowned obstetrician, Dr. Isaac Baker Brown (see E. Sheehan's excellent article), founder and director of his own hospital ("The Surgical Home for Women"), elected President of the Medical Society of London in 1865, published, the following year, a book with a strange and frightening title: *On the Curability of Certain Forms of Insanity, Epilepsy, Catalepsy, and Hysteria in Females*, in which he proposed clitoridectomy as the cure for every one of these afflictions, on the basis that their principal cause was masturbation. He specified that he had been successfully practising such operations in his hospital since 1858. The book became very popular when published and gave rise to few criticisms, such that clitoridectomy spread as a practice across all of Victorian England and also in America. Several months were required for a reaction to set in, for the physicians and surgeons to protest and for female patients to contest the effectiveness of Dr. Brown's so-called miracle cure (a cure that, what's more, had to be paid up front with "hard cash"). Finally, a coalition of young gynaecologists denounced Dr. Brown's practices and obtained that an enquiry of the Obstetrical Society be made into clitoridectomies (December 1866). In February 1867, the Board of the Obstetrical Society, ruling on the notoriety and debates surrounding Brown's use of clitoridectomy, recommended his expulsion from the society. A strong majority voted in favour of the expulsion in April of the same year, and Brown resigned not only from the Society he had presided but also as director of the hospital he had founded.[43] But the harm was done, and clitoridectomy remained a recommended practice in books written only a few years before by renowned gynaecologists, particularly in America, so that it was only abandoned progressively.[44] A new blow was struck in relation to the usefulness of the clitoris in 1884, with the discovery of the microscopic process of procreation by the famous Belgian embryologist, **Edouard van Beneden**.[45] His work on fecundation showed that embryos result from the sperm's penetration of the ovum and that the latter is a living cell that makes its way from the maternal ovary without any help from some substance secreted by the clitoris. It was the end of all the legends claiming that cyprine[46] was secreted by the clitoris and played a role in the fecundation process. Henceforth, the clitoris's "physiological depreciation" was sealed, and even gynaecologists began to think of the clitoris as an apparatus presenting no interest and thus forgot the pertinent pages written by Colombo and later by Kobelt.

Very surprisingly, it was **Sigmund Freud**[47] who, by seriously depreciating the clitoris, gave the final blow. To serve his own purposes, Freud appropriated the theory proposed by his colleague Dr. Krafft-Ebing, the famous Viennese psychiatrist, that there are two erogenous zones that succeed one another in the construction of the woman individual, the clitoris for virgins and the vagina after being deflowered. And in his "Three Essays on Sexual Theory", which was published in 1905, Freud integrated into his grand theory of Oedipal

[42] Thus, today we are puzzled when consulting Bourgery and Jacob's magnificent book from 1866 to 1867, in which an isolated clitoris is described, without the spongy bulbs (which are described later in the book, under their ancient name of "retiform plexus").

[43] Brown died miserably, paralysed after several heart attacks and had to be financially assisted by a "Baker Brown Charitable Fund" contributed to by physicians and rather non-vindictive former patients, which had been operated.

[44] Some cases of "therapeutic" clitoridectomies, thankfully exceptional, were published until 1940.

[45] Edouard Joseph Louis Marie van Beneden (1846–1910), a renowned embryologist, was a Professor of Zoology, Anatomy and Comparative Physiology and Embryology at the University of Liège. He provided evidence not only for the fecundation process but also for the meiosis and mitosis processes.

[46] It is recalled that previous physicians referred to the liquid, which moistens the vagina during sexual excitation, "cyprine". They believed it to be secreted by the clitoris and that this secretion played a role during fecundation.

[47] Sigismund Schlomo Freud (1856–1939) simplified his name in 1877 and was thereafter called Sigmund Freud. As neurologist and physician, and student of Meynert and Charcot, he became Professor and Director of Neurology at the "Clinique des Enfants malades" (Sick Children's Clinic). He then settled in Vienna, which he only left in 1938 for London, after the Nazi invasion of Austria. He was the founder and uncontested Master of psychoanalysis. He was first diagnosed with oral cancer in 1923 and eventually died despite going through at least thirty different operations.

evolution the postulate according to which clitoral orgasms represent an infantile stage, while normal adult orgasms are vaginal. This was as much as to say that there were two types of women, the "clitoral", who had remained sexually immature, and the "vaginal", with a flourishing adult sexuality. Although contested, Freud's theory made enough of an impact that the clitoris's role in female pleasure was forgotten. Despite the Master's theories, his student, patient and faithful friend, Marie Bonaparte, was going to provide food for thought in relation to the clitoris, for she remained convinced of this organ's importance, though adding another limiting factor: the distance between the glans clitoridis and the vagina.[48] She therefore believed that the frigidity she suffered from was due to a "too great a distance" between these two areas and decided to be operated to reposition her clitoris. Sadly, the operation (though performed by the famous Austrian surgeon, **Joseph von Halban**) was unsuccessful.

1.1.5 The Return to Favour

Freud's ideas were also to be discussed, with many authors, who referred back to Kobelt's work and recalled that the "vaginal canal is not the sensory apparatus". In 1953, the Freudian theory lost its halo when the Kinsey Report "Human Behaviour in the Human Female"[49] stated that "vaginal orgasms do not exist", but this report was also criticised for several reasons. Finally, the vaginal and clitoral theories concerning female orgasms remain valid to this day, with the clitoris's predominance being based on the vagina's poor innervation, compared to the significant innervation of the clitoris. Scientists in favour of the vaginal predominance found new arguments with the discovery of the extremely mediatised G-spot.

The second half of the twentieth century was rich in events that altered people's sex lives (sexual revolution of the 1960s and 1970s, introduction of oral contraceptives, development of sexology, particularly with the books written by Masters and Johnson,[50] feminist movements) and succeeded in putting an end to a certain number of taboos, thus allowing women to live more fulfilling sexual lives.

In this liberal climate, the clitoris's rehabilitation begun, after being ignored for years, during which it was not considered as an actual part of women's bodies. It is spoken of more freely. Media (including television) have begun to address this topic. New scientific studies have been devoted thereto. But the road ahead is long, and at the beginning of the twenty-first century, conservatism still predominates, thus preventing the public, and particularly the male public, from truly learning about this "extraordinary jewel" of the female body.

It is for this reason that we have decided to publish this study.

---

[48] Relationships of the bulbo-clitoral organ, connections with the Urethra, Chap. 13, p. 114–120.

[49] Dr. Alfred Charles Kinsey, Professor of Entomology and Zoology, published two books on human sexual behaviour with an enormous impact: *Sexual Behaviour in the Human Male* in 1948 and *Sexual Behaviour in the Human Female* in 1953. The latter was translated into French in 1954 as *Comportement sexuel de la femme*.

[50] William Howell Masters, a gynaecologist, and Virginia Eshelman Johnson, a psychologist, are known for their remarkable work on human sexuality in 1957–1965. Most notably, they described the various phases of sexual intercourse (see Chap. 15) and studied treatments for sexual disorders and dysfunctions.

2 Formation of the Clitoris

The formation of the clitoris, a component of the external female genital apparatus, is complex and cannot be dissociated from the organisation of the internal female genital apparatus. This organisation not only involves gonosomes (sexual chromosomes), hormonal factors and growth factors but also specific genes, the homeotic genes[1] regulating the embryonic organisation plane.

The first development stages of the male and female genital apparatuses are similar and the external genitals of the two sexes resemble each other until the third month of development.

In order to facilitate the reader's understanding, we will successively study the internal phenomena and then the external phenomena before specifying what induces the said phenomena.

2.1 Internal Phenomena

The initial situation can be considered by examining the caudal end of the embryo on the 34th day of the development[2] (Fig. 2.1). This end is occupied by a cavity, the cloacal cavity or cloaca, covered by an entoblastic embryonic layer.[3] This cloaca is closed at its end by a didermic membrane, i.e. with double layer (endoderm lined with ectoderm), the cloacal membrane.

On the ventral surface of the caudal end, a growth can be observed: the cloacal tubercle consisting of mesoderm covered by ectoderm.

The development in the caudal direction of the partition initially separating the primitive intestine and the allantois, the urorectal fold (which itself consists of mesenchyme coated with endoderm), will divide the cloaca, in its cranial two-third, into two ducts: a ventral duct (the urogenital sinus) and a dorsal duct (the posterior intestine). The embryologists still refer to this fold as the "fold of Tourneux".[4] The cloacal tubercle is at the front of the urogenital sinus. The cranial half of this tubercle will become the infra-umbilical anterior abdominal wall (pelvic wall). As for its caudal half, it will be the site of complex phenomena leading to the formation of the urogenital sinus: it will be colonised by a budding of the ventral endodermal wall of the cloaca, the cloacal entoblastic plate, ("cloacal entoblastic lamina of Solère" for French authors). This lamina progresses towards the top of the cloacal tubercle, which it reaches around the 36th day of development (Fig. 2.2).

As of the 37th day (Fig. 2.3), apoptosis phenomena will occur within the cloacal entoblastic plate towards the top of the cloacal tubercle, with formation of vacuoles whose convergence will perform of a sort of tunnelling process. This hollowing phenomenon occurs within the cloacal entoblast, the ventral extension of the primitive urogenital sinus, which will be the phallic portion of this sinus. The urorectal fold, which has come closer to the perineum by progressing towards the cloacal tubercle, is at present referred to as the perineal urorectal septum. This septum will end up by merging with the remainder of the endodermal lamina,[5] as of the

[1] The homeotic genes are genes which determine the organisation plane of the individual: among these genes, there is the **Hox** gene, which determines the respective positions of the organs and the **sonic hedgehog** gene, which regulates morphogenesis.

[2] The embryonic stages are expressed in numbers of days or weeks of development (it is necessary to add 2 weeks or 14 days to each of these figures to obtain the number of weeks of amenorrhoea: thus the 6th week of development corresponds to the 8th week of amenorrhoea).

[3] The human embryo consists of 3 layers:

- The ectoblast (epiblast and neurectoblast), represented on the diagrams by a blue line
- The mesoblast, represented on the diagrams by a red line or red crosses
- The entoblast, represented on the diagrams by a yellow line.

[4] F. Tourneux is the author of an Atlas of Embryology concerning genito-urinary organs, published in 1892 in *Travaux et Mémoires des Facultés de Lille* (*Works and Memories of the Faculties of Lille*).

[5] Certain embryologists refer to converging paramedian folds, the Rathke folds, instead of the entoblastic lamina of Solere. These folds are linked at the Tourneux fold to partition the cloaca. We preferred the process generated by the development of the single entoblastic lamina, which is more compliant with our observations of embryos.

V. Di Marino, H. Lepidi, *Anatomic Study of the Clitoris and the Bulbo-Clitoral Organ*,
DOI 10.1007/978-3-319-04894-9_2,

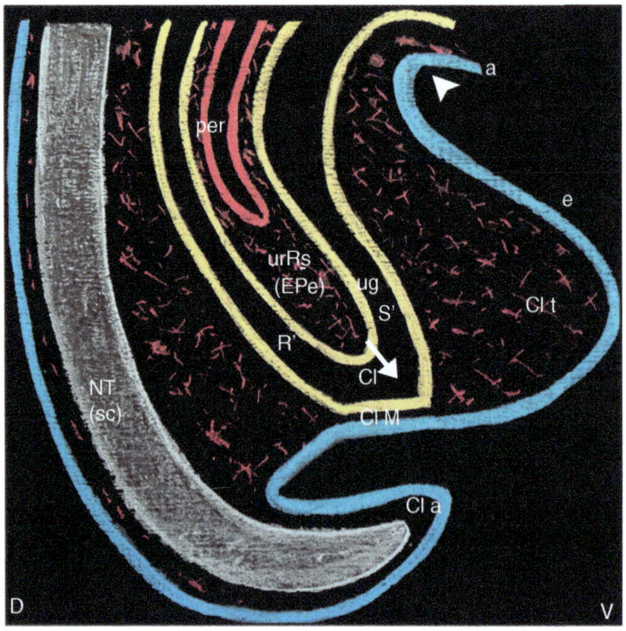

Fig. 2.1 Median sagittal section of the pelvic region (drawing of a micrograph female human fœtus; 34th day of development) (The ectoblast is drawn in *blue* (*a* amnios, *e* epiblast, *NT* neurectoblaste of the neural tube). The mesoblast is drawn in *red*. The entoblast is drawn in *yellow*). *D* dorsal, *V* ventral, *Cl* cloaca, *Cl a* cloacal appendix, *Cl M* cloacal membrane, *Cl t* cloacal tubercle, *NT* neural tube (*sc* spinal cord), *per* peritoneum, *R′* future rectum, *ugS′* future urogenital sinus, *urRs* urorectal septum (eperon perineal of French anatomists), *oblique white arrow* it shows the direction of urRs′ progress towards the cloacal membrane, *white arrowhead* it indicates the point where the epiblast follows the amnios (which covers the umbilical cord)

Fig. 2.2 Median sagittal section of the pelvic region (drawing of a micrograph female human fœtus; 36th day of development). *Cl* cloaca, *Clepl* cloacal entoblastic plate (lame entoblastique cloacale of French authors), *Cl M* cloacal membrane, *Cl t* cloacal tubercle, *urRs* urorectal septum, *oblique white arrow* it shows the direction of urRs′ progress towards the cloacal membrane

42nd day of development (Fig. 2.4), thus determining two ducts, the ventral urogenital duct and the dorsal intestinal duct, as well as two didermic membranes, i.e. a ventral membrane, the urogenital membrane and a dorsal membrane, the anal membrane. At the same time as the development of the urogenital duct, which has become the urogenital sinus, four ducts are also developing in the embryonic mesenchyme (mesoblast). They will end at this sinus and thus divide it into a cranial, urinary, vesicourethral section (future urinary bladder + future proximal urethra) and a caudal genital section: the pelvic section of the urogenital sinus. These four ducts are the two mesonephric ducts: Wolff channels and the two paramesonephric ducts (Muller channels).[6]

On the last day of the embryonic stage, i.e. on the 56th day, at the end of the 2nd month of development (Fig. 2.5), all is ready so that the last modifications can occur. Following these modifications, it will rapidly be possible to identify the sex of the foetus. The 56-day-old embryo shows, on the median sagittal section, two endodermal ducts: a dorsal duct corresponding to the posterior intestine and its caudal expansion, the rectum, and a ventral duct corresponding to the low urinary tracts with an expansion, which is the bladder, with an extension forming the three parts of the urogenital sinus, — from top to bottom, the future urethra (Sg-u1 portion of the urogenital sinus) whose external orifice will be located just above the vaginal orifice,[7] the Sg-u 2 portion (pelvic portion of the urogenital sinus) and the distal portion (phallic portion of the urogenital sinus). The two endodermal ducts are blind and closed by the anal membrane and the urogenital membrane, respectively, between which the future perineum will be positioned. The junction between the pelvic and phallic portions of the urogenital sinus comes in contact with the utero-vaginal channel of Mullerian origin, at the end of which an endodermal budding (sino-vaginal bulbs) has developed and which will grow hollow to form the distal third of the vagina.[8] Between the uterine body and the rectum, the peritoneal pouch (Douglas pouch) has inserted itself between the uterine body and the rectum, thus forming the bottom of the peritoneal

[6]The mesonephric ducts will form, in foetuses of both sexes, the vesical trigone and the ureters. Furthermore, they will develop into spermatic ducts in male foetuses. On the other hand, the paramesonephric ducts will form, in female foetuses, the two fallopian tubes, the uterus and the top part of the vagina (utero-vaginal channel) by distal fusion on the centre line (in male foetuses, these ducts will involute following the secretion of an "anti-Mullerian" hormone, secreted by the precursors of the Sertoli cells of the male gonad).

[7]In the male foetus, the pelvic portion of the urogenital sinus (Su-g 1) will become the prostatic urethra portion overlying the seminal colliculus (at the top of which ends the prostatic utricle, a Mullerian residue of the male sex!).

[8]This hollowing process will be nearly complete, except for the area of junction with the urogenital sinus, where the hymeneal membrane (which will become the hymen) persists.

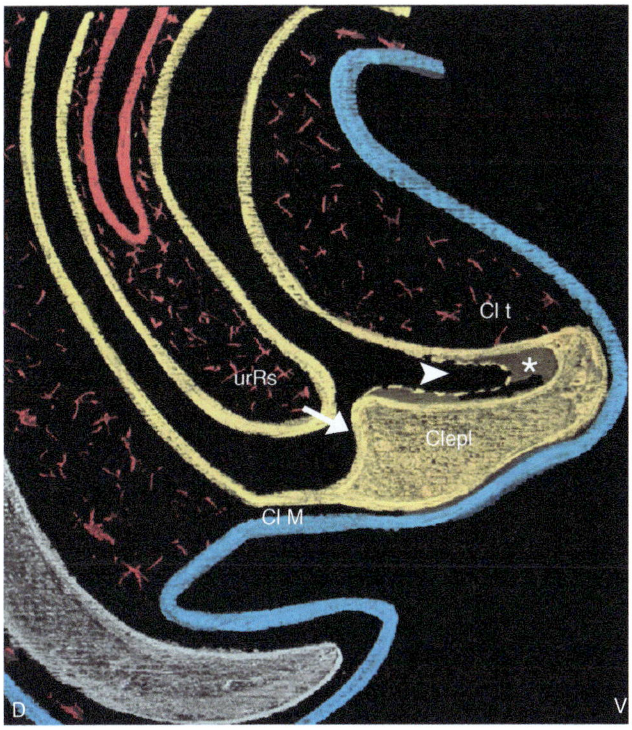

Fig. 2.3 Median sagittal section of the pelvic region (drawing of a micrograph female human fœtus; 40th day of development) (While the urRs is progressing toward the Clepl, which is digged (*asterisk*) toward the top of the cloacal tubercle (phenomenon explained by apoptosis and convergent vacuolisations)) *Clepl* cloacal entoblastic plate (lame entoblastique cloacale of French authors), *Cl M* cloacal membrane, *Cl t* cloacal tubercle, *urRs* urorectal septum, *Oblique white arrow* it shows the direction of urRs's progress toward the Clepl (observe that the urRs is quite in contact with the Clepl), *White arrowhead* it shows the digging of the Clepl which will result in the formation of the phallic part of the future urogenital sinus

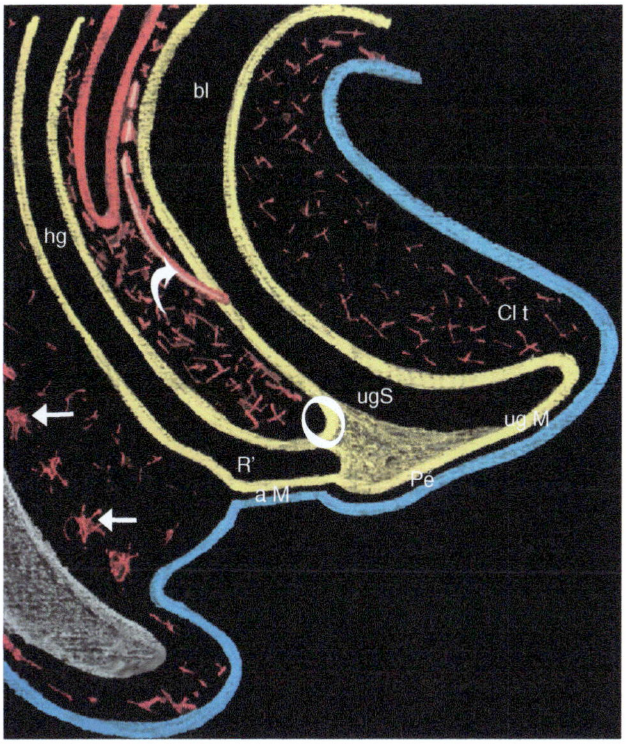

Fig. 2.4 Median sagittal section of the pelvic region (drawing of a micrograph female human fœtus; 42th day of development) (The fusion of the urRs with the Clepl (*white circle*), which occurred at the end of the 6th week of development, divides the primitive cloaca into two ducts: the urogenital sinus, ventral and the hindgut, dorsal. The primitive perineum separates the urogenital membrane from the anal). *a M* anal membrane, *bl* bladder, *Cl t* cloacal tubercle, *hg* hindgut, *Pe* perineum, *ug M* urogenital membrane, *ug S* urogenital sinus, *R′* future rectum, *curved white arrow* it shows one of the two paramesonephric ducts opening into the urogenital sinus, *white right arrows* mesenchymatous condensations which will form the vertebras

pouch: the utero-rectal pouch. At the same time and as of the 7th week, the top of the old perineal tubercle has been modelled so as to form a spherical end: the genital tubercle located in front of the distal end of the phallic urethra. It is also at this stage that are now distinguished, from a ventral viewpoint, the pubic bones and the pubic symphysis, which have developed from the pre-sinus mesenchyme, and, from a dorsal viewpoint, the sacral and coccygeal vertebrae, which have developed from the dorsal mesenchyme and which surround the distal part of the neural tube (distal end of the spinal cord). As from the 1st day of the 3rd month of development, the embryo becomes a foetus. It is still not possible to externally identify if the foetus is male or female.

2.2 External Phenomena (Fig. 2.8)

At the beginning of the 6th week of development, epithelial folds have developed: they are the cloacal folds. They border the cloacal membrane on either side. Further outwards, other folds have formed: the labio-scrotal swellings.

The partitioning of the cloaca will cause the scission of the cloacal folds so that, at the end of the 6th week, two urogenital folds bordering the urogenital membrane and two anal folds bordering the anal membrane are to be observed. All these folds are ectodermal. It is to be noted that these two types of folds are separated by the superficial part of the perineal formation or primitive perineum, resulting from the posterior apposition of the urogenital folds and from the coalescence of the posterior section (dorsal) of the labio-scrotal swelling.

As observed previously, it is during the 7th week that an epithelial growth appears on the old cloacal tubercle: the genital tubercle or primitive phallus, which will become, via modelling morphogenesis phenomena, towards the end of the 2nd month, the glans clitoridis in women and penian glans in men. During the formation of the genital tubercle, in a region above the latter, the following is observed: a double cartilaginous draft of mesenchymal origin (two pubic bones connected by a pubic symphysis) and, at a higher level, the formation of the infra-umbilical ventral wall.

The perineum, which is now well represented, increases and progresses, in continuity with the mesenchymal

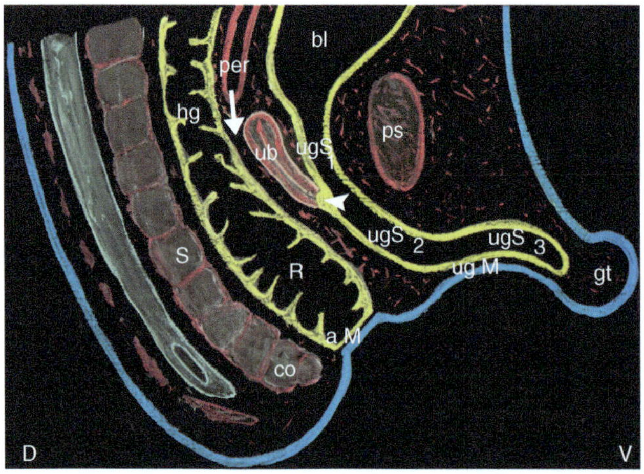

Fig. 2.5 Median sagittal section of the pelvic region (drawing of a micrograph female human fœtus; 56th day of development) (The uterus body and the superior part of the vagina are formed by the fusion of the two paramesonephric ducts (ducts of Muller) before opening into the primitive urogenital sinus. At this stage (end of the 2nd month), it is not possible to specify the embryonic sex). *a M* anal membrane, *bl* bladder, *co* coccyx, *gt* genital tubercle, *hg* hindgut, *per* peritoneum, *ps* pubic symphysis, *R* rectum, *S* sacrum, *ub* uterus body, *ug M* urogenital membrane, *ugS1* urethral part of the urogenital sinus (future urethra), *ugS2* pelvic part of the urogenital sinus, *ugS3* phallic part of the urogenital sinus, *right white arrow* it shows the descent of the peritoneal pouch which will become the recto-uterine pouch, *white arrowhead* it shows the sino-vaginal bulbs which form the distal part of the vagina (the breaking down of their epithelium will connect the utero-vaginal canal with the urogenital sinus)

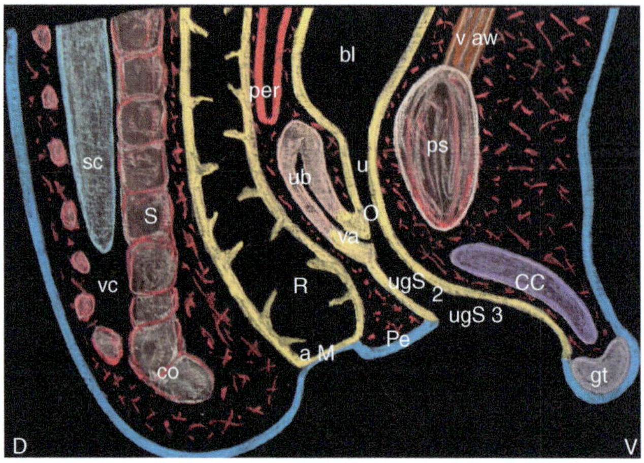

Fig. 2.6 Median sagittal section of the pelvic region (drawing of a micrograph female human fœtus; 63th day of development) (During the 9th week of development: The erectile bodies develop in the clitoral "draft". The urogenital membrane disintegrates opening the phallic part of the urogenital sinus to the exterior and finally creating the vulvar cleft, (no secondary closure happening for the female fœtus)). *a M* anal membrane, *bl* bladder, *CC* corpus cavernosum clitoridis, *co* coccyx, *gt* genital tubercle, *Pe* perineum, *per* peritoneum (recto-uterine pouch), *ps* pubic symphysis, *R* rectum, *S* sacrum, *sc* spinal cord, *u* urethra, *ub* uterus body, *ugS2* pelvic part of the urogenital sinus, *ugS3* phallic part of the urogenital sinus, *v aw* ventral abdominal wall, *va* vagina, *vc* vertebral canal, *white circle* external urethral orifice

partitions, which perfectly separate the genito-urinary and anorectal compartments.

At the start of the 9th week, the ventral growth of the genital tubercle will stretch the urogenital membrane. This membrane will break up a few days later, with the first rupture occurring at its anterior end. The anal membrane remains, but a small dip in its ectoderm appears and corresponds to the start of the formation of the future ectodermal anal canal.

From now, the anal membrane separates the ectodermal and endodermal parts of the anal canal. At the end of the 9th week, 63rd day (Figs. 2.6 and 2.7), the urogenital membrane has almost completely disappeared,[9] thus exposing the endodermal ceiling of the phallic part of the urogenital sinus, in the male foetus as well as in the female foetus.[10] However, major modifications will occur at the level of the external apparatus according to the sex of the individual: while in a male, the groove will secondarily close again and form the penian urethral cylinder through the fusion of the urethral folds and the bursa or scrotum will form through the fusion of the labio-scrotal swellings on the centre line,[11] the groove will persist in the female sex, thus opening the vestibule onto the vulvar cleft with, on either side, the labio-scrotal swellings, which are still separate and will become the labia majora (Figs. 2.8 and 2.9). The pelvic portion of the female urogenital sinus (Su-g 2) now becomes the vestibule, into which emerges, at its cranial part, the external orifice of the urethra and directly below, the external orifice of the vagina. The genital tubercle protrudes below the ventral commissure of the labia. Inside of the labia majora, the labia minora, which are connected to the bottom surface of the genital tubercle, are hardly visible (Fig. 2.9). The epithelium which covers their medial surface comes, such as the vestibule, from the phallic portion of the urogenital sinus. It is thus of endodermal origin.

The erectile bodies are already visible at this stage: They developed in the mesenchyme of the old cloacal tubercle. These are, first and foremost, the two corpora cavernosa whose attachments to each descending branch of the pubes (still cartilaginous) are quite visible (transverse sections:

[9]The anal membrane will only disappear during the 10th week.

[10]It is not possible to specify the sex of a foetus until the 3rd month because the external genitals are similar until this date of development. This is why it is only during the second echography (morphological echography of the 2nd quarter), performed between 18th and the 24th week of amenorrhoea (16 to 22 weeks of development), that it is possible to inform a mother of the sex of the child she is carrying, if she wishes to know.

[11]The closure of the penile cylinder and that of the bursa with the formation of their median raphe will only be completed during the 14th week of development.

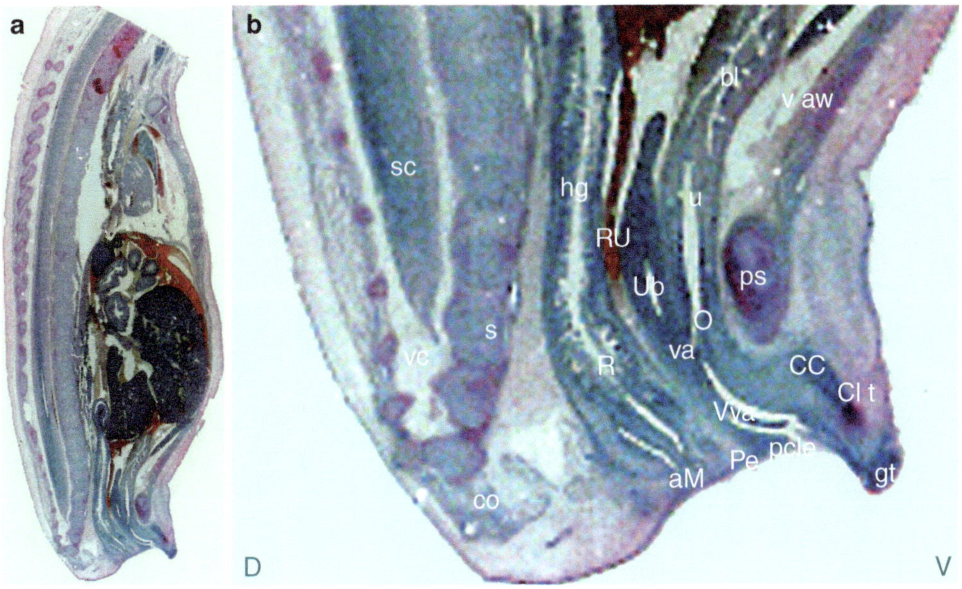

Fig. 2.7 (a) Photomicrograph of a sagittal section of a female human foetus 53 mm, (b) Photomicrograph of the pelvic region (magnification of the caudal end: x13), (Compare this sagittal section of a female human foetus with the picture of the 63th day of development). Note the remainders of the urogenital membrane, still present. As the anal membrane, it will break down at the end of the 10th week, establishing a continuity between the anal pit and the rectal part of the hindgut). *D* dorsal, *V* ventral, *aM* anal membrane, *bl* bladder, *CC* corpus cavernosum, *Cl t* cloacal tubercle, *co* coccyx, *gt* genital tubercle, *hg* hingut, *pcle* pudendal cleft, *Pe* perineum, *ps* pubic symphysis, *R* rectum, *RU* recto-uterine pouch, *s* sacrum, *sc* spinal cord, *u* urethra, *Ub* uterus body, *v aw* ventral abdominal wall, *va* vagina, *vc* vertebral canal, *Vva* vestibule of the vagina, *white circle* around the external urethral orifice

Fig. 2.10). These corpora cavernosa follow a development comparable to that of the male foetus. They have extended towards the genital tubercle and progress parallel to the cranial part of the genital formation (Figs. 2.7 and 2.11). At their end, they come closer in a V shape with the top pointing towards the genital tubercle (Fig. 2.12). The two spongious bodies initially develop in the mesenchyme, which surrounds the vulvar cleft and in the mesenchyme which occupies the genital tubercle (Fig. 2.10). At the level of the glans, the connections of this spongious body with the two corpora cavernosa (in preparation of the future disposition in adults) are perfectly visible. However, the development of these female spongious bodies is very particular. Due to the absence of a "penile" urethral cylinder, these spongious bodies have no periurethral future and remain separated by the vestibule, thus forming vestibular bulbs on the right and left. Moreover, while, in the male foetus, the urethral corpus spongiosum connects to that of the glans to form a single spongy flow, in the female foetus, the distal extension of the perivestibular spongy formation, despite the phallic urethra, which acts as a support stake, involutes and provides what is referred to as the "residual spongy part" without any significant erectile capacity.

The following external stages will perfect the development of the external apparatus: They correspond to the formation of the preputial clitoral hood and the modelling of the clitoris.

The formation of the preputial clitoral hood is similar to that of the male penile prepuce.[12] It starts as of the 8th week of development. A slight lamellar proliferation, referred to as the ectoblastic preputial plate, penetrating more or less deeply, from the surface epithelium, develops at the ectoblastic epithelial layer, which covers the genital tubercle (Fig. 2.12). Vacuoles appear in this lamina following apoptosis phenomena.[13] The convergence of these vacuoles creates a space around the glans, which ends up separating, on the periphery, the glans and the portion of the prepuce which envelopes it.[14] The preputial chamber of the glans clitoridis is thus created. It is bordered by the same epithelium

[12] In the male fœtus, the formation of the prepuce is moreover related to the development of the terminal urethra (at the level of the glans).

The preputial plate is the origin not only of the formation of the prepuce but also of that of the glandular urethral cylinder and the small navicular inside the glans: It has invaginated in the depth of the latter from its top and up to the level where the primitive penile urethra ends. The balanic (or glandular) urethra (with its small navicular), thus created, is placed in communication with the phallic urethra, resulting from the closure of Su-g3 to form the final urethra, which thus emerges at the top of the penis.

[13] For other authors, the preputial chamber does not result from apoptosis phenomena but from a simple invagination of the epithelium of the glans surface.

[14] The separation of the ectoblastic preputial plate is often incomplete. It then forms, for the glans, a free part and a buried, and thus hidden, part.

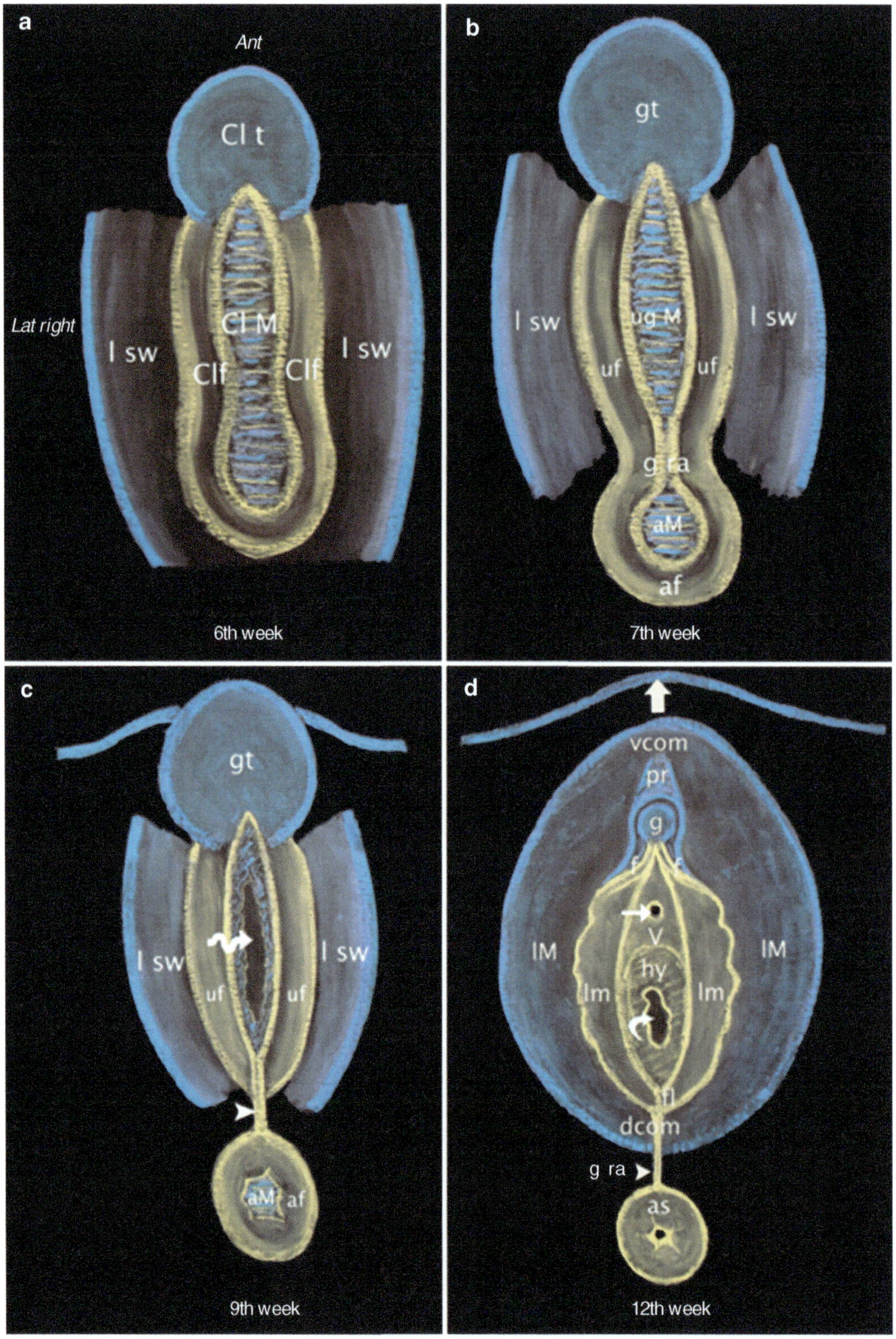

Fig. 2.8 Differentiation of the external genitalia of the female human foetus. (**a**) 6th week: the cloacal membrane closes up the cloaca (caudal part of the hindgut); (**b**) 7th week: ventrally, the genital tubercle appears at the top of the cloacal tubercle. Dorsally, a partial fusion of the two urethral folds (creation of the genital raphe) separates the cloacal membrane into two parts, urogenital (ventral) and anal (dorsal). (**c**) 9th week: the urogenital membrane disintegrates, opening the phallic part of the urogenital sinus to the exterior. (**d**) 12th week: the urethral folds have become the labia minora, the labial swelling has become the labia majora and the hymen hides the vaginal orifice. The clitoris is in a shaping phase and the prepuce is already formed; *af* anal fold, *aM* anal membrane, *as* anal sphincter, *Cl M* cloacal membrane, *Clf* cloacal fold, *Cl t* cloacal tubercle, *dcom* dorsal commissure, *f* frenulum, *fl* frenum of the labia minora, *g* glans, *g ra* genital raphe, *gt* genital tubercle (it appears at the 7th week, on the top of the cloacal tubercle), *hy* hymen, *lm* labium minus, *lM* labium majus, *l sw* labial swelling, *pr* prepuce, *uf* urethral fold, *ug M* urogenital membrane, *V* vestibule, *vcom* ventral commissure, *white arrowhead* perineal raphe, *thick white right arrow* mons pubis, *fine right arrow* external urethral orifice, *curved white arrow* hymeneal ostium, *waved arrow* disintegrating urogenital membrane

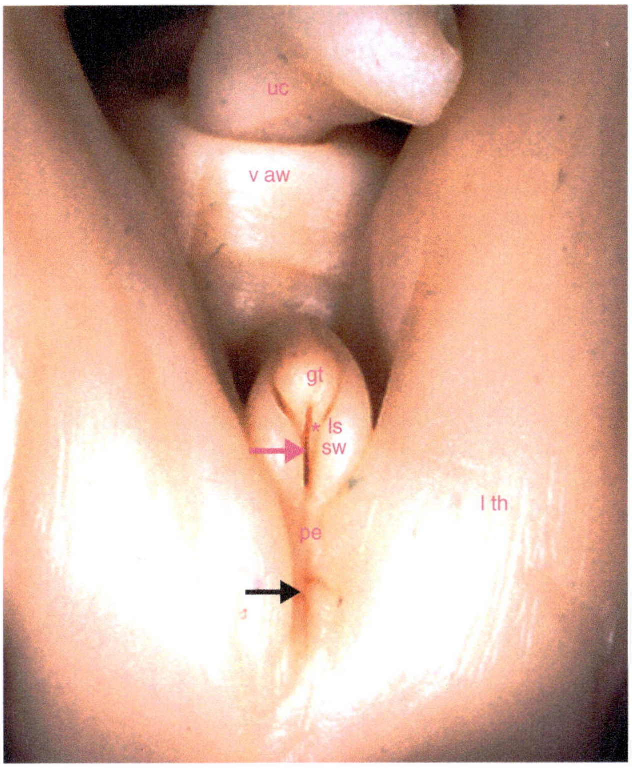

Fig. 2.9 Perineal view of the external genitalia of a female human fœtus (beginning of the 11th week of development). The fœtus is in gynaecological position. *gt* genital tubercle, *ls sw* left labio-scrotal swelling (future left greater lip), *l th* left thigh (posterior face), *pe* superficial perineum, *uc* umbilical cord, *v aw* ventral abdominal wall (infra-umbilical part), *magenta asterisk* left urethral fold (future left lesser lip), *magenta arrow* urogenital (or urethral) groove (future vulvar cleft), *black arrow* anus (the anal membrane is not completely disintegrated). At this stage, it is still difficult to know by a basic examination of the external genitalia, which is the sex of the foetus

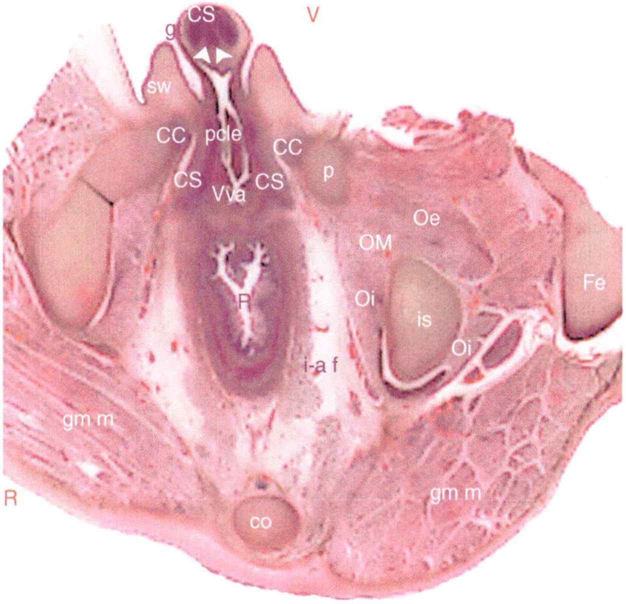

Fig. 2.10 Transverse section of the pelvic region (photomicrograph) female human fœtus; 11th week of the development. *R* right, *V* ventral, *CC* corpus cavernosum, *CS* corpus spongiosum, *co* coccyx, *Fe* femur, *gm m* gluteus maximus muscle, *gt* genital tubercle, *i-a f* ischio-anal fossa, *is* ischium, *l sw* labial swelling, *Oe* obturator externus muscle, *Oi* obturator internus muscle, *OM* obturator membrane, *p* pubis, *pcle* pudendal cleft, *R* rectum, *Vva* vestibule of vagina, *white arrowheads* they locate the two corpora cavernosa converging towards the future glans clitoridis, behind the corpus spongiosum. Observe: The corpora spongiosa, on either side of the vulva. They will become the bulbs. The corpora cavernosa (crura clitoridis), attached to the ischio-pubic rami

corresponding to the internal layer of the hood and, on the other side, to the surface covering of the glans.

The clitoris will also undergo a reduction, thus modelling its shape and reducing its size (which makes us forget that at certain stages of its development, it was longer and more voluminous than the penis at same stage). Thus, in a 4-month-old female foetus (Fig. 2.13), and in relation to a normal development,[15] the clitoris will have already acquired its final aspect: minuscule formation covered by the prepuce and whose end, the glans, is hidden under the hood.

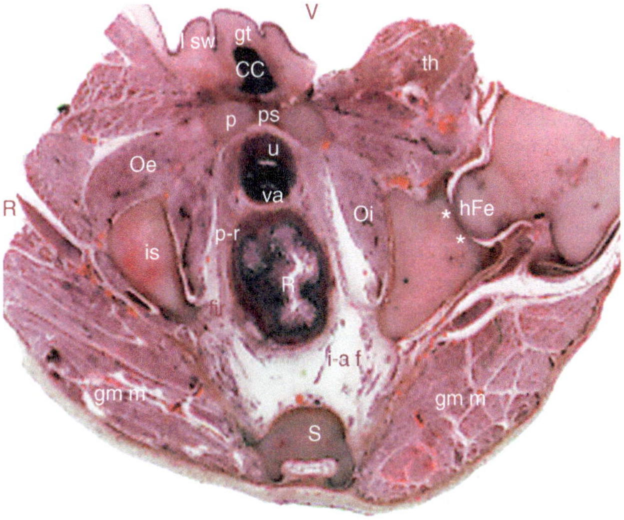

Fig. 2.11 Transverse section of the pelvic region (photomicrograph) female human fœtus; 11th week of development (section at the level of the middle third of the pubic symphysis). *R* right, *V* ventral, *CC* corpora cavernosa, *gm m* gluteus maximus muscle, *gt* genital tubercle, *hFe* head of femur, *i-a f* ischio-anal fossa, *is* ischium, *l sw* labial swelling, *Oe* obturator externus muscle, *Oi* obturator internus muscle, *p* pubis, *pr* pubo-rectalis (bundle of levator ani muscle), *ps* pubic symphysis, *R* rectum, *S* sacrum, *u* urethra, *th* thigh, *va* vagina, **acetabulum. The present section, located above the previous one, is at the level of the pubic symphysis. It concerns, on the *left side*, the hip joint

[15] Very rare cases in which there is no clitoris have been reported in literature. It seems that, in certain young girls presenting this "agenesis", the development of the embryonic clitoris actually did occur but that then an apoptosis gene intervened and caused its disappearance (gene comparable to the Bmp4 gene whose activation causes the disappearance of the penis in many birds including the cockerel).

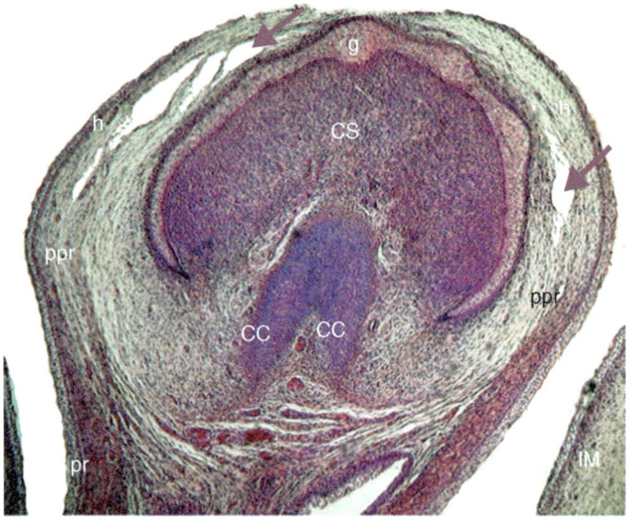

Fig. 2.12 Transverse section of the genital tubercle (photomicrograph showing the formation of the hood on a 43mm female human fœtus). *CC* corpus cavernosum, *CS* corpus spongiosum occupying the top of the genital tubercle, *g* top of the future glans, *h* future hood, *lM* labium majus (greater lip), *ppr* epithelial preputial plate (preputial lamina), *pr* prepuce, *mauve arrows* they show the vacuoles resulting from apoptosis phenomena

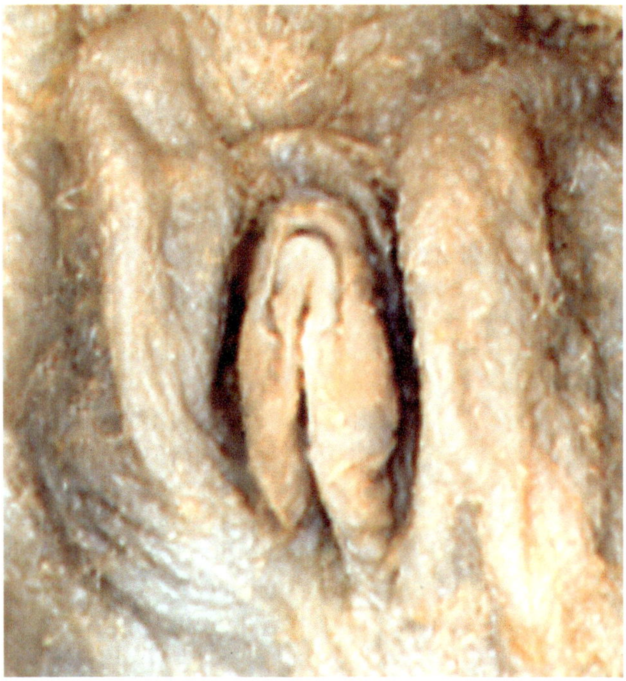

Fig. 2.13 Female human fœtus at the end of the fourth month (18 weeks of amenorrhea). Macrophotography of the external genitalia Biometry of the fœtus; length vertex-coccyx: 14 cm. Weight: 220 g. All structures (greater and lesser lips, clitoris, frenula, commissures) are in place, normally developed. The genital tubercle declined, turning into clitoral glans. Compare with the genitalia of the fœtus at the 11th week

2.3 Determinism of Sexual Differentiation and Related Pathologies

We shall now examine how the sex of an embryo is determined and how it can be affected by pathological phenomena.

2.3.1 Basic Concepts

It is known that each somatic cell of each individual contains 23 pairs of chromosomes including 22 pairs of autosomes (44 autosomes) and a pair of sexual chromosomes or gonosomes (2 gonosomes). It is the latter which determine the sex that the embryo, and then the foetus, will have. If the two gonosomes are XX, the sex of the foetus will be female. If the two gonosomes are XY, the sex of the foetus will be male.[16] As regards the sexual cells, i.e. the gametes (ovocytes in the ovaries and spermatozoa in the testes), there are only 23 chromosomes (one of each pair). The pairs are reformed in the egg cell (once the ovocyte has been fertilised by the spermatozoon).

2.3.2 Modern Concepts

It was necessary to wait for the second half of the twentieth century (1959) for the determinants of masculinity and femininity to be specified. Geneticists (especially P.A. Jacobs and J.A. Strong) managed to establish in an absolute manner that sexual determinism is related to gonosome Y: If this sexual chromosome is present, the foetus will be male. If it is absent, the foetus will be female.[17] Seven years later, more progress (work performed by P.A. Jacobs and A. Ross) has been made and made it possible to locate the area determining the sex, in the short arm of chromosome Y. Finally, since 1987, the progress made in genetics (work performed by D.C. Page et al.) has made it possible to specify the area carrying the masculinity gene: this gene is referred to as gene SRY: the Sex-Determining Region of the Y chromosome is located in a small area of the short arm of Y, region 1 (work of M.S. Palmer et al. in 1989 and A.H. Sinclair et al. in 1990).

SRY is a trigger for genetic phenomena, which will generate, from the undifferentiated gonad, not only the forma-

[16]This could be affirmed by the study of two subjects with a chromosomal anomaly: the Turner syndrome (karyotype: 45, X and female phenotype) and the Klinefelter syndrome (karyotype: 47, XXY and male phenotype).

[17]Beyond this simplified diagram, it is also necessary to recall the major role of the hormones produced by the autosomes!

tion of the testes and the secretion of the male hormone but also the secretion of the anti-Mullerian hormone, whose role is to cause the degeneration of the paramesonephric ducts of Müller. In the absence of the SRY gene, the ovaries develop and female hormones are produced.[18]

Later work showed that, in spite of its significance, the SRY gene is not alone and that there actually exists a gene cascade involved in sexual determinism.[19]

2.3.3 Pathologies of Sexual Differentiation

They can be of genetic or hormonal origin.

- Pathologies of **genetic origin** are the consequence of gonosome anomalies. Firstly, they include the **Klinefelter** and **Turner** syndromes (see below).

  This also concerns cases of **true hermaphroditism**, i.e. cases in which gonads have testicular tissue and ovarian tissues (the standard case being uni- or bilateral the "ovotestis"). These cases are not very frequent. Karyotype anomalies are variable and all is visible, including "mosaic" cases with, in the same individual, cells comprising a Y chromosome and cells which do not contain this chromosome (karyotype example: 46, XX/46, XY). The phenotype is often interpreted at birth as male due to the presence of a penis.
- Pathologies of **hormonal origin** are due to a deficit in androgens (DHT or dihydrotestosterone[20] and testosterone) or to anomalies affecting the receptors for androgens (feminising testicle), as regards male subjects. On the other hand, it is an early exposure to androgens[21] which is responsible for pathologies in women. These hormonal anomalies are the cause of pseudo-hermaphroditism cases:
  - Male pseudo-hermaphroditism: normal male karyotype (46, XY), normal gonads (normal testes but often ectopic testes), **ambiguity of the genital tracts and external genitals** (hypospadias, at a minimum, especially deficits in DHT; external female genitalia, especially in feminising testicles!)
  - Female pseudo-hermaphroditism: normal female karyotype (46, XX), normal gonads (ovaries), **ambiguity of the genital tracts and external genitals**. All cases are possible, from the **peniform clitoris** associated to hypospadias and an empty scrotum to a male aspect of the external genitals but with however no testes in the bursa.[22]

2.3.4 The Revenge of Chromosome X

The most recent discoveries in the field of genetics have modulated the rigorous diagram referred to previously and the prominence of the SRY gene in the world of mammals (placental and marsupial mammals). Researchers (F Veyrunes, CNRS, et al., Institute of Genomics of the IRD) have discovered (December 2009), by studying the African pygmy mouse, Mus minutoides (close to the domestic mouse), that the Y chromosome does not always define the sex.

First of all, these researchers have noted, in the population of the above-mentioned mice, a majority of fertile females carrying XY gonosomes.

In order to understand, they then studied the "normal" females of this population, i.e. females carrying standard XX gonosomes. They then discovered that these two X sexual chromosomes could be differentiated morphologically and decided to call them (X) and (X*). They then turned their attention back to studying females carrying a Y chromosome and observed that the gonosomes of these females were in fact X*Y and not XY. The molecular and cytogenetic analyses, which they carried out at the same time on the Y chromosome of these females, were used to affirm that there was no mutation on the SRY gene and that there was no anomaly on this gene which could explain its non-expression. It must then be concluded that the chromosome (X*) carries a **mutation**, which is still unknown and which is **capable of inhibiting the expression of the powerful SRY and thus cause a sex reversion**. Extraordinary options are therefore now open to research and researchers are certainly in store for new surprises.

---

[18] In certain relatively rare cases, it is possible for male subjects to have an XX gonosome. This is explained by an abnormal crossing over which, at the time of the meiosis, makes it possible for an X chromosome to incorporate a fragment of the short arm of the Y chromosome, bearing the region where the SRY gene is located.

[19] The following have already been identified: SF-1 and WT1 expressed upstream of SRY and necessary for the development of the undifferentiated primitive gonad; DAX1 necessary for ovarian development; SOX9 necessary, with SRY, for testicular development; and SF-1 (steroidogenic factor-1) necessary for the secretion of testosterone by the Leydig cells of the testis and for the secretion of the anti-Mullerian hormone by the Sertoli cells of this testicle. Other genes are still being identified!

[20] Deficiencies in DHT are generally due to a mutation of an enzyme, 5-alpha-reductase.

[21] This early exposure can be due to:

- A tumour of the mother's adrenal gland
- Foetal congenital adrenal hyperplasia
- Hormonal treatment with androgens during the pregnancy.

[22] We can imagine the serious civil status mistakes resulting from the male aspect of a newborn baby!

3 Terminology and Related Difficulties

Few organs have involved as many difficulties in relation to their designation and name than the bulbo-clitoral organ and its components. The first difficulties appeared when referring to the clitoris and then the bulbs, within the external female genitalia, that previous anatomists had named, without any hesitation, "shameful parts of the woman". However, later on, equally significant difficulties arose when anatomists finally understood that it was necessary to integrate the female cavernous and spongy bodies into a single organ and that a name should be given to the organ in question.

3.1 The First Names

One of the first names was the **woman's penis** (adopted by Latin scientists: virga vel penis muliebris). Then Hippocrates[1] provided his own names: **columella** (or small pillar) or the extremely imaged **uvula**.[2] Aristotle, for his part, called it "**coles feminarum**" (quotation from A. Portal). Aetius and Paul from Aïgina, considering that "this small formation is hidden under the labia like young brides under their veils" called it the **nymph**.[3] Avicenna gave it the name of **al bathara** (or **el bathr**), i.e. penis. Albucasis, closer to physiology, was to call it "**tentigo**" (which is placed under tension) but also "**softness of love**", a term that Colombo was going to reuse and claim as his own a few centuries later. For the Romans, and if we are to believe Rufus of Ephesus (see Chap. 1), born in c.98, who wrote a treatise entitled "On the Names of the Parts of the Human Body", the "woman's shameful parts" chapter contains the definition and the various names still used during his era in relation to the clitoris: it actually is a "**muscular wattle**, which hangs in the middle of the opening of the cleft". It is generally referred to by means of three terms, **numoé** (which means "the veiled component"), **murton** (i.e. the bilberry)[4] or **hypodermis** (the organ which is under the skin).

However, all the terms, which we have just seen, slowly disappeared and only two names remained, one for the Latin world, **landica**, and the other for the Greek world, **clitoris**.

3.2 The Two Major Names

3.2.1 Landica

This term (landica, ae) seems to have appeared relatively late, towards the end of the second century, and to have initially been part of "good classical Latin" (E. W. Fay). However, very quickly, this term was regarded as obscene (for reasons which we are unaware of) and was rarely used in the current language.

It can thus be found on drawings representing the Priapea (78.5.), a collection of obscene poems dedicated to Priape, the phallic god of fertility. It was even engraved on a sling-shot projectile during the siege of Perugia (Perusia in Latin), as the attackers and besieged exchanged stones on which were engraved obscene insults (fulmen peto landicam Fulviae/culum Octavia)![5]

More interesting still, the term "landica" was written by Cicéron. In a famous letter, this author refers to word games which can be made, voluntarily or not, by associating perfectly innocent syllables and, which when combined, are pronounced like an obscene word,[6] in this case "landica".

[1]Hippocrates, Lib. II, de naturà muliebris (quoted by A. Portal).

[2]According to Galen, quoted by J. Riolan, the name employed by Hippocrates explains the role of the clitoris: "to protect the matrix against the cold: "The uvula is for the pharynx. It (the clitoris) plays the same role as for the matrix, as it covers the orifice of the matrix and closes, by the same means, the passage of cold things, which could affect it".

[3]Aetius calls (which is very appropriate according to J. Riolan) the labia minora "the winged parts".

[4]For the Romans, the myrtle is related to Venus, such as laurels are related to Apollo, the oak to Jupiter and the olive tree to Minerva.

[5]Hallett J.P.: Fulvia, Mother of Iullus Antonius, new approaches to the sources on Julia's adultery at Rome, Helios, 2006, 33, No. 2.

[6]Cicero, Epistulae ad familiares 9.22 (ad Paetum amicum):

(Memini in senatu disertum consularem ita eloqui: "Hanc culpam maiorem an il**lam dicam**?" Potuit obscenius? "Non", inquis, "non enim ita sensit").

V. Di Marino, H. Lepidi, *Anatomic Study of the Clitoris and the Bulbo-Clitoral Organ*,
DOI 10.1007/978-3-319-04894-9_3, © Springer International Publishing Switzerland 2014

The term "landica" is also found in the master work of Soranos from Ephesus concerning "the diseases of women", the first actual treatise on gynaecology.

Landica gave rise to the adjective "landicosa", which was also rarely used and meant "who has a large clitoris".[7]

But where does the word landica come? Few answers concern this etymology. There is, however, one sentence, which seems interesting and which was provided by E.W. Fay: the origin of the term "landica" could in fact be the term "glandica", derived from the glans referring to the distal end of the male penis. The "g" would have been progressively lost due to the pronunciation difficulty for Latin people (their tongue retains the g), which has led to the disappearance of this letter.

However, as it has not been greatly used, the "landica" of the Romans rapidly gave way to the Greek word "clitoris", which successive authors had widely introduced to Rome, such as testified, according to J. Riolan, by the texts written by Rufus of Ephesus, Pollux and Suidas. On the other hand, the term "landica" was still used in languages of Roman origin, including the old French language ("old French") or Romanian. Thus, in "old French", landica became landie or landye. And quite naturally in the historical dictionary of the old French language, which is the glossary of the French language from its origin until the century of Louis XIV, a definition for landie is provided: natural parts of the woman (thus with an extension of the term, which now includes the entire external genitalia). On the other hand, there is no trace of the term "clitoris" in this dictionary, although this word had previously invaded the Latin language.

3.2.2 Clitoris

The etymology of the word "clitoris" is still prone to discussion. However, there are numerous other possibilities even if they remain uncertain. It should also be noted that the term has often been written in different ways: **kleitoris** for Rufus, **kletoris** (for Pollux, during the second century), **klitoris** for Hesychius (during the sixth century) and finally **clitoris** for Suidas (during the eleventh century).

The term "clitoris" only appeared in France during the seventeenth century. It appeared in the dictionary of R. Cotgrave in 1611 (p. 203)[8] with the following definition: "A woman's Priuities" (or a woman's "priuities"). It was soon going to spread in all of Europe, especially via Italian and English anatomists, and as a result make the words derived from "landica" disappear forever.

[7]"Laxa et landicosa": inscription written on the house of a prostitute from Pompeî (undoubtedly written by a malevolent neighbour!).

[8]"Landie" and "landies" have also been included in Cotgrave's dictionary (p.570).

Among the multiple origins which have been proposed for the etymology of the term "clitoris", we will retain the following:

- Derived from the verb cleitorizein or kleitorizein, which means touch and tickle or titillate lasciviously
- Derived from Kleitor, a city founded by Clitor and which had a famous fountain whose water tasted like wine
- Derived from Kleo, verb meaning "to celebrate, praise" or from kleitos, adjective in the richest senses: glorious, prominent, famous and superb
- Derived from Kleitor or Klitora, city of Arcadia (mountainous region in the South of Greece) located on a hill and acting as a lock for several valleys
- Derived from Klitoris, a dark coloured stone, which can be found on the banks of the Indus (Plutarch also used this term to refer to a black stone, which can be found on Mount Lilée)
- Derived from Clitoris, a legendary daughter of Myrmidon
- Derived from Kleio, verb meaning "I close", or from kleis, name meaning "what is used to close (key or lock)"
- Derived from Kleitus, which means side or slope of a mountain or hill, or even a small hill

Linguists seem to want to adopt the last two etymologies (especially the Kleio etymology, which appears in several dictionaries[9]). However, the discussion remains open, especially as certain eminent specialists, such as M. Cohen[10] have found similarities between the term clitoris and word formulations belonging to the living language of modern Ethiopia.

The terminology induced from "clitoris" is also interesting. The following can be mentioned:

Clitoral	which refers to the clitoris
Clitorism	extended and painful erection (pathological) of the clitoris (it is identical to priapism in men)
Clitorise to	lasciviously touch the clitoris
Clitoridectomy	resection of part or all of the clitoris
Clitoroplasty	plastic surgery or repair surgery of the clitoris
Clitoromegaly	large-sized clitoris
Clitoresque[11]	which evokes the clitoris
Acomoclitism[12]	fetishism related to hairless pubes

[9]e.g.: JB Morin and Ansse de Villoison: etymological dictionary of French words derived from Greek (1809).

[10]M. Cohen: The mysterious origins of the word "clitoris" in The classic clitoris; historic contributions to scientific sexuality, Th. Power, Lowry. Publisher: Nelson hall nh, Chicago.

[11]see: "Possible words and existing words, the case of derivatives in esque", M. Plenet, ERSS, Univ. Toulouse Le Mirail.

[12](**a** private + **come**, hair + **clit**, from clitoris) thus literally: "clitoris without hair"!

3.3 The Word "Clitoris": Evolution of Its Definition Through the Centuries

- The first dictionary citing the word "clitoris" was that of R. Cotgrave,[13] A Dictionnarie of the French and English tongues (1611) which defined so "A woman's Priuities". That definition was included in the Godefroy's dictionary, "Dictionnaire de l'ancienne langue française et de tous ses dialectes du IXème au XVème siècle" (1881–1902),[14] but with a little modification: "A woman's Priuities" becoming "A womans privities", more comprehensible!

  Godefroy gives also his own definition: "Petit organe charnu à l'entrée de la vulve" (small fleshy organ at the entrance of the vulva).
- In the dictionary of P. Richelet, "Dictionnaire françois" (1680),[15] the definition is intended anatomical: "clitoris, Terme d'Anatomie: chair qui est en haut et entre les lèvres de la matrice" (anatomic word: flesh that is above and between the lips of the matrix).
- It is the same for the "Dictionnaire Universel" (Universal Dictionary) of A. Furetière (1690),[16] the author based on findings of G. Fallope (see Chap. 1): "Clitoris, Terme d'Anatomie: C'est un nom que Fallope a donné à une petite caroncule qui est au devant de la vulve. Elle a 2 ligaments et quatre petits muscles et une glande couverte d'une peau déliée, comme d'un prépuce. Quelques-uns l'appellent verge féminine. On la retranche quelquefois par opération de Chirurgie quand elle sort trop en dehors". (Clitoris : Anatomical term. It is a name that Fallope gave to a small wattle in front of the vulva. She as two ligaments and four small muscles and a gland covered by an untied skin like a foreskin. Some call it, female penis. It is sometimes subtracted by operation surgery when she goes outside too").
- In "Le Dictionnaire des Arts et des Sciences" of Th. Corneille (1694)[17] (The Dictionary of Arts and Sciences), copy of the two previous definitions is clear: "Clitoris, Terme d'Anatomie. Petite caroncule qui est au haut et entre les lèvres de la matrice. Elle a 2 ligaments, quatre petits muscles, et une glande couverte d'une peau fort déliée. Ce mot vient du grec kintoron qui signifie en latin pudenda mulieris" (Clitoris, Anatomical term. Small wattle that is above and between the lips of the matrix. She as two ligaments, four small muscles and a gland covered by a very untied skin. This word originates from Greek "kintoron" which means "Pudenda mulieris"). Note that "kintoron" does not appear in Greek language! As against the position of the clitoris is much more accurate than that given by the definition of Furetiere.
- The first edition of the French Academia[18] (founded by cardinal Richelieu) appears the same year (1694) followed by numerous following editions.
  - So, in 1762 (4th edition) the clitoris is defined: "terme d'Anatomie. Petite partie de chair ronde qui est dans l'endroit le plus élevé des parties naturelles de la femme" (Anatomical term. Small part of round flesh which is in the highest place of the natural parts of woman).
  - In 1798 (5th edition), same definition.
  - In1835 (6th edition), "Le clitoris: terme d'Anatomie. Petit organe charnu, de forme ronde et allongée, qui est placé à l'endroit le plus élevé des parties naturelles de la femme et de toutes les femelles d'animaux quadrupèdes" (The clitoris: Anatomical term. Round and elongated shaped fleshy small organ, which is placed at the highest point of natural parts of women and all female animals quadrupeds). Note the reappearance of the word "organ" already used by Godefroy and the reference to the animals.
  - In 1879 (7th edition), same definition for the word clitoris than in the previous edition.
  - In 1932–1935[19] (8th edition), "Le clitoris: terme d'Anatomie. Petit organe charnu, de forme ronde et allongée qui chez la femme, est placé à l'entrée de la vulve" (The clitoris: Anatomical term. Round and elongated shaped fleshy small organ, which, in the women, is placed at the entrance of the vulva). Unfortunately, it is clear that this definition is a return to imprecision! For completeness, we also include the Universal Dictionary (Dictionnaire Universel) published in 1854.[20] The author, M. La Châtre, contemporary of G.L. Kobelt, indeed gives a long definition of the clitoris, containing some anatomic considerations: "Petit corps rond et long , situé dans l'endroit le plus élevé des parties naturelles de la femme et des femelles de mammifères. Il est semblable au membre viril par sa structure et se termine

[13]R. Cotgrave: A dictionnarie of the French and English tongues. London, Adam Islip, ed. (1611).

[14]Godefroy Fr., Dictionnaire de l'ancienne langue française et de tous ses dialectes du IX e au XVe siècle, Paris, Vieweg F. Bouillon E., 10 tomes, 1881–1902.

[15]Richelet P., Dictionnaire françois, J-H. Widerhold, Genève, 1680.

[16]Furetière A. , Dictionnaire Universel, A et R. Leers, La Haye et rotterdam, 3 Tomes, 1690.

[17]Corneille Th., Le Dictionnaire des Arts et des Sciences, Veuve J.B. Coignard, Paris, 1694.

[18]Dictionnaire de l'Académie Française, Ière ed, 1694.

Dictionnaire de l'Académie Française, 4ème ed., Veuve de B. Brunet, Paris, 1762.

Dictionnaire de l'Académie française, 5ème ed, J.J. Smith et Ce, Paris, 1798.

Dictionnaire de l'Académie française, 6 ème ed, Firmin Didot frères, Paris, 1835.

Dictionnaire de l'Académie française, 7 ème ed, Firmin Didot et Cie, Paris, 1879.

[19] Dictionnaire de l'Académie française, 8 ème ed, Librairie Hachette, Paris, 1932–1935.

[20]Le Dictionnaire Universel, M. La Châtre, Administration de Librairie, Paris, 1854.

de même par un petit gland qui diffère de celui de la verge en ce qu'il n'est pas percé. Il en diffère aussi par son volume qui est de beaucoup moindre" (Small round and long body located in the highest place of the natural parts of the woman or female mammal. It is similar to the male member in structure and ends even by a small glans that differs from that of the penis in that it is not pierced. It also differs in its volume which is much less.).

Consulting the recent French dictionaries shows that the current definition of the word clitoris has not really progressed. So in the Larousse dictionaries 2013[21] (Petit Larousse, Larousse de la Langue Française), "Le clitoris: Petit organe érectile situé à la partie supérieure de la vulve" (The clitoris: small erectile organ located in the upper part beyond vulva). However, the erectile nature of the clitoris is henceforth mentioned.

In the Hachette dictionary 2013,[22] same definition as in Larousse dictionary.

It is not the same with the English recent dictionaries in which finally appears the clitoral function. So in the English dictionary 2013 of the Cambridge University,[23] the definition is as follows: "Clitoris, a small organ above the vagina that can give a woman sexual pleasure when is touched". Unfortunately, the location is inaccurate and erectile character is not mentioned. By cons, to our great satisfaction, the definition of the clitoris in the dictionary of the Oxford University[24] **"a small, sensitive, erectile part of the female genitals, at the anterior end of the vulva"** is by finally, almost perfect (it lacks only the word "organ"). In the dictionary of contemporary English,[25] there is also a good definition: **small organ at the front of the vulva, that is a centre of sexual sensation in women** (it lacks only the "erectile" function). The ideal would be to combine the two previous definitions. Note that more than four centuries have been necessary to achieve a complete and accurate definition of the clitoris!

3.4 Names for the Bulbs

R. de Graaf and Swammerdam were the first to mention the bulbs in their respective work, published the same year (1672) (see Chap. 1). Each of the two authors wanted to claim the anteriority of the discovery. However, if we follow the remarkable demonstration of L. Kobelt in his chapter on the "reservoir for venous blood and its muscle", it is de Graaf, who was the first, as he had already shown his friends, 18 months earlier, images in which these plexuses already appeared. Furthermore, still according to Kobelt,[26] it is de Graaf who "was the first to describe and represent this part exactly" as the **retiform plexus** or **reticular plexus**. As for Swammerdam, he did not find any other name for "his" discovery than that of **crus clitoridis interna**, thus creating certain confusion. This was also the case for Taberranus, who, in his "observationes anatomicae", published in Lucques in 1753, called them **plexus cavernosus** or even for Santorini, who named them **corpus cavernosum**. It is only later that the term of **bulbs of the vagina** was going to appear in French books. This term was then modified by Taylor to become **semi-bulbs** and then by Kobelt himself: to become **bulbs of the vestibule**. The exact name of the bulbs is still subject to discussion and we will address this issue in the "Relationships" chapter (refer to this chapter).

The other current topic of discussion concerns the entire unit, which forms the cavernous and spongy parts.

3.5 A Holistic Name for the United Cavernous and Spongy Parts

At present, the aim is no longer to separate the spongy and cavernous parts and in particular the clitoris and the bulbs. Such as demonstrated by Kobelt and repeated by H.E. O'Connell et al., anatomists have to deal with, in women (as in men), a "unified structure", which must be provided with a name, such as has been the case for men.

But which name is to be chosen? Kobelt had already thought about this, especially insofar as he wanted to treat the structures of the external genitalia of man and those of woman in the same manner. Finally, it is their physiology which inspired him and he thus described them as "**the apparatus of the genital sense**" of man and woman.

Since then, the terminology has not changed too much and the words corresponding to this beautiful physiological entity are missing. H.E. O'Connell concluded: "There is appeal in using a simple term, the clitoris, to describe the cluster off erectile tissues responsible for female orgasm". For our part, we believe that "clitoris" or even "clitoral complex" (also proposed by H.E. O'Connell) would be too restrictive and rapidly lead to forgetting the bulbs again. Moreover, we still have to name a true organ, with its own vessels and nerves and its specific functions. It is therefore a "cavernous-spongy organ" which must be named. So let us take that step and propose **bulbo-clitoral organ**, with verbal terms and groups which are easy to pronounce in Latin as well as English.

---

[21]Larousse 2013:
- Petit Larousse
- Dictionnaire de la Langue Française.

[22]Dictionnaire encyclopédique Hachette 2013.

[23]Cambridge English dictionary.

[24]Oxford English dictionary.

[25]Longman Dictionary of Contemporay English. Longman Group UK Limited.

[26]Kobelt reports that "Verheyen was reproached by Morgagni for omitting the retiform plexus in his edition of 1699". This author also reported that "in Germany, the bulb was forgotten due to the fact that the retiform plexus was considered as a simple venous network". Similarly for Tiedemann, "in his report on Duverney's gland, the retiform plexus is a simple venous plexus, a vascular network which enfolds the vagina (confusion with the vaginal venous plexus)".

4 External Morphology of the Clitoris

The frequent lack of knowledge of the anatomy of the clitoris, even among women considered as "experts", is explained by the fact that the major part of this essential component of the external female genital apparatus is hidden and therefore cannot be observed.

Only the distal end of the clitoris can be visualised during a vulvar examination, subject to slightly separating the folds consisting of the labia minora or nymphs, operation which also draws aside the labia majora.[1] A median, convex and hemi-cylindrical protrusion is observed at the ventral end of the vulva. It occupies the dip formed by the labia majora before they meet at their ventral commissure. This protrusion corresponds to the body of the clitoris and the skin covering the clitoris is the clitoral prepuce. A cutaneous and nearly circular fold is present at the loose end of this prepuce: the clitoral hood. This hood overhangs and surrounds part of a more or less spherical and central formation, the glans clitoridis. Very often, the glans remains hidden at the bottom of the "preputial cylinder" and is not immediately visible. It can only be observed once the hood has been gently pulled back.

From an anatomical viewpoint, the clitoris belongs to the vulva[2] (Fig. 4.2). It is thus located at the level of the anterior perineum,[3] behind the ventral commissure of the labia majora (their point of union with the mons pubis).[4] In order to complete the topographic arrangement, it is essential to recall that the glans clitoridis hangs over the external ostium of the urethra.

The hood of the clitoris is a cutaneous fold, which covers the glans clitoridis over the 2/3 of its periphery (top surface and lateral surfaces). It has various shapes (Fig. 4.3). The most common shape is a rounded hood resembling a roman

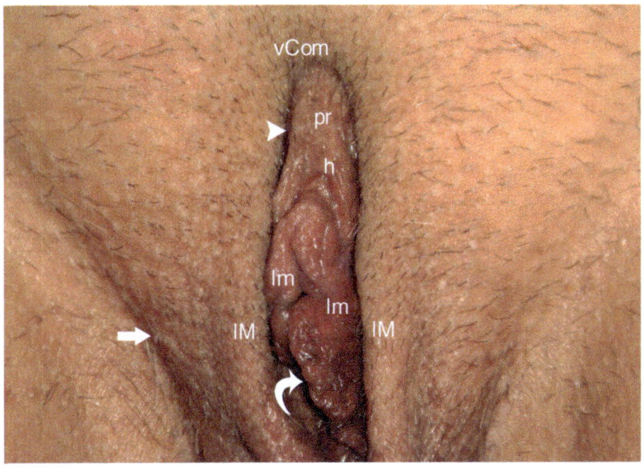

Fig. 4.1 Morphological aspect of the female pudendum (vulva) (after shaving). *h* hood of the glans, *lM* labium majus (greater lip), *lm* labium minus (lesser lip), *pr* female prepuce (praeputium clitoridis) covering the clitoral shaft, *vCom* ventral commissure (anterior commissure) of the lips, *white arrowhead* "labio-preputial" sulcus (continuation of the sulcus between majus et minus labia), *white curved arrow* penetrating into the pudendal cleft in order to accede to the vaginal vestibule, *right white arrow* genito-femoralis sulcus. NB: The labia minora are visible because the subject observed is a multipare female. Usually, closer young women, the labia minora are concealed within the pudendal cleft and the labia majora are approximated

[1]If the vulvar lips are not separated, it is impossible to visualise the entire external female genital apparatus: The only visible elements laterally are the labia majora and, in the centre, the ventro-dorsal vulvar cleft (anteroposterior vertical alignment, in a gynaecological position), bordered by the protruding part of the generally irregular edge of the labia minora. Above the vulvar cleft, the mons pubis (mons veneris), the ventral commissure of the labia majora and the clitoral prepuce are only visible if the hairy pubic triangle is shaved or plucked (Fig. 4.1).

[2]The vulva refers to the entire external genital apparatus, i.e. the mons pubis, the labia majora and minora and the vestibule of the vagina.

[3]The perineum consists of all the soft parts, which close at the base of the pelvic excavation, below the levator ani muscles. The external female genital apparatus is located in the anterior part or urogenital part of this perineum (the posterior part of the perineum is called the anal perineum).

[4]Under the teguments, the clitoris extends beyond the ventral bone boundary of the anterior perineum, as its median attachments rise and cover the bottom ¼ of the ventral surface of the pubic symphysis (see Chap. 13).

V. Di Marino, H. Lepidi, *Anatomic Study of the Clitoris and the Bulbo-Clitoral Organ*,
DOI 10.1007/978-3-319-04894-9_4, © Springer International Publishing Switzerland 2014

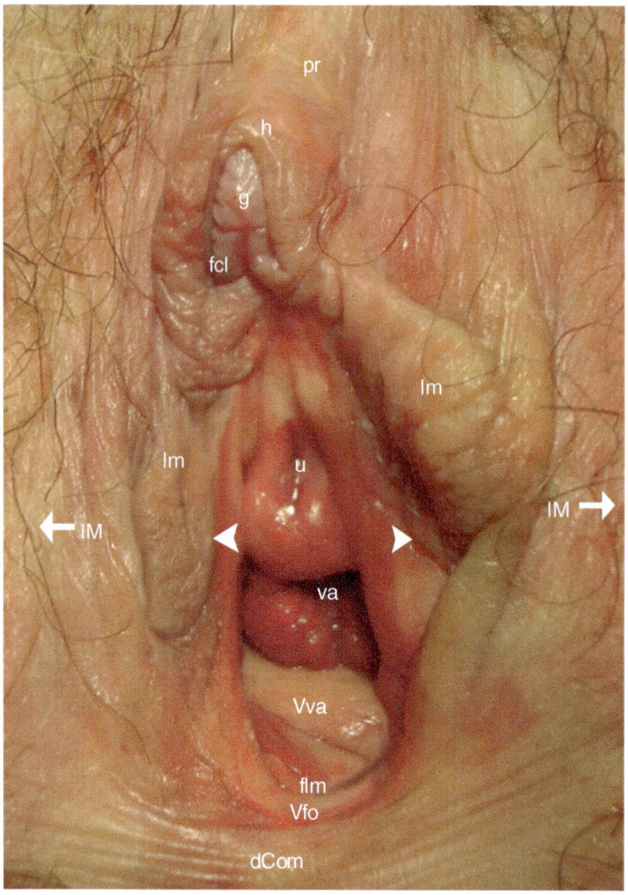

Fig. 4.2 View of the opened pudendum (vulva) showing the external urethral orifice and the external genitalia. Each labium majus is tracted laterally (*white arrow*). *dCom* dorsal commissure of the labia, *fcl* frenum (frenulum clitoridis), *flm* frenum of the labia minora (frenulum labiorum pudendi), *g* glans clitoridis, *h* hood, *lm* labium minus (lesser lip), *lM* labium majus (greater lip), *pr* female prepuce (praeputium clitoridis), *u* external urethral orifice, *va* vaginal orifice (introïtus), *Vfo* vestibular fossa, *Vva* vestibule of the vagina, *white arrowheads* they show the Hart's line (line of junction between labium minus inner skin and vestibule's epithelium)

arch. The hood may also have an ogival shape. It can even be shaped as a dihedron with a top edge (roof-shaped hood) (Fig. 4.9). It extends the skin covering of the body of the clitoris. The **clitoral prepuce** consists of this skin covering and the hood. The hood often has several folds. In most cases, the hood is not very thick but it can, even in young women, thicken and adopt a dysmorphic aspect: This hood is called by the French authors a crassiform hood (from the Latin word "crassus" meaning "fat") (Fig. 4.4). The hood can slide along the surface of the glans, so that the latter can be retracted (such as with the male sexual apparatus). A cavity is therefore formed between the glans and the internal surface of the hood, the preputial chamber (Fig. 4.5). The bottom of this chamber is an arc-shaped groove, which stops, such as the hood, at the level of its inferior surface (due to the presence of the frenula): it is the neck of the glans clitoridis (equivalent to the balano-preputial groove of the male penis). It is at the level of this neck that the external epithelium of the glans is reflected on the internal coating of the hood (Fig. 4.6). The external surface of the prepuce is covered by a keratinised squamous epithelium. In the case of young girls, a mucous layer is generally present on the internal surface of the hood. In the case of women of an age to procreate, this layer is gradually replaced by a non-keratinised squamous epithelium.[5] The presence of sebaceous glands (preputial glands) is observed at the neck of the glans, which explains the accumulation of smegma,[6] especially if the hood is fully covering or if the level of hygiene is not satisfactory. With age, the prepuce tissues tend to undergo a ptosis and distend. This causes the formation of a sort of withered sheath, inside which the glans is often completely hidden (Fig. 4.3). The lateral parts of the hood extend beneath the glans and are gradually incorporated into the lateral surface of the labia minora. The fusion with the labia minora can be precocious (high fusion). It can also occur at a lower level, at the top-third to mid-third junction of the labia minora (low fusion). This case corresponds to a lateral extension of the hood, sliding on the lateral surface of the labium minus, before merging with the latter.[7] Generally, there is no symmetry between the right and left levels of the junctions between the lateral extension of the hood and the homolateral labium minus.

The glans is the best known part of the clitoris due to its surface location and accessibility. Moreover, at present, it is rightly considered as the ideal site for female pleasure.

The glans clitoridis challenges neophytes by its resemblance to the glans of the penis, and at first sight, an observer will view the clitoris as a simple miniature penis with a glans whose aspect recalls that of a small-sized male glans! However, an accurate analysis rapidly shows the specific characters and multiplies differences contradicting this point of view.

[5]The epithelia of the prepuce and hood include many immunising cells (dendritic cells, lymphocytes, NK-T cells, etc.). The internal epithelium of the hood, which is not keratinised, does not oppose any effective "mechanical barrier" to the possible penetration of HIV, which the above-mentioned cells will therefore incorporate!

[6]The smegma is a white substance produced in the preputial chamber: it is comprised of desquamed epithelial and sebaceous cells (secretion product of the sebaceous glands, which lubrify the epithelia of the glans and hood and enable the hood to slide). Certain bacteria of the genital flora can generate an unpleasant smegma smell.

[7]In exceptional cases, this protrusion can, on one side, remain independent of the corresponding labium minus (F.L. Jayle).

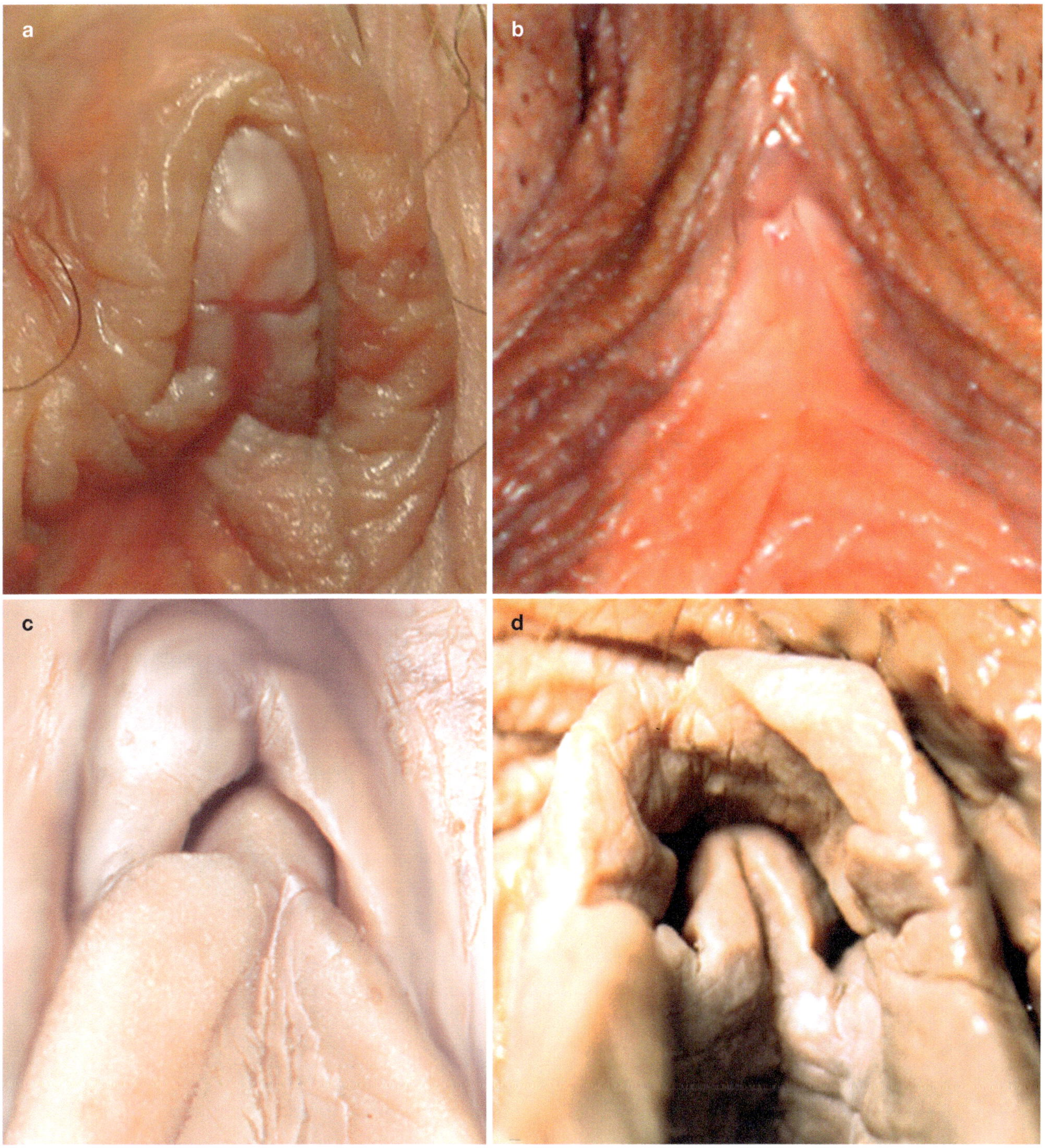

Fig. 4.3 Variations of the hood and their consequences on the cover of glans clitoridis. (**a**) Protruding glans; (**b**) barely apparent glans (ptosis of the preputial tissues in an old woman); (**c**) hidden glans (thick hood and large frenula); (**d**) hidden glans (only the frenula are visible)

The glans clitoridis is a protruding tubercle, which is more or less rounded (convex in all directions), and oblique rearwardly. It is often compared to a berry: the bilberry or a tiny acorn (with different colours). However, a slight flatness of the lateral surfaces and a dome-shaped top section are often observed, with a reduction in the diameters from the base to the end, such that the geometrical figure, which best represents the surface of the glans clitoridis, is that of an

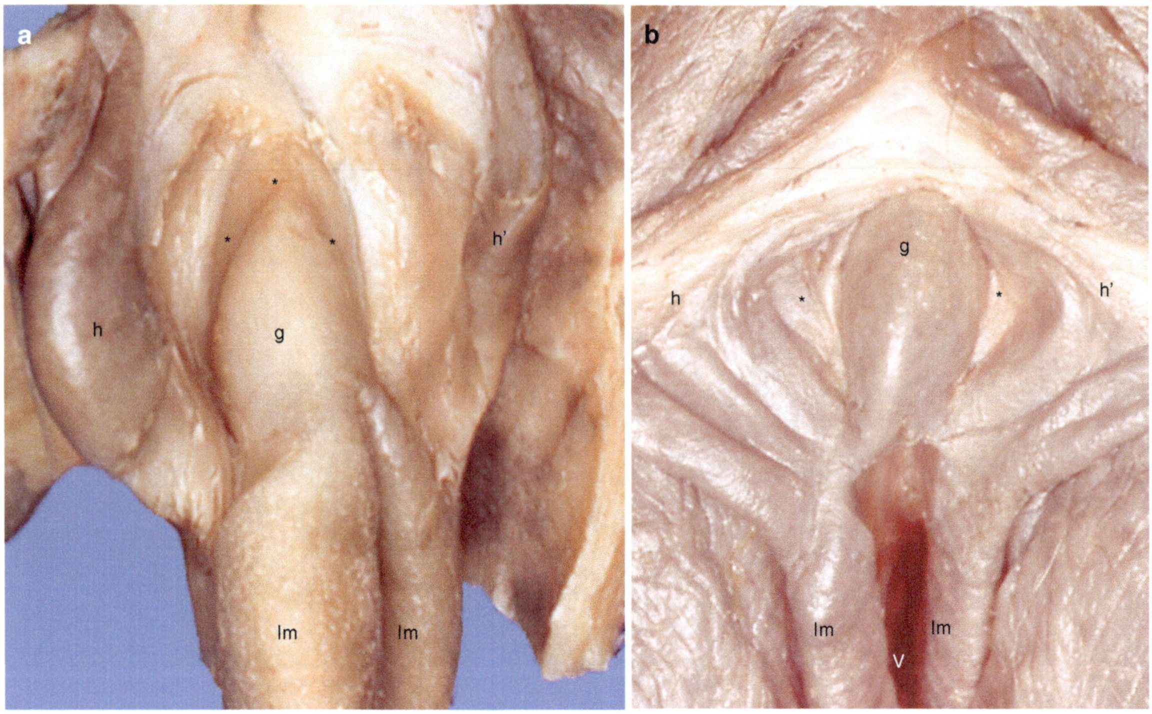

Fig. 4.4 Aspects of two glans (**a**, **b**) after sagittal sections of their hoods. These sections allow to see the preputial chamber and the free part of each glans (**a**, **b**) better. *g* glans clitoridis, *h* clitoral hood (right half), *h′* clitoral hood (left half), *lm* labium minus, ***asterisks locating the preputial chamber on each specimen, *V* vestibule. NB: On specimen **a**, the labia minora seem to be born directly from the glans (no frenula!)

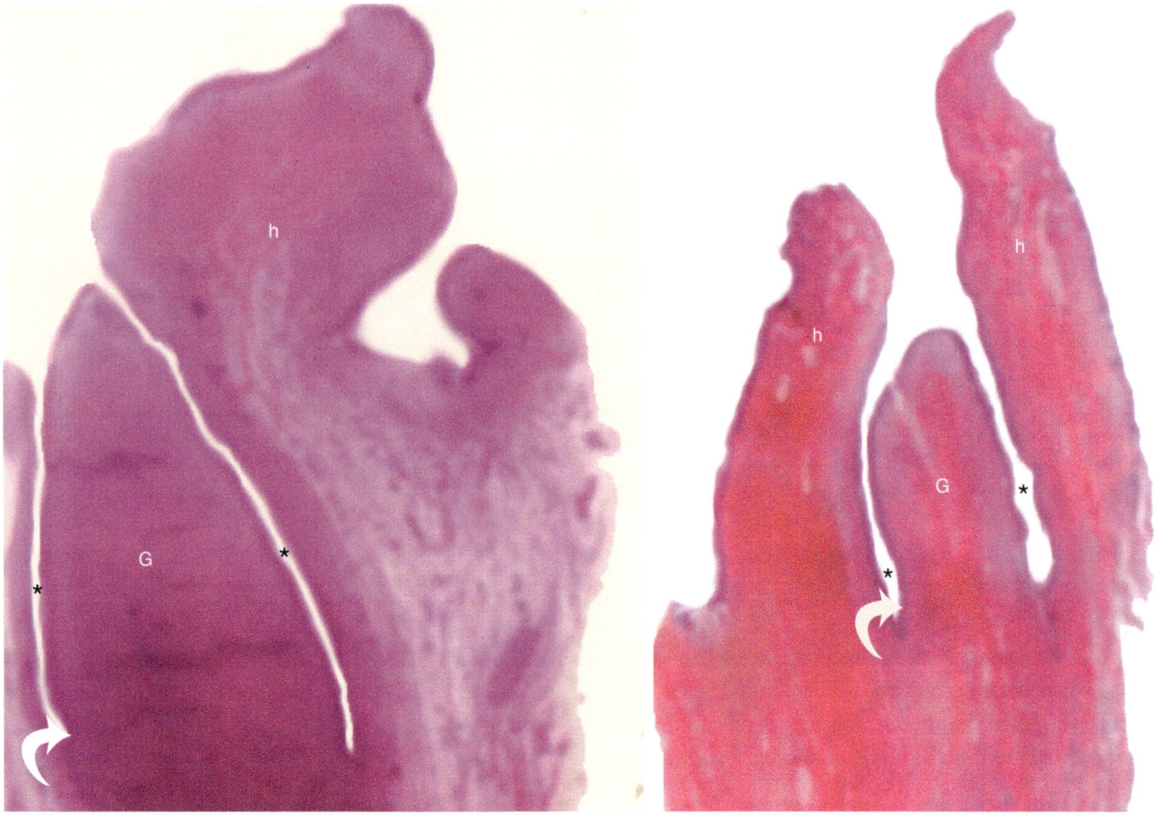

Fig. 4.5 Relations of the glans to the hood (end of prepuce) on histological transverse sections, in the axis of the glans. *G* glans of the clitoris (glans clitoridis), *h* hood (end of the prepuce), *asterisk* preputial chamber (clitoro-praeputial sulcus) (The female preputial chamber is analogous to the male preputial sulcus), *white curved arrow* neck of the glans

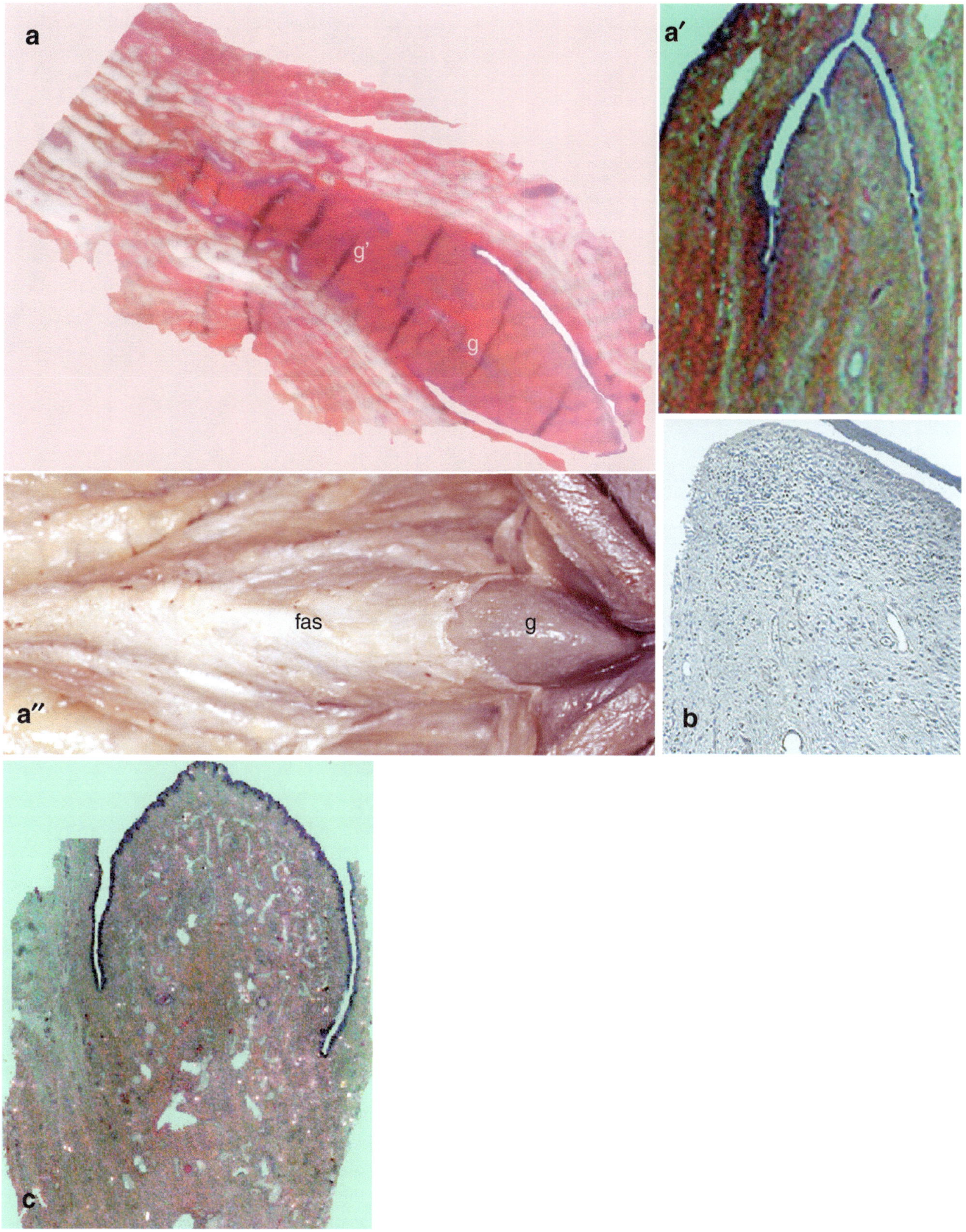

Fig. 4.6 The three principal morphological aspects of the glans clitoridis. (**a**, **a**′, **b**, **c**) Histological transverse Sections. (**a**″) Superior view of a dissected clitoris, after resection of the prepuce (skin of the clitoral shaft + hood of the glans). (**a**, **a**′, **a**″) Conoidal appearance; (**b**) paraboloidal appearance; (**c**) bulbous appearance. *g* free part of glans, *g*′ hidden part of glans, *fas* fascia of the clitoral body

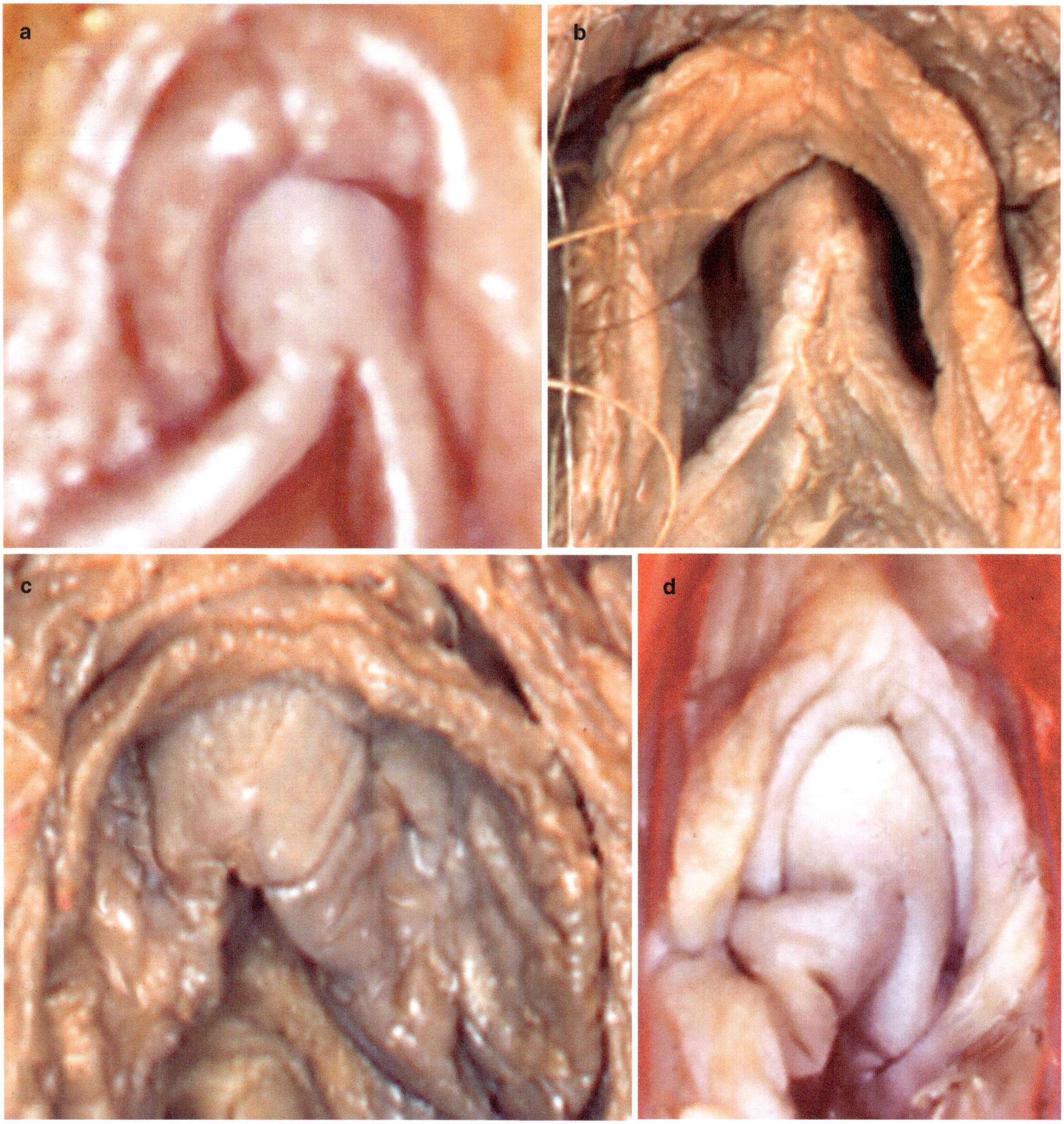

Fig. 4.7 Some morphological aspects of the glans clitoridis. (**a**) Cubic glans with asymmetrical tubular frenula. (**b**) Oblong glans with symmetrical tubular frenula. (**c**) Glans head fish shaped with large implantation of frenula. (**d**) Glans with flattened end and angular frenula

elliptic paraboloid (ends of a miniature missile or a mini-suppository). It can also have an oblong or cone-type aspect (with a rounded end or, on the other hand, a frankly acuminate end). Finally, it can have a levelled and flat end, shaped as a fish snout (Figs. 4.7, 4.8 and 4.9).

The prepuce detaches at the neck or base of the glans and becomes the hood. This way it frees the end of this formation. It is rare to observe, at the level of the neck, a basal crown such as is usually observed at the glans of the penis. When this type of crown exists at the glans clitoridis, it is not very marked, except in certain cases (see the glans c on Fig. 5.12). The glans is smooth except on the lower third of its circumference. Its surface is usually uniform; however, a thin white centre line may run along the glans and seem to divide it into two parts. It can, in some cases, lose the regularity of its curves and take on a polyhedral aspect with

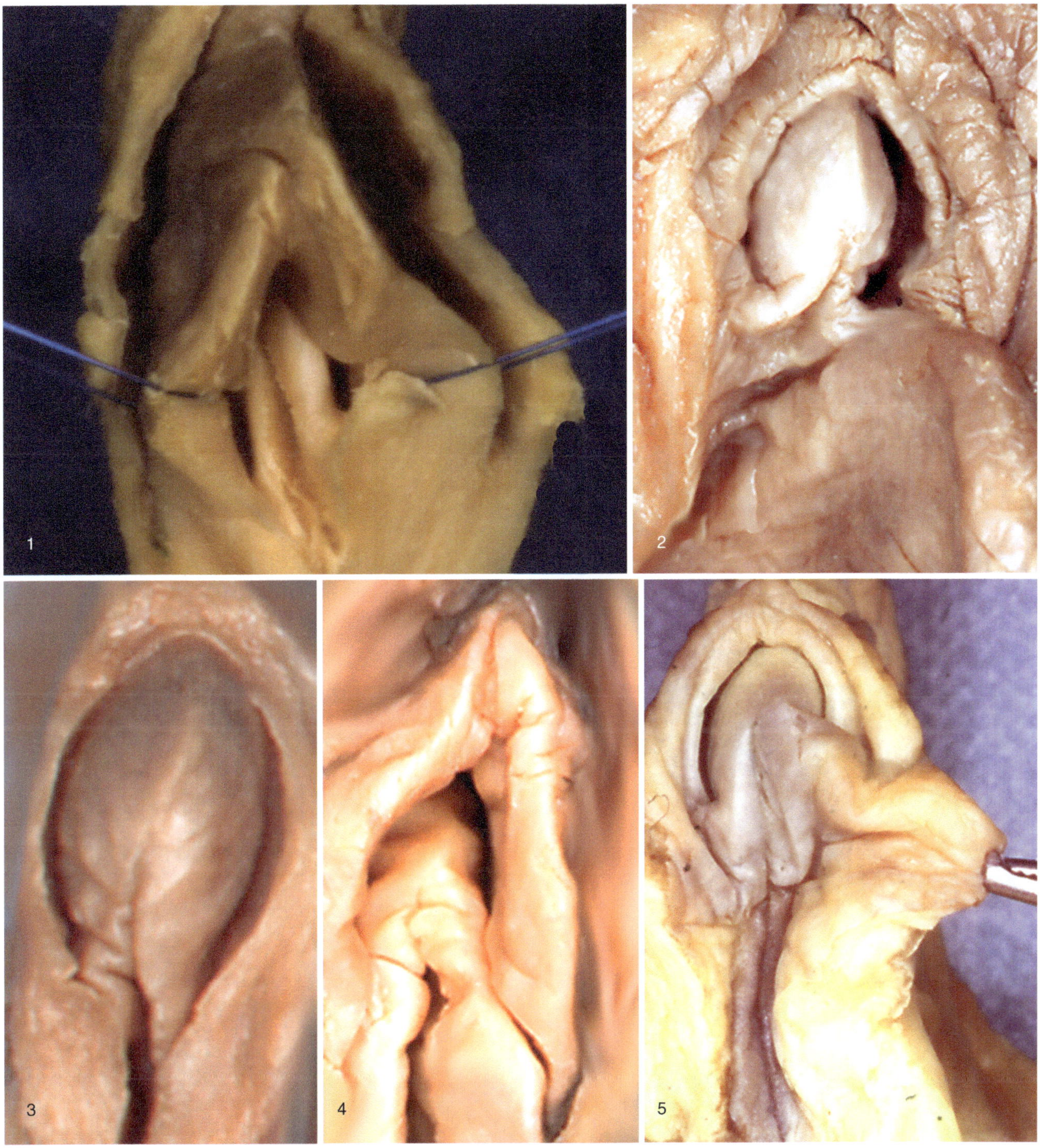

Fig. 4.8 Some variations of the glans' anatomy. *1* rod-shaped frena (rod-shaped frenula clitoridis), *2* glans with several folds of the hood, *3* glans with visible septum, *4* exceptional feature of an accordion-shaped glans, *5* glans with large left frenum arising from its peak

several facets ("accordion-shaped" glans): this is referred to as the grape currant (Fig. 4.8). The colour of the glans is variable, delicate pink in young women, a more reddish tint (with a tonality identical to that of the medial surface of the labia minora) in women who are not menopaused, and generally a duller colour in menopaused women. It may even have blue-grey aspects in aged women. Its colour actually varies with the intensity of its capillary and arteriolar circulation.

The dimensions of the glans have paradoxically been the subject of in-depth studies performed in vivo by various authors. It is easy to imagine the difficulties involved in such measurements and the discomfort of the patients who had to undergo them!

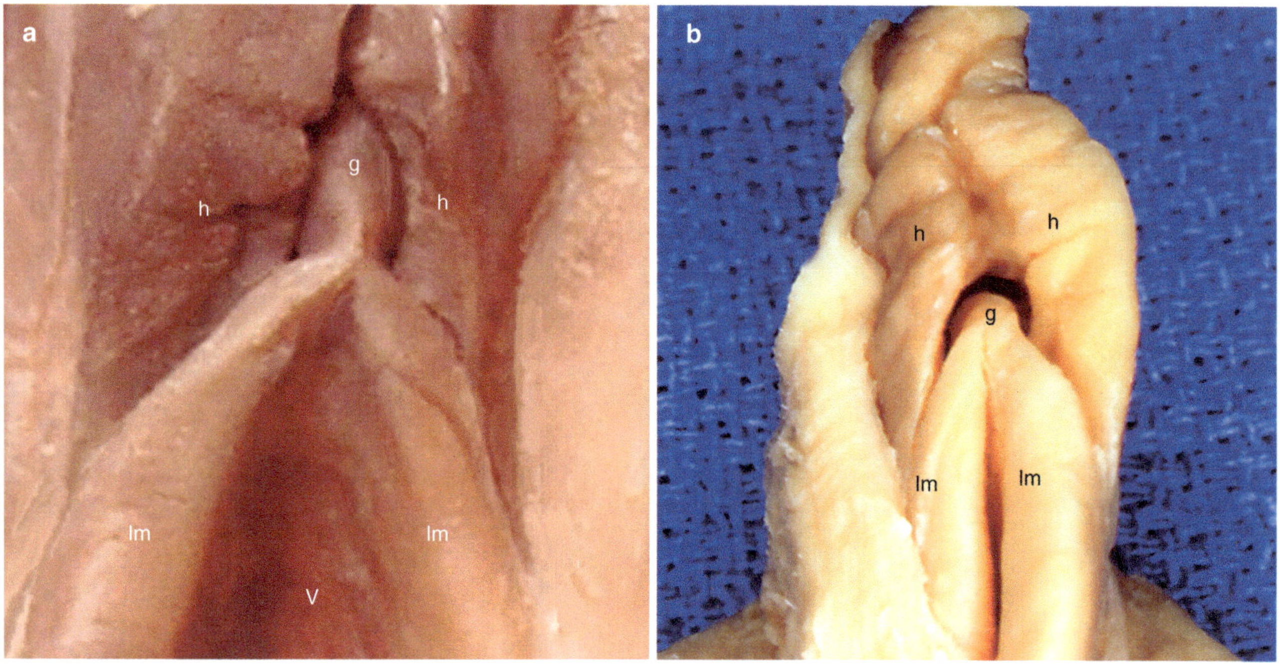

Fig. 4.9 Two examples of "crassiforme" hoods. (**a**) On the *left*: the glans, apparent, is flattened laterally. It seems that there is a single frenulum getting lost in a sort of ventral commissure of the labia minora. The top of the vestibule is visible (labia minora moved away). (**b**) On the *right*: the gland is spherical and barely visible. The very nearby labia minora close the vulve. These two labia minora seem to be born directly from extremity of the glans (frenula very short, almost non-existent!). *g* glans clitoridis, *h* hood, *lm* labium minus, *v* vestibule

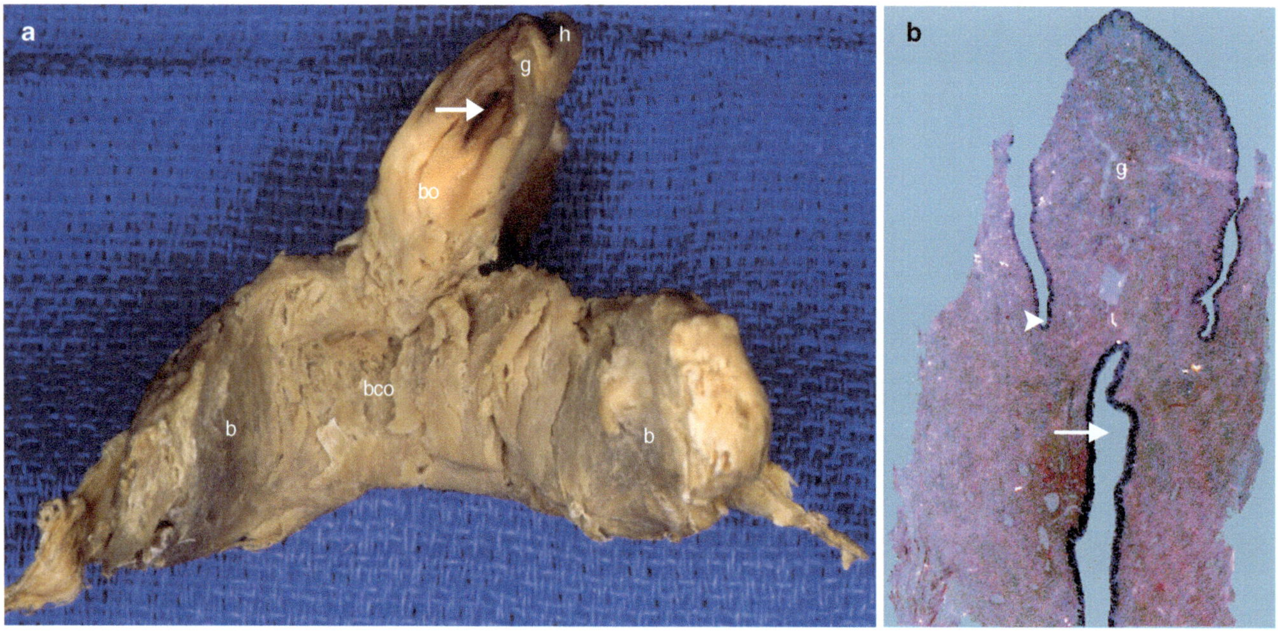

Fig. 4.10 Examples of "clitoral canal". (**a**) *Left picture*: inferior view of a "perforated" clitoris. (**b**) *Right picture*: section of a "clitoral canal". *b* bulbe, *bco* bulbar commissure, *bo* clitoral body, *g* glans of the clitoris, *h* clitoral hood, *white arrows* they show the one-eyed canal (embryologic residue of the phallic portion of the urogenital sinus), *white arrowhead* neck of the glans. Remark on the histological section, the epithelium covering the aborted urethra

In his remarkable book on Gynaecology, published in 1931, F.L. Jayle, after having carefully examined and measured the (glans of the) clitoris of 36 successive patients (table attached, Fig. 4.10), found the following average length 5.5 mm (extreme lengths: 2.5 and 10 mm) and as an average width (diameter) 3.5 mm. He concluded that the glans is more long than large! 15 cases are above average including 10, between 7 and 8 mm long. He noted that the width is proportional to the length and classified the large (glans of the) clitorises (7–8 mm long for 4–5 mm large) and small (glans of the) clitorises (4–5 mm long for 2.5–3 mm large).

A more recent American study (1992) of B.S. Verkauf et al., based on measurements carried out during 200 gynaecological examinations, provided practically identical results with an average width (diameter) of 3.4 mm ± 1.0 (extreme diameters: 2 and 5 mm) and an average length of 5.1 mm ± 1.4 (extreme lengths: 3 and 8 mm). The authors did not observe any significant difference in the measurements performed on the glans under study, according, on the one hand, to the size and weight of the women subject to examination and according, on the other hand, to the use or not of oral contraceptives. However, all the women who have undergone one or more pregnancies provided significantly greater glans measurements.

There are also significant measurement differences between ethnic groups, such as has been demonstrated by a comparative study performed by Mr. Phillip et al., who measured the external genitals of 570 children, newborn babies born at term, from two different populations (the first included 221 children and the second 349).

Other measurements were used to study the variation of the measurements of the glans clitoridis[8] from birth to adulthood.

It is Oberfield et al. who provided the measurements of newborn baby girls born at term (82 cases): the average length is of 4 ± 1.24 mm. The average width (diameter) is of 3.3 ± 0.78 mm.

It is Sane K. and Pescovitz OH. who studied the size of the clitoris from childhood to adolescence (80 studied cases). For an average age of 7 years old, they found a sagittal length average of 5 mm and an average transverse diameter[9] of 3.3 mm. According to these authors, there is no significant change until puberty. At the time of puberty, the clitoral growth is moderate and is especially correlated with the increase of body surface. These same authors have also defined a measurement index: the CI or clitoral index, expressed in mm^2 and obtained by multiplying the highest sagittal length, expressed in mm, by the transverse diameter, also expressed in mm. Thus, the average CI varies with age: for a newborn baby, it is 13.3 mm^2; for an average age of 7 years old, it is 16.7 mm^2; from 13 to 18 years old, i.e. for normal postpuberty girls, the average IC is of 20.7 mm^2.

GE. Tagatz et al. have studied the clitoral index among adult women, in relation to a bio-essay studying the effects of overdosing with androgens. The results of this study show that for a normal adult woman, the IC does not exceed 35 mm^2. On the other hand, beyond 35 mm^2, there exists, in half of the cases observed, at least another clinical sign of an excess of androgens. This shows the significance of the IC in endocrinological practice. For our part, it was possible to perform measurements on adult women in the best conditions possible, due to the fact that the glans measurements, carried out on anatomical subjects, during dissection operations, were supplemented by micrometric inspections performed on the histological sections from the dissected parts. Among the 10 cases, in which this double measurement could be carried out, we obtained an average of 5.1 mm for the length (extreme lengths: 2.5 and 6 mm) and of 3.5 mm for the largest diameter (extreme diameters: 3 and 4.5 mm), which corroborates the in vivo measurements carried out by the above-mentioned authors.

On the basis of these various measurements, it is possible to conclude that a normal glans clitoridis is a very small anatomical formation.

What about the rare cases in which dimensions are greater?

Most of the time, these cases are anomalies resulting from severe hormonal disturbances involving the secretion of androgens.[10] Clitoridomegaly or clitoromegaly refers to cases in which the dimensions of the glans exceed 10 mm for the length and 7 mm for the diameter.[11] Due to their exceptional dimensions, certain hypertrophies have even been the subject of publications or other works: we can mention the treaty of the surgeon, Ambroise Paré, "Des monstres et des prodiges" (Monsters and prodigies) (1573) and that of the not less famous anatomist, Jean Riolan[12] "Discours sur les hermaphrodites" (Speech about hermaphrodites) (1614) who mentions clitorises whose length and diameter are that of a finger!

In such cases (pseudo-female hermaphroditism), the clitoris can actually resemble a penis: it is then referred to as a "peniform clitoris". We can easily imagine the resulting personal, family and social problems and the repercussions on the sexual life of women having this hypertrophy, given that it also involves alterations of the secondary sexual characters and even, in some cases, a coalescence of the labia majora.

The external morphology of the glans varies not only according to its dimensions but also according to its connections with the clitoral hood! Such as has already been

[8]It is to be noted that in their publications, the authors express their measurements as the "length of the clitoris", whereas they measure only the glans, i.e. the accessible part while reclining the hood (such as specified).

[9]The authors never specify which transverse diameter is used for their measurements: Is it the diameter of the base of the glans or the largest transverse diameter (the latter is not always the diameter of the base of the glans, according to the clitoral morphology)?

[10]Female pseudo-hermaphroditism occurs following a hypersecretion of androgens by the foetal adrenal gland (endogenous origin). The same virilisation phenomena can occur following the administration of anabolic steroids (exogenous origin)!

[11]If a clitoromegaly is detected in a woman, any sign of androgenic hypersecretion must be researched.

[12]Riolan had already considered that these hermaphrodites had two sexual potentialities. He classified them among what the Greeks called "tribades" and the Romans "fricatrices" (on this topic, read the excellent chapter written by K. Park: The Rediscovery of the Clitoris, French Medicine and the Tribade, 1570–1620).

proposed by Jayle, four categories of glans clitoridis have been identified:

- Protruding glans, which is immediately visible without needing to separate the labia minora. They are generally very long glans.
- Apparent glans, which is easily visible as soon as the labia minora have been separated.
- Hooded glans, which is easily visible by reclining the hood.
- Hidden glans, which is very difficult to externalise, due to the length of the hood and its many folds.

Regardless of its morphological dimensions and aspects, the glans clitoridis shows two characteristics, which clearly differentiate it from the glans of the penis:

- It does not have an external urethral orifice.
- Its bottom surface is entirely occupied by a double frenulum.

The absence of any orifice at the end of the glans clitoridis is the most common aspect. However, there exist clitorises provided with an orifice opening onto a blind channel, from 4 to 6 mm long, which can be catheterised with the mandrel of a trocar for fluid puncture. We observed this phenomenon three times on the specimen under study. This type of channel is quite visible on histological preparations, where it is bordered by an ordinary squamous epithelium.

Two cases may arise:

- Either the external ostium of such a channel is positioned at the top of the glans clitoridis and a comparison with the external orifice of the male urethra is required. Knowing that in men it is the same epithelial layer, which, while becoming hollow, forms the prepuce of the glans and the terminal urethra with its small fossa navicularis (urethral portion, which will communicate with the phallic urethra from the urogenital sinus), we believe that a similar hollowing process may occur in women. However, this process is abortive since it does not encounter a channel with which it can communicate.
- Or the external ostium of this channel is positioned underneath the neck of the glans (Fig. 4.11), a few mm from the apex of the glans. The abnormal persistence of the terminal segment of the phallic portion of the urogenital sinus may be mentioned as an embryological explanation.

In any event, this may explain certain old texts and the resulting errors, such as the belief according to which "female semen" was released by a clitoral channel (see Chap. 1). It should be remembered that even if a small orifice and a ductal residue are sometimes present at the glans clitoridis, these are not functional urethral structures!

The lower surface of the glans clitoridis is very particular. It may even be said that, in most cases, it does not exist as it is nearly entirely occupied by the double frenulum of the labia minora or nymphs, which we prefer to refer to as **frenula clitoridis** (Fig. 4.12). The frenula therefore resemble two protrusions of the lower surface of the glans, which each diverge backwards and outwards to join the top edge of the counterpart labium minus. The aspect of each frenulum varies and a right-left symmetry is not very frequent. Each frenulum may be flat (webbed frenulum) or shaped as a cylinder (tubular frenulum) (see Fig. 4.9). There exist short and long frenula exceeding 1 cm in length. The end of each frenulum may vary: In most cases, the frenulum extends with the internal surface of the homolateral labium minus. It may also merge with the extension of the hood and thus form the ventral end of the corresponding labium minus. The inter-frenulum angle is important and plays a considerable role during the coitus. The frenula have a colour similar to that of the upper edges of the labia minora. The frenula may be straight, curved, wrinkled or bent (Fig. 4.8). If the glans has a certain degree of binary constitution (glans with a visible septal line), it may give the impression that each of the two sections extends into the homolateral frenulum. In these cases, the two frenula can be parallel and run alongside each other over ¾ of their length (dual rod frenula) (Fig. 4.9). It will be also noted that in cases where the glans has an orifice (opening onto a short blind channel), the formation of each frenulum starts in contact with this orifice. The two frenula can sometimes delimit a small area between them or a small cavity ("fossa") in the bottom surface of the clitoris, which is referred to as the infra-glandular cavity. The frenula, such as the medial surface of the labia minora, are covered with a squamous epithelium. This epithelium covers the relatively dense conjunctive tissue, which occupies the internal part of each frenulum. Overall, each of the labia minora (or nymphs) is formed at its juxta clitoral section, by the contribution of two homolateral components: infero-lateral part of the clitoral hood and clitoral frenulum.[13] Therefore, there exists a functional nympho-clitoral unit whose physiological importance will be demonstrated in another chapter.

Upstream of the glans, there is another part of the clitoris, visible via the protrusion it forms in the preputial groove, which separates the two labia majora, immediately behind their ventral commissure: This is the **protrusion of the clitoral body**,[14] covered by the prepuce (M12). On either side of

[13]This contribution is variable:

- The labium minus can be formed through the fusion of the frenulum and the lateral extension of hood.
- The labia minora can be formed from a single frenulum, while the extension of the hood gets lost on its lateral surface.
- More rarely, it is the lateral extension of the hood which forms the labium minus, which will be joined, on its medial surface, by the frenulum.

[14]Still referred to as the "clitoral shaft".

Fig. 4.11 Measurements of the glans clitoridis carried out by Dr. F.L. Jayle on 36 of his successive patients. Extract from "La Gynecologie", 1918. Chap: L'anatomie morphologique de la femme

Longueur en millimètres.	*Largeur en millimètres.*	*Age et enfants.*
10	5	69 ans, 0 pare.
8 1/2	4	32 ans, 1 pare.
8	6 1/2	52 ans, 0 pare.
8	5	26 ans, 0 pare.
8	5	32 ans, ovariectomisée.
8	4	32 ans, 0 pare.
7 1/2	4	30 ans, 3 pare.
7 1/2	3 3/4	30 ans, 1 pare.
7	5	26 ans, 0 pare.
7	4	48 ans, 1 pare
7	3	26 ans.
7	2 1/2	21 ans, 0 pare.
6 1/2	4 1/2	34 ans, 0 pare.
6	4	29 ans, 4 gest., 2 pare.
6	4	33 ans, 2 pare.
5 1/2	3	27 ans, 1 pare.
5	4	45 ans.
5	3 1/2	24 ans, 0 pare.
5	3 1/2	26 ans, 1 pare.
5	3	33 ans, 2 pare.
5	2 1/2	27 ans, 1 pare.
5	2 1/2	43 ans, 2 pare.
4 1/2	3	32 ans, 1 pare.
4	3 1/2	28 ans, 5 pare.
4	3	25 ans, 0 pare.
4	3	53 ans, 1 pare.
4	3	30 ans, 0 pare.
4	3	43 ans, 3 pare.
4	2 1/2	36 ans, 3 pare.
4	2 1/2	26 ans, 0 pare.
4	2 1/2	34 ans, 0 pare.
4	2 1/2	30 ans, 1 pare.
3 1/2	3	18 ans, 0 pare.
3 1/2	2 1/2	22 ans, 0 pare.
3	1 1/2	20 ans, 0 pare.
2 1/2	2 1/2	39 ans, 1 pare.

this protrusion, a more or less deep labio-preputial groove extends the interlabial groove[15] up to the ventral commissure of the labia minora.

The epithelium of the prepuce is laterally in continuity with that of the labia majora: it is thus also a keratinised squamous epithelium, with high content of sebaceous glands. However, it is less thick and has a lower content in hair follicles, which contrasts with the significant hairiness of the nearby labia majora and especially that of the mons pubis (mons veneris) which is above it. The cutaneous covering of the body of the clitoris plays the same role as the hood at the level of the glans: This consists of a protection role dedicated to the entire clitoral prepuce. For the body of the clitoris, this protection is increased by the presence of a more or less deep groove in which it is located. As for the hood, the protection, which it provides for the glans clitoridis, has been compared by certain anatomists to that of the eyelids for the eye bulb.

From a morphological viewpoint, the body of the clitoris, beyond the glans, is cylindrical. The prepuce covers it like roof tile. As it is necessary to release this clitoral body by dissection in order to study all of its sections, we will describe it in the following chapter with the hidden structures, which are not visible from outside.

[15]The interlabial sulcus separates each labium minus from the homolateral labium majus.

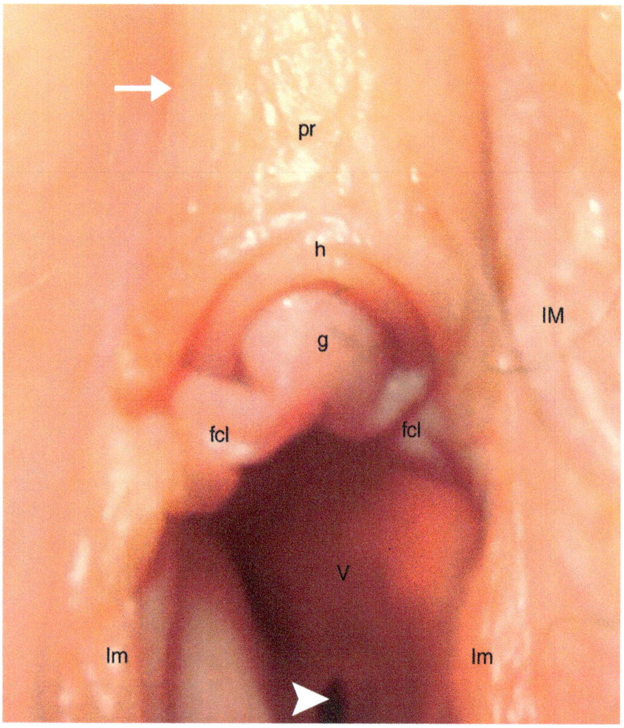

Fig. 4.12 Details of the glans' external aspect (perineal view). *fcl* frenum (frenulum clitoridis), *g* glans clitoridis, *h* hood, *lm* labium minus (lesser lip), *lM* labium majus (greater lip), *pr* female prepuce (preputium clitoridis) covering the clitoral shaft, *V* vestibule, *white arrowhead* external urethral orifice, *white arrow* labio-preputial sulcus

5 Structure of the Bulbo-clitoral Organ

5.1 Terminology Difficulties

The description of structures referred to as "clitoral" represents a true semantic problem. When referring to the anatomical nomenclature, the old nomenclature or the recent international nomenclature, the clitoris and the vestibular bulbs are described separately.

When considering anatomic facts, i.e. what is observed and what is revealed through dissection, the clitoris and vestibular bulbs are closely related and form a single entity, specific to the external female genital apparatus. However, difficulties arise in relation to the name of this entity.

O'Connell H.E. et al. solved the problem by referring to the two above-mentioned structures with a single term: **the clitoris**. However, if this is true from an anatomical observation viewpoint, it is inaccurate from a terminological viewpoint.[1]

The question becomes even more complicated when considering the glans, which is qualified (as is customary) by the adjective clitoral (due to the fact that, from a topographical viewpoint, it represents the free and exteriorised part of the clitoris), while, at the same time, it is embryologically derived, for its major part, from the primitive corpus spongiosum.

In order to solve all these difficulties, we propose to call this complex unit, derived from the female embryonic genital tubercle, the "**bulbo-clitoral organ**".[2] This unit was referred to by Testut et al. in their famous book on Anatomy, which has been published many times, as **erectile organs of the vulva**, which Köbelt had referred to by using a holistic term: **genital sense's apparatus** in female.[3]

The bulbo-clitoral organ, such as it has always been observed, includes:

- Cavernous structure parts: the crura and the body of the clitoris
- Spongy structure parts or parts derived from the foetal corpus spongiosum: the vestibular bulbs, the commissure of the bulbs and the infra-corporeal residual spongy part
- A part ensuring the junction and the communication between the cavernous and spongy vascular structures: Köbelt's pars intermedia

This organ resembles a dihedron (Fig. 5.3) whose top edge comprises a slanted extension directed downwards and backwards. This dihedron caps a truncated urogenital pyramid consisting of the vaginal canal surmounted by the urethra. Overall, through dissections, the considerable significance of the buried parts of this bulbo-clitoral organ is rapidly observed. The vision of the glans clitoridis, alone free and visible part of the clitoris, only provides a minimalist overview of the said apparatus.

While the architecture of the bulbo-clitoral organ is constant, this is not the case for its global morphological aspect and a significant range of various aspects can be encountered, not only due to the various lengths, thicknesses and shapes of its components but also due to the variety of connection angles of these same components. At first sight, as soon as an anatomical sample has been collected and even after the most careful dissection of this specimen, the bulbo-clitoral organ has a massive aspect, which is always astonishing, and sometimes even a little monstrous (Fig. 5.1).

[1]The Unicist Theory, rehabilitated by O'Connell, corresponds to reality (see the histological dissections and sections…). However, the fact of referring to the two structures as the clitoris does not correspond to "nomina anatomica" and does not make sense; it is therefore a source of confusion.

[2]We chose the "bulbo-clitoral" adjective ("bulbo-clitoridien" in French) rather than the "clitorido-bulbar" adjective ("clitorido-bulbaire" in French), as it is easier to pronounce and translate into English.

[3]L. Köbelt had grouped the cavernous and spongy units under the same term "Genital Sense's Apparatus in Female", a consensual term which is too far removed from descriptive anatomy.

V. Di Marino, H. Lepidi, *Anatomic Study of the Clitoris and the Bulbo-Clitoral Organ*,
DOI 10.1007/978-3-319-04894-9_5, © Springer International Publishing Switzerland 2014

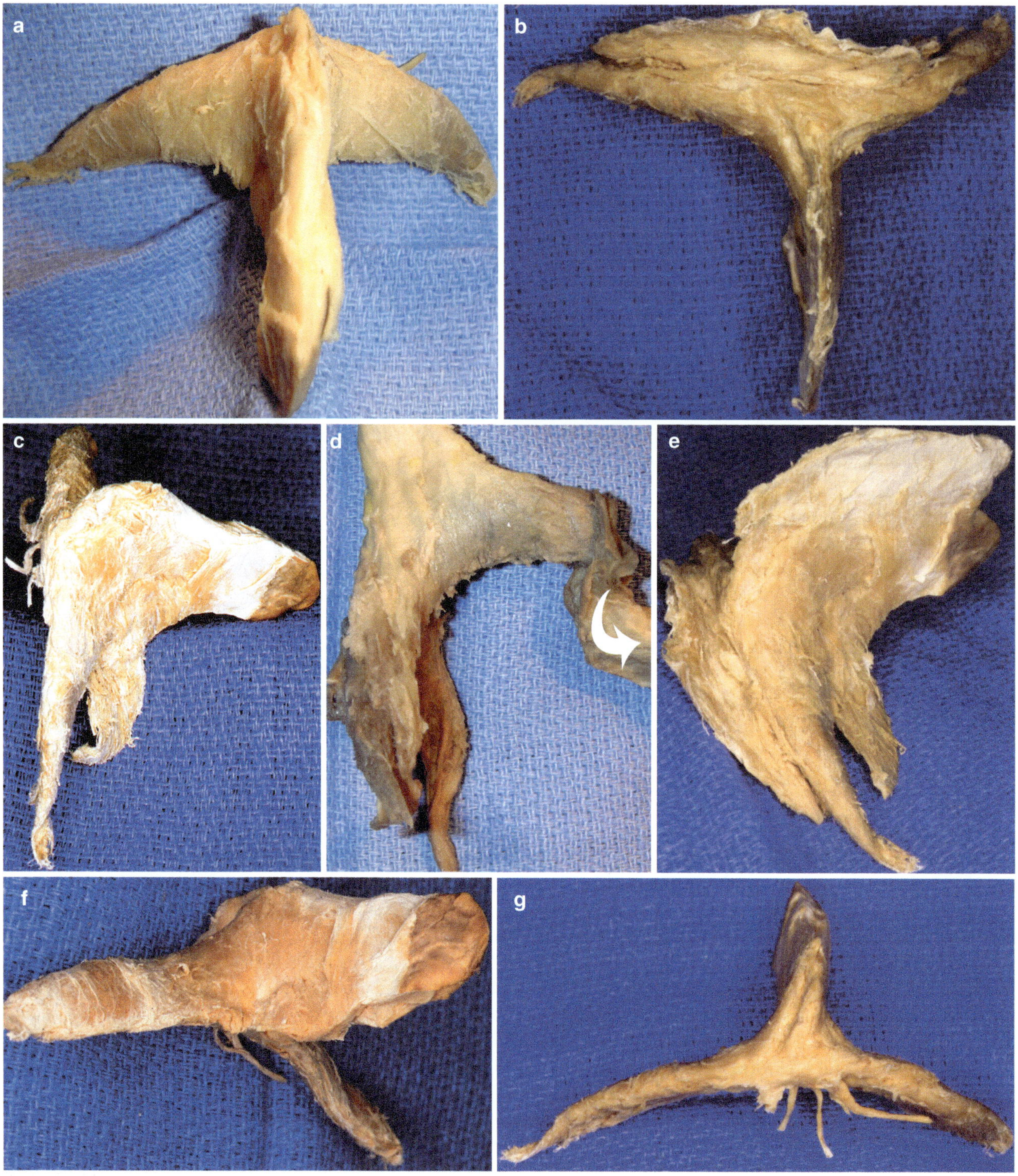

Fig. 5.1 Morphology of the clitoral parts of bulbo-clitoral organs after resection of bulbs and spongy tissue. (**a**) Antero-superior view, usual aspect; (**b**) superior view, tapered aspect; (**c**) right lateral view, usual aspect; (**d**) right lateral view, usual aspect (the spongy part dissected, but always attached to the distal clitoris, is tilted forwards (*curved white arrow*)); (**e**) right lateral view: bulging aspect; (**f**) right infero-lateral view: usual aspect; (**g**) posterior view (with neurovascular hilum and very long crura)

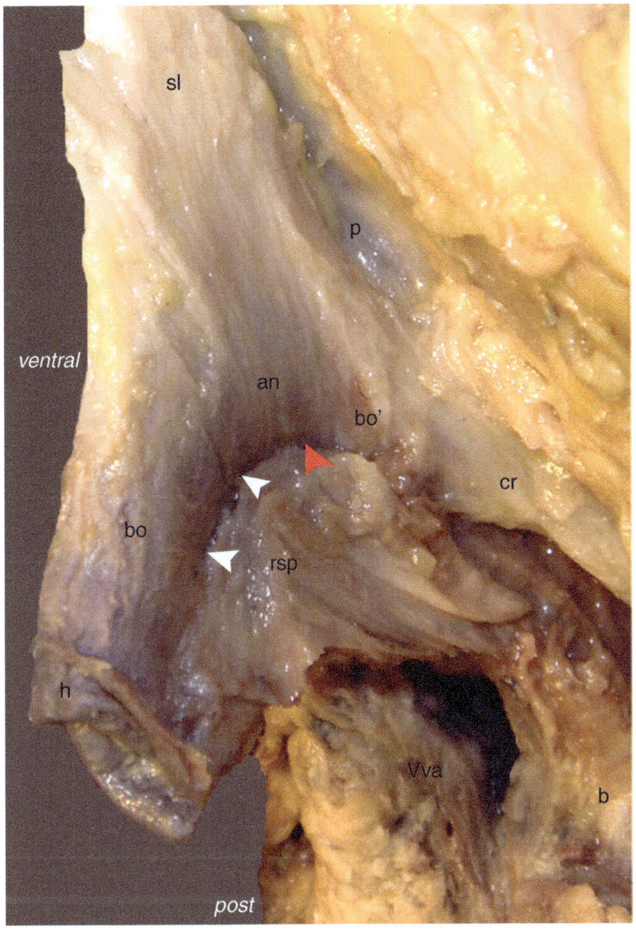

Fig. 5.2 Lateral aspect of a dissection showing the structure of the bulbo-clitoral organ. *bo* body of the clitoris, descending part, *bo′* body of the clitoris, ascending part, *cr* crus clitoridis, *an* angle (or elbow) of the clitoral body, *h* hood, *p* pubis, *rsp* residual spongy part, *Vva* vaginal vestibule. *Arrowheads*: They show the area of the intermediate network of Köbelt. Note: The *red arrowhead* indicates the part where this network's communications are the most dense: between the commissura bulborum and the angle of clitoral body

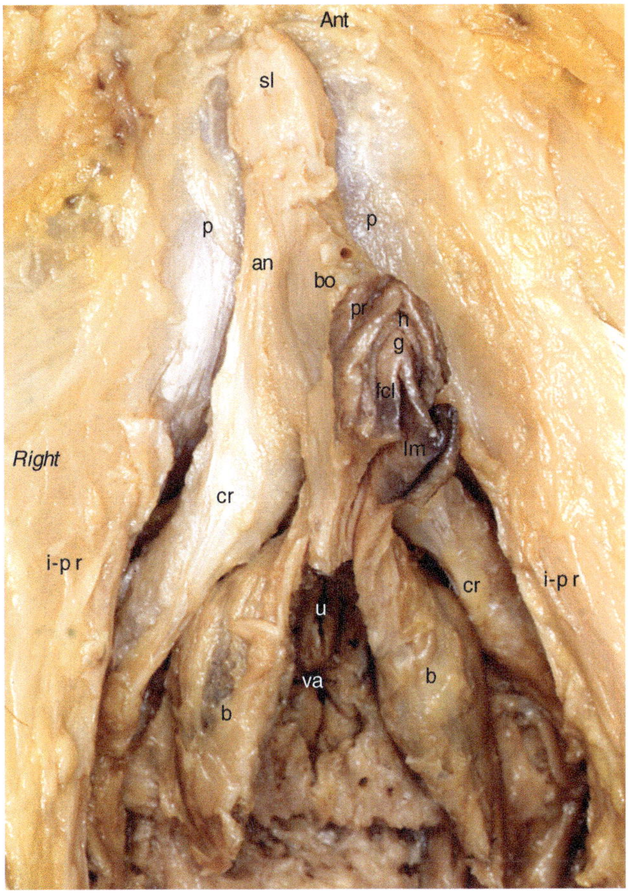

Fig. 5.3 Dissection of the urogenital female perineum showing the bulbo-clitoral organ in situ. *an* angle (or elbow of the clitoral body), *b* bulb (spongiosus), *bo* body of clitoris, *cr* crus clitoridis, *fcl* frenum (frenulum clitoridis), *g* glans clitoridis, *h* hood, *i-p r* ischio-pubic ramus, *lm* labium minus (lesser lip), *p* pubis (anterior surface), *pr* prepuce, *sl* suspensory ligament, *u* external urethral orifice, *va* vaginal orifice (introïtus). The photography is taken in slightly ¾ right, in order to see the clitoral body

Furthermore, this apparatus occupies the three dimensions (Fig. 5.1). This triplanar characteristic, which could only be demonstrated by 3D reproductions, requires, in relation to the pictures intended for the pages of a book, the use of artifices, such as contrast and light variations, or unusual shots such as right or left ¾ photographs (Fig. 5.2).

5.2 Parts of Cavernous Structure

They are the crura and body of the clitoris.

5.2.1 The Crura

The **crura clitoridis** (international nomenclature), or **pillars** or **roots**, are the two cavernous structures, shaped as a long cornet, attached, via the ischiocavernosus muscles, to the ischio-pubic rami of the two coxal bones, which will meet frontwards and from within to form the body of the clitoris (Fig. 5.3). These crura are located at the anterior perineum in the superficial perineal region,[4] in contact with the lateral

[4]The superficial perineal region lies between the superficial fascia of the perineum (fascia of Colles) and the lower fascia of the urogenital diaphragm (perineal membrane).

Fig. 5.4 Dissection of a bulbo-clitoral organ showing the relationships between the different components (ventral view of an ex situ dissection). *b* bulb (spongiosus), *cr* crus clitoridis, *f* frenum (frenulum clitoridis), *g* glans clitoridis, *lm* basis of the labium minus, *n* dorsal nerve of clitoris, *pr* prepuce

edges of the subpubic angle. In order to reach them, it is necessary to cross the superficial fascia of the perineum, to almost completely free the labial fatty body,[5] contained in the labia majora[6] and to push this formation towards the ventral surface of the pubis.

When the crura are still in place and not detached from their insertions, it is all the more difficult to apprehend their shape as they are hidden by the muscle fibres of the ischiocavernosus muscles, which are pushed against their lower and medial parts. Moreover, these crura are attached to the internal surfaces of the ischio-pubic rami. The lower edges of these rami partially hide the crura. Overall, it is actually only possible to study the crura when they have been released from their attachments and therefore only parts analysed ex situ (Fig. 5.4). There exists a right-left symmetry but when the two crura are examined on the sampled parts, a few discrete variations in length, thickness and sometimes, morphology, are observed. Each crus is initially tapered (top of the cone) and becomes gradually wider to reach its maximum diameter in front of the distal part of the ventral surface of the pubis, before joining the opposite crus on the centre line, in front of the pubic symphysis. The crura of the corpora cavernosa, with the ischiocavernosus muscles which cover them (which themselves are sheathed in their envelope fascia), make up structures solidly attached to the lateral edges of the pelvic outlet via bone and fibrous attachments. The bone insertions are located on the middle part of the internal surface of the ischio-pubic ramus, under the sickle-shaped crest[7] and slightly above the lower edge of the bone. The fibrous insertions are located on the juxta-bone part of the lower fascia of the urogenital diaphragm (perineal membrane). At this level, the fascia clearly becomes thicker. We propose to call this thickening "attaching lamina of the corpora cavernosa".

The dimensions of the crura vary, we have already observed this! In our research, the average length of the crura (measured up to the level where the joining between the left and right crura starts) is of 37 mm (30 to 35 mm for L. Testut, 30 mm for P. Kamina). The average diameter (measured at mid-length of the crus) is of 9 mm for the parts which we have measured.

The structure of the crura is simple: a thick fibro-elastic or **albuginea** envelope, surrounding and protecting the cavernous tissue. It is the albuginea one, which provides the crus with its firm consistency. Each crus receives a feeding artery, a branch of the superficial perineal artery.

The two crura have a prepubic converging direction and their axes determine, between them, a connection angle, shaped as a reversed V, of variable amplitude, of 85° on average (extreme angles: 50 and 95 degrees)[8] with, from an architectural viewpoint, a relatively ogival disposition. This connection is not a unitive convergence such as seems initially apparent. By examining the back of the sampled parts, the impression is given that there is a triangular commissure (Fig. 5.5), which merges together the terminal portion of the two crura. Actually, the dissection and sections (Fig. 5.6) are used to affirm that there is no fusion of the initial crura into a single corporeal element. There is no coalescence or fusion but only a joining of the two pillars, which will merge to form a single corporeal cylinder. As for the pseudo-commissure of the pillars, which it is better to call retro-crural fascia (old

[5]The labial fatty body, a considerably vascularised fatty formation, which can be perfectly mobilised and can maintain an excellent vascularisation, if its upper pedicle is preserved, is used successfully, in surgery, for curing vesico-vaginal fistula (operation of Martius).

[6]The labia majora, due to their contents, were called "fatty bags" by the anatomist Marie Philibert Constant Sappey (1810–1896).

[7]The sickle-shaped extension of the sacro-tuberal ligament inserts itself into this sickle-shaped crest.

[8]In one case, among the 30 that we studied, the 2 crura had a hardly ascending and nearly horizontal route, and a connection angle exceeding 150°. This case was excluded from the previous average.

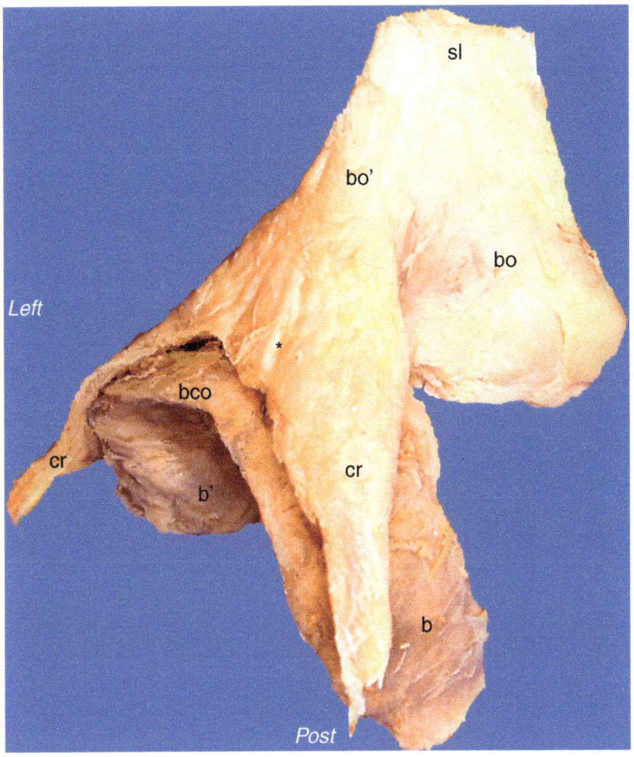

"ligament of the inter-thigh"), it is only formed by a single dense fibrous layer, which merges more or less with the fibrous coating, which coats the ventral surface of the pubic symphysis and from which it can be detached to unstick the clitoris from the symphysis and slightly rotate it backwards and downwards. This pseudo-commissure (Figs. 5.5, 5.7 and 5.8) covers a trapezoidal or triangular surface (isosceles or equilateral triangle), of variable height (between 1.3 cm and 2 cm). The base of the commissural triangle borders the lower edge of the pubic symphysis. This fibrous structure is not only a unitive collagen lamina coating the dorsal surfaces of the crura. It is an actual protective tunic whose major role is to cover the nerves and vessels, before they reach the cavernous structures. This fascia delimits a loose conjunctive region, the retro-crural space (Fig. 5.9), inside

Fig. 5.5 Three-quarter view of an ex situ dissection of the bulbo-clitoral organ showing the reciprocal positions of clitoral and bulbar parts. *b* right bulb, *b′* left bulb, *bco* bulbar commissure, *bo* clitoral body, post-angular part (descending part), *bo′* clitoral body, pre-angular part (ascending part), *bco* bulbar commissure, *cr* crus clitoridis, *sl* suspensory ligament, * the asterisk locates the retro-crural fascia

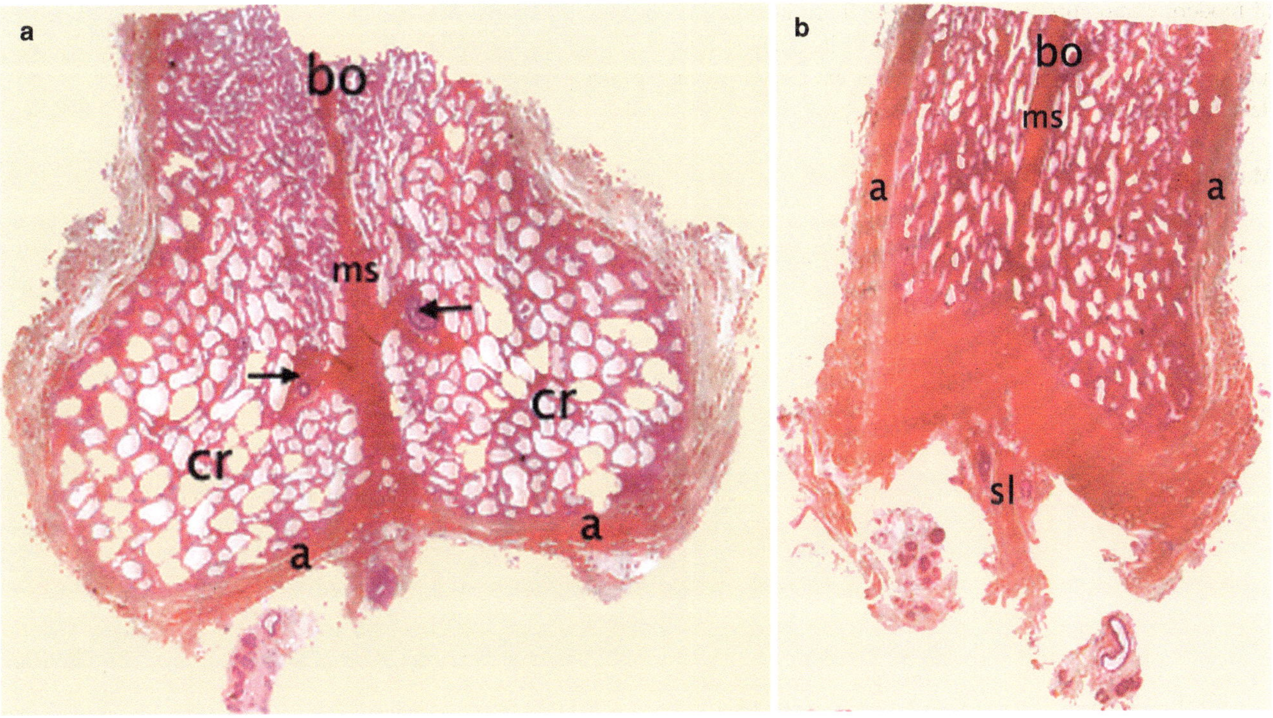

Fig. 5.6 Photomicrographs of two transversalis sections of the corpora cavernosa into the clitoral body. (**a**) This section concerns the elbow (or angle) of the clitoral body. The CC of each crus is continuous with the CC of the corresponding half clitoral body. Note the round aspect and the great diameter of the cavernous gaps into the crura and the tight longitudinal meshes of clitoral body. The median septum thins towards the end of the post-angular body. (**b**) This section concerns only the post-angular portion of the clitoral body. The cavernous gaps only have tight longitudinal meshes. The 2 CCs separated by a median septum are incorporated in a single clitoral cylinder. *a* albuginea, *bo* clitoral body (post-angular part), *cr* crus clitoridis, *ms* median septum, *sl* suspensory ligament, *black arrows* they locate the cavernosal arteries (deep clitoral arteries)

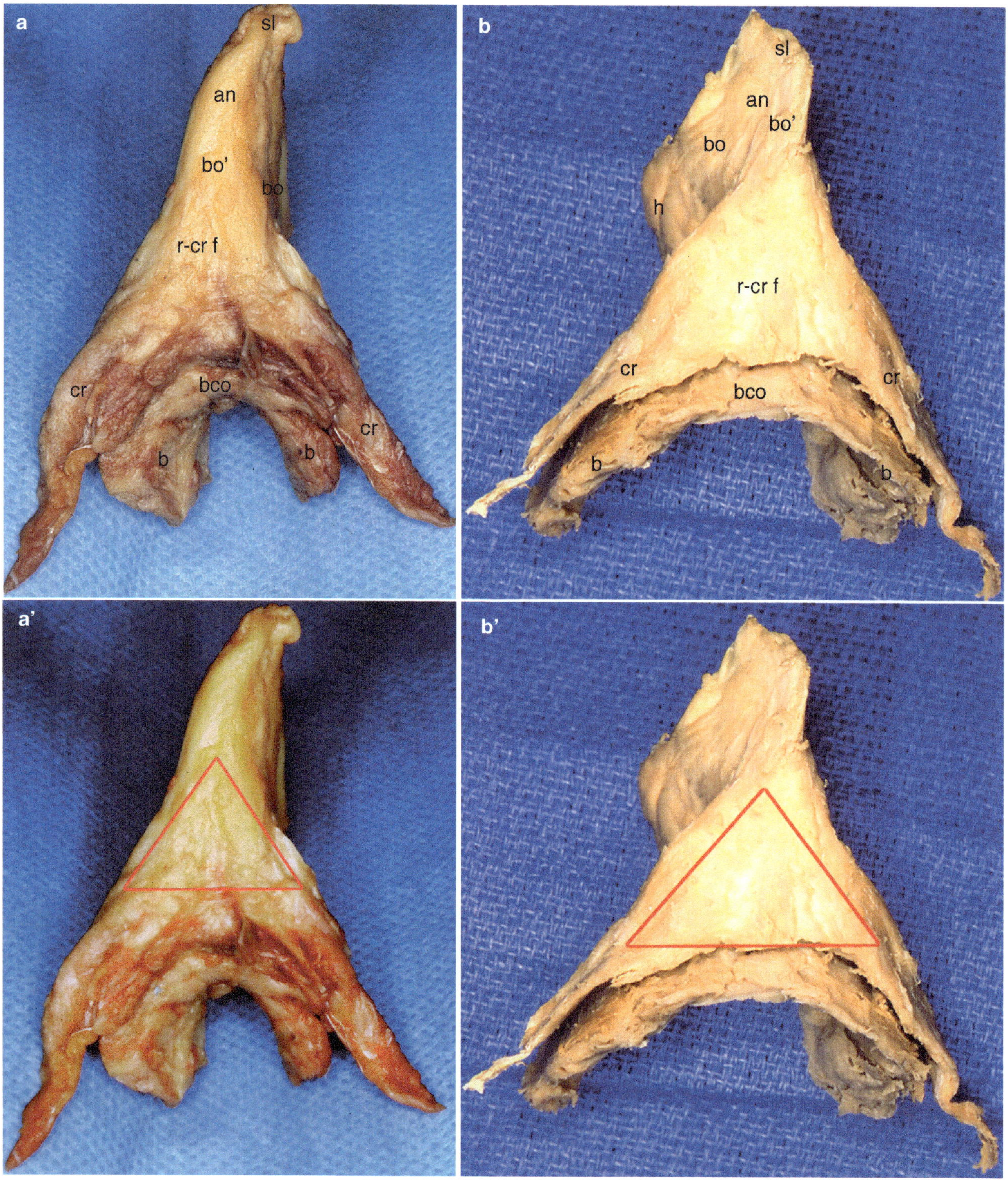

Fig. 5.7 Dorsal aspects of 2 bulbo-clitoral organs (**a**, **b**), showing the retro-crural fascias with their triangular surface. *an* angle, *b* bulb, *bco* bulbar commissure, *bo* clitoral body, descending part (clitoral shaft), *bo′* clitoral body, ascending part, *cr* crus clitoridis, *h* hood, *r-cr f* retro crural fascia, *sl* suspensory ligament, On the pictures (**a′**, **b′**), the triangular surface of each retro-crural fascia has been delimited by *red lines*

which are routed small arteries, veinlets, nerve rami with, at their contact, many Pacini's corpuscles. In this region, a median fibrous partition (aligned on the median septum of the clitoral body), separating the right and left vasculo-nervous contingents, is often observed.

5.2.2 The Body

Such as we have just observed, it is the joining of the two crura which will form the body of the clitoris. At the start, this body continues the direction of the crura, in front of the pubic symphysis. It then has a large diameter, representing the sum of the diameters of each clitoral crus. The joining of the medial surfaces of the two albugineae (Fig. 5.10) creates a true fibrous partition: the septum of the crura (pectiniform partition of Testut), which will extend into the corporeal clitoral cylinder, by the median septum. When it reaches mid-height of the ventral symphysis surface, the body of the clitoris will suddenly change directions and leave the bone plane and protrude downwards and backwards into a sagittal plane. This change of direction and plane corresponds to the

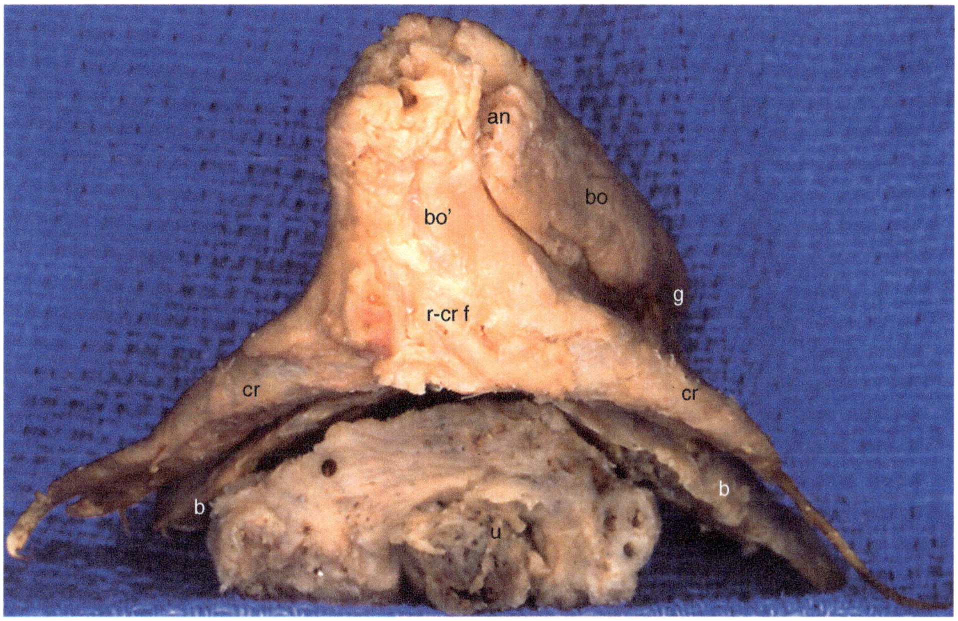

Fig. 5.8 Relationships between the crura clitoridis, the bulbs and the urethra (dorsal view of an ex situ dissection of a bulbo-clitoral organ removed with the underlying urethra). *an* angle, *b* bulb, *bo* clitoral body, post-angular part (descending part), *bo′* clitoral body, pre-angular part, (ascending part), *cr* crura clitoridis, *g* glans, *r-cr f* retrocrural fascia. The bulbar commissure draws an arch which creeps between the joined crura and the distal urethra's dorsum

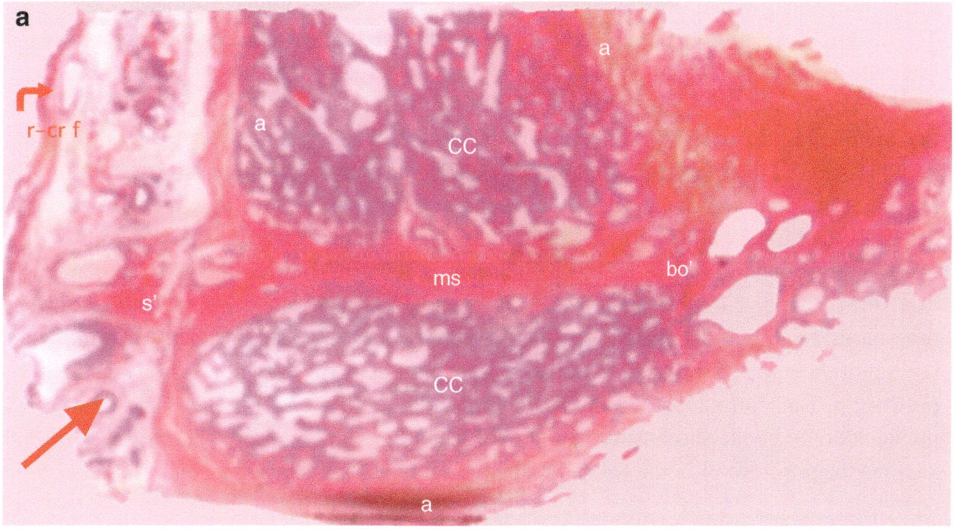

Fig. 5.9 The retro-crural fascia and the retro-crural space. (**a**) Photomicrograph of a section of the joined crura of the clitoris showing the retro-crural space; (**b**) microscopic aspect (standard staining by HEAS); (**c**) microscopic aspect (PSA staining). *a* albuginea, *CC* corpus cavernosum, *ms* median septum, *N* bundle of the dorsal nerve of clitoris, *r-cr f* (with *curved red arrow*): retro-crural fascia, *s′* septum of the retro-crural space, *long thin red arrow* it shows the retro-crural space, *short thick red arrow* it shows the microscopic aspect of the retro-crural fascia

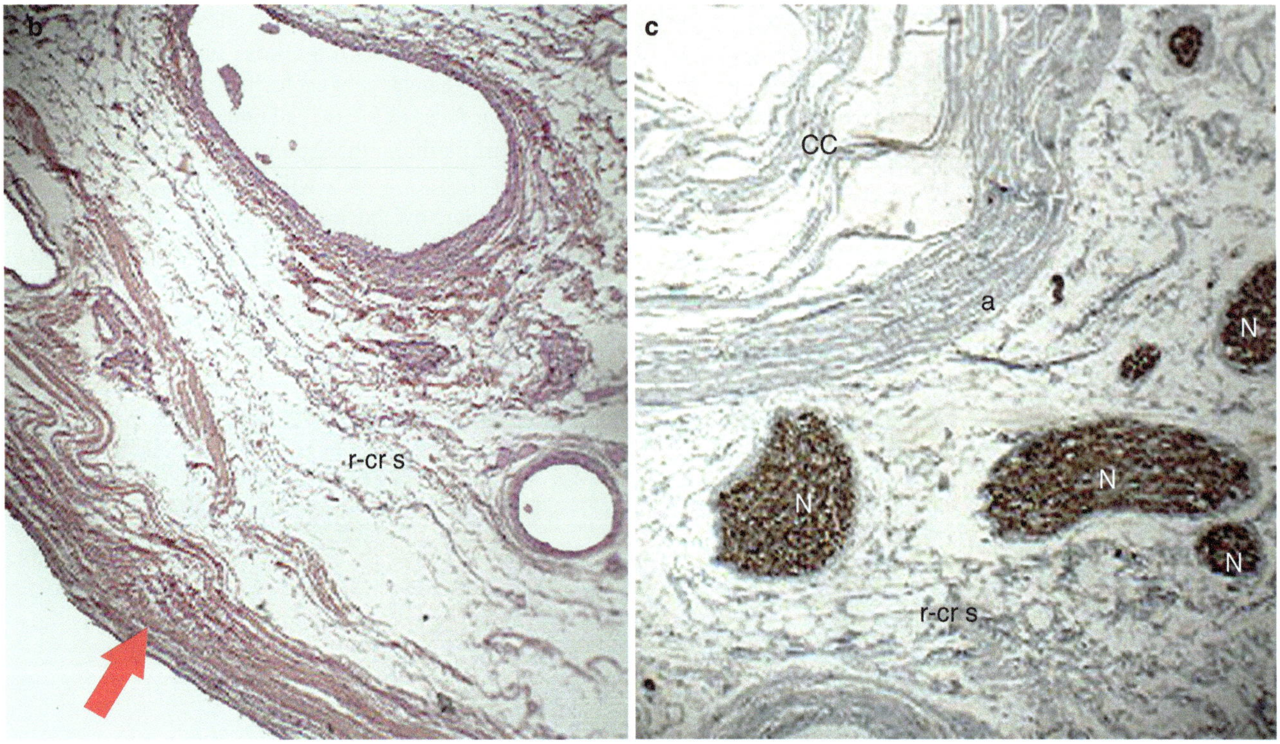

Fig. 5.9 (continued)

angle or **elbow of the clitoris**. This elbow therefore belongs to the clitoris body, which it separates into two segments: a short segment, against the pubis, and a long segment, arranged on a sagittal plane, perpendicular to the symphysis plane. The entire body of the clitoris thus has, via lateral vision, a bent aspect, with a high convexity (Fig. 5.5), which sometimes recalls a sort of bump.

The **angle** or **elbow** (that L. Köbelt referred to as the **knee**) deserves to be studied with care, in addition to the fact that it corresponds to a significant change in direction, it has two major characteristics: It is the area to which is attached a major portion of the suspensory ligament (see relevant chapter). It is also the area from which the corporeal structure acquires its final volume and its cylindrical shape. This elbow corresponds to an angle, the Köbelt angle of inflection, 90° on average (extreme angles are of 85 and 150°). In most cases, the top of the angle is rounded. Sometimes, even the top of the angle is replaced by an arch or a small dented protrusion. This clitoral elbow is attached very solidly to the pubis, not only by the suspensory ligament of the clitoris but also via the retro-crural fascia backed to the prepubic fibrous coating.

The **ascending segment** (ascending part or pars ascendens) of the body of the clitoris, located upstream of the elbow, is the **pubic segment** (or **pre-angular** segment) **of the body**. It results from the joining of the 2 corpora cavernosas of the crura clitoridis. This segment is relatively short and does not exceed 2 cm (average length: 1.28 cm). Initially positioned against the pubic symphysis, it then slightly detaches therefrom while maintaining its slanted ventro-cranial orientation (Fig. 5.5). In this segment, the 2 corpora cavernosas of the crura still retain their cylindrical shape. They are tangents via the medial edges of their albuginea. However, the medial parts of the 2 albugineae will imperceptibly merge, as if forceps were applying lateral tensions and crushing the 2 cavernous cylinders against one other (Fig. 5.10). The 2 cylinders will take the shape of 2 tangent hemi-cylinders via their medial parts, which have become straight. Firstly, a true fibrous commissure is created, the **cavernous commissure**. Then, the albuginea will merge and cause the coalescence of the medial fibrous parts, which have become rectilinear. Thus are formed the double **corpora cavernosa of the clitoral body** and its septum, **the median septum**. It should also be noted that the relatively large cavernous cavities at the crura will tend to be packed inside the corpus cavernosum of the clitoral body and thus become narrower (Fig. 5.6) and occupy a smaller surface whereas their number has more or less doubled as it includes the cavities of the two crura! This cavernous structure is quite visible in the **preputial segment** or **post-angular segment of the clitoral body** (Fig. 5.11). This segment, with a descending direction (descending part of the clitoral body), is directed downwards and to the rear. The ¾ distal sections of this part of the clitoral body can be felt under the skin of

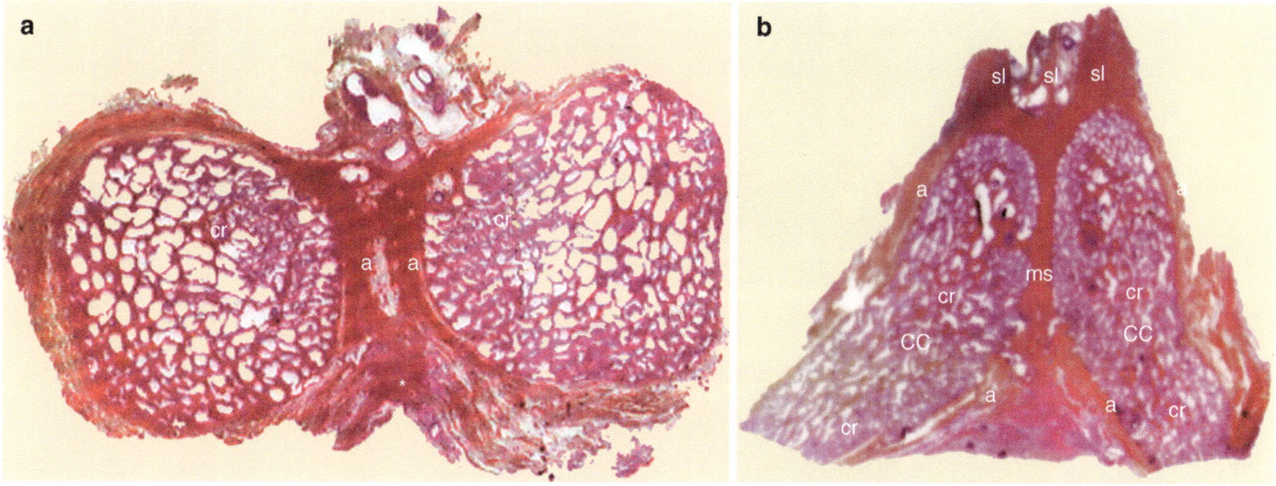

Fig. 5.10 Photomicrographs of two sections of the crura clitoridis. (**a**) Transversal section just before backing (the medial parts of the tunicae albugineae are still separated and form the cavernous commissure). (**b**) Coronal section of the joined crura in the pre-angular part of the clitoral body (ascending part), showing the formation of the clitoral body (the medial part of the 2 tunicae albugineae has merged in a single median septum). *a* albuginea (tunica albuginea), *CC* corpora cavernosa, *cr* crus clitoridis, *ms* median septum, *sl* suspensory ligament, * retro-crural fascia

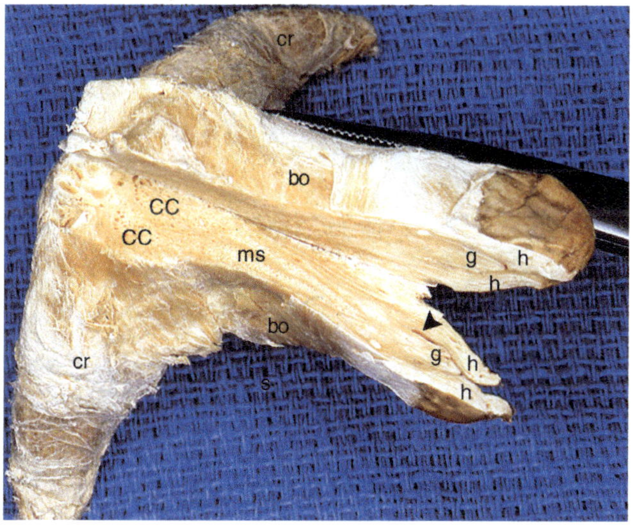

Fig. 5.11 Transverse section of a bulbo-clitoral organ showing the structure of both the clitoral body and the glans (the superior half of the body is slightly tilted upwards by a forceps, in order to see the inferior section's surface). *bo* clitoral body (descending part), *CC* joined corpora cavernosa, *cr* crus clitoridis, *g* glans clitoridis, *h* hood, *ms* midline septum, *black arrowhead* neck of the glans. Note the gradual reduction of the joined corpora cavernosa's diameter which become very sharp at their distal end

the prepuce, at the interlabial groove. It is this part which is extended, as we have already observed, by a free end: the glans. This part of the body of the clitoris is sometimes referred to as the **clitoral shaft**. Anatomists, as well as gynaecologists, have not conducted many studies on the length of the clitoral body. During our work, we observed an average length of 1.28 cm for the pubic part and 3.24 cm for the shaft (the glans being excluded). If the glans is associated to the shaft, we obtain 3.75 on average for the entire descending part! Even less knowledge is available concerning the diameter of the clitoral body. It is true that the definition of this diameter is difficult and represents an immediate problem: Is the clitoral body to be measured with all its envelopes? Is it necessary to include the skin of the prepuce in the analysis of the diameter of the descending portion? Is it necessary to include the latero-corporeal fibres of the suspensory ligament when making measurements? Finally, we chose the least debatable type of measurement: that of the diameter of the clitoral body, completely stripped of its envelopes and therefore simply covered by its albuginea. The average diameter is then of 1.10 cm for the posterior third, which confirms the size reduction occurring after the elbow. Otherwise, as each root has an average diameter of 0.9 cm, we should have found for the body: 0.9 × 2, i.e. 1.8 cm. This diameter decreases further because the transverse diameter of each of the corpora cavernosa progressively decreases and the average diameter of the distal third of the shaft's pars cavernosa is only of 8 mm (see comparative sections). Then, a very slight widening of the distal part of the body occurs; this corresponds to the transition towards the anatomical glans. Due to the fact that the insertion of the hood only rarely occurs at the junction of the corporeal cylinder and of the base of the anatomical glans, in most cases, the visible part of the glans (or free part) only corresponds to the end of this anatomical glans (see Fig. 5.12). The structure of the shaft results from its formation mode: merging of 2 crura and thus of 2 corpora cavernosas, which is perfectly visible on the sections perpendicular to the axis of the descending part of the body—2 corpora cavernosas, surrounded by a single peripheral albuginea, resulting from the merging of

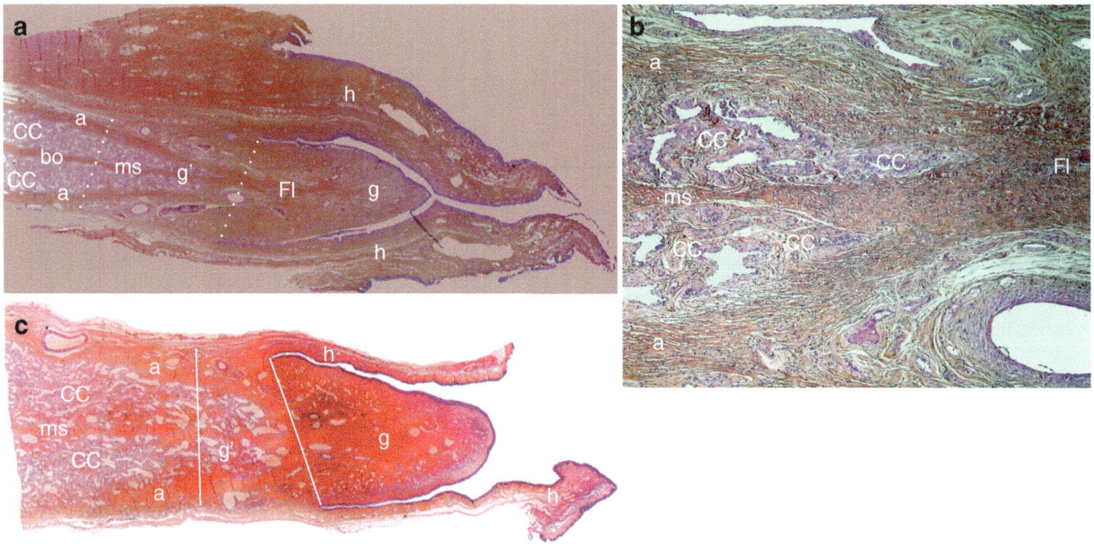

Fig. 5.12 The end of clitoral body and the glans. (**a**) Photomicrograph of a transverse section of the end of the body's descending part. (**b**) Microscopic aspect of g′ (hidden part of the glans or end of the body for other authors). (**c**) Photomicrograph of another specimen (section similar to **a**). *a* tunica albuginea, *bo* clitoral body, descending part, *CC* corpora cavernosa, *Fl* Forster's ligament, *g* glans (free end), *g′* glans (hidden part), *h* hood, *ms* median septum. The *dotted* and *bound white lines* delimit what is considered by many authors as the hidden part of the glans. Note also the peniform morphology of the glans **c**

the 2 albugineae of the crura, are observed. These 2 corpora cavernosas, shaped as quarter moons, with rounded angles, are separated by **the median septum** ("pectiniform partition" of the previous authors) became the cavernous septum, resulting from the fusion or coalescence of the joining area of the albuginea from the two crura. This septum, such as the albuginea, is essentially formed by a dense collagen fibre network. It is an incomplete septum, allowing, without any difficulty, the communication of the blood flow between the 2 corpora cavernosas (Fig. 5.13). The 2 cavernous arteries are connected to this septum and arranged symmetrically thereto. We will study in detail, further in this document, the tissue of the corpora cavernosa in order to gain a better understanding of its operating mode whose finality is a clitoral erection. We can firstly say that many rami are located within the cavernous tissue but that larger nerve branches, as well as a great number of vessels, are routed on the lateral walls of the albuginea, in the direction of the glans. The cavernous cylinder and the vessels and nerves, which are routed along the cylinder, are surrounded by a thin conjunctive envelope: the clitoral fascia (Fig. 4.7), analogous to the deep fascia described by Buck for the male penis. The neurovascular formations are not the only components, which can be found in abundance, around the corpora cavernosa. The body of the clitoris has a high content of sensory corpuscles, especially mechanoreceptors of Vater-Pacini (see Chap. 9). These corpuscles are located in a site in contact with the neurovascular bundles, which is not surprising as each mechanoreceptor contains a sensitive nerve fibre.

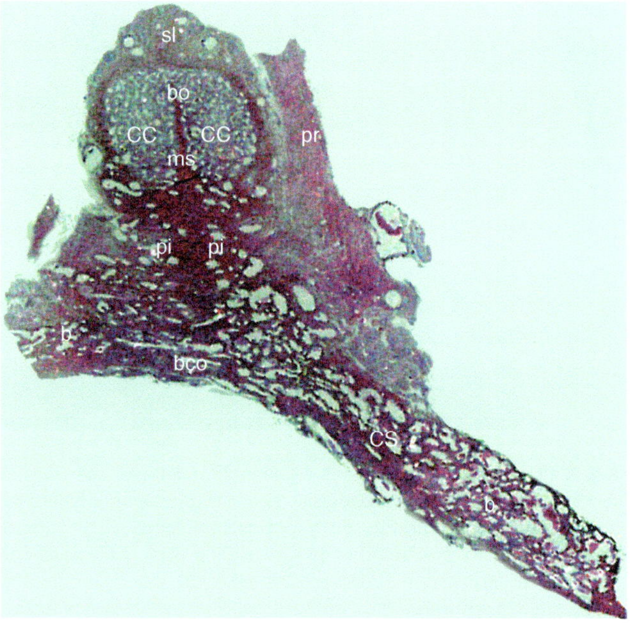

Fig. 5.13 Photomicrograph of a bulbo-clitoral organ's frontal section at the angle of clitoral body. *b* bulb (corpus spongiosum), *bco* bulbar commissure, *bo* clitoral body, *CC* conjoined corpora cavernosa, *CS* corpus spongiosum, *ms* median septum, separating the two conjoined corpora cavernosas, *pi* pars intermedia of Köbelt, *pr* prepuce. NB: on this section, it is evident that the two constituent parts of the female organ (corpora cavernosa and corpora spongiosa) form an unified structure: the bulbo-clitoral organ

- The "top" part of the corporeal cylinder has key connections: The dense and most ventral fibres of the suspensory ligament of the clitoris are attached to its proximal part, beyond the elbow, such as they are attached to the latter. This is also the case of, when they exist, the fibres of the fundiform ligament of the clitoris, of which many pass on lateral surfaces of the body (see Chap. 9). Its distal half is superficial and subcutaneous: It is the back of the clitoral shaft, which can be felt beneath the ventral commissure of the lips, covered by the skin of the prepuce. Between the preputial subcutaneous tissue and the clitoral fascia, some neurovascular structures are observed: the superficial vessels and nerves of the clitoris (Fig. 12.3). A microscopic study also shows, at this level, some smooth muscle fibres similar to those of the dartos of the penis, as well as large tact corpuscles.

- The "bottom" part of the corporeal cylinder is in direct connection with the commissure of the bulbs as well as with the infra-corporeal residual spongy part. These spongious structures are connected to the body of the clitoris via the network or pars intermedia described by Köbelt.

5.3 Parts of Spongy Structure

These structures are the "vestibular" bulbs joined together by the commissure of the bulbs and the infra-corporeal residual spongy part.

5.3.1 "Vestibular" Bulbs

It is undeniable that the **two bulbs** belong to the bulbo-clitoral organ, as opposed to what can be read in certain books on anatomy, even the best known works,[9] while at the same time Köbelt thought that he had definitely demonstrated this connection since 1851. After the work published by Köbelt, other vicissitudes awaited the bulbs, whose name changed from one author to another over time, from the "plexus retiform" terminology to that of "vestibular bulbs", which has been employed by the commissions of the anatomical nomenclature, including terms such as "vaginal bulbs" (Testut) or "urethral bulbs" or even "urethro-vaginal bulbs", which is better suited to anatomical facts (see Chap. 13).

The 2 bulbs are generally represented as two bulky, long, curved and convex formations, which are convex on their bottom surface, via which they are discovered during a perineal dissection (Fig. 5.3), once the superficial perineal fascia (fascia of Colles) has been broken down and the **labial fatty body**, which hides them, has been dissected and folded forwards. This is why, the bulbs have been graphically compared to "2 leeches full of blood" (Köbelt), positioned against the urethro-vaginal pyramid (see Fig. 5.3), laterally bordering the external orifice of the urethra and the vaginal orifice (introïtus), arranged diagonally and directed frontwards and inwards. The curved aspect may sometimes concern the bottom surface as well as the top surface: In such a case, the bulb takes the aspect of drop-shaped pendant, determining a sort of small latero-vaginal cavity. The bulbs may also have a more or less flattened shape, which is hardly convex on its bottom surface and is comparable to a flattened dry fig (Fig. 5.4), with a tapered antero-medial end and a large swollen posterolateral end. The dimensions of the bulbs are extremely variable: 3.15 cm on average with extreme measurements of 1.5 and 4.2 cm. The greatest width, generally located near the posterolateral end, is on average of 2.2 cm on average with extreme widths of 1.3 and 3.2 cm. The thickness varies on average between 3 mm (flattened bulbs) and 11 mm (bulging bulbs). The 2 bulbs are rarely perfectly symmetrical. Similarly, in most cases, the dimensions measured on the right and on the left vary considerably. The **top surfaces** of the 2 bulbs are in a supero-medial plane and include two insertion areas, which are sometimes separated by a small edge: an insertion on the lateral edges of the distal ends of the urethra and vagina and an insertion on the bottom fascia (perineal membrane) of the neighbouring urogenital diaphragm. The **lower surfaces** of the bulbs are in a slightly oblique, infero-lateral plane and are hidden by the fibres of the bulbospongiosus muscles applied to their albuginea. For each bulb, a **lower** medial **edge** (urethro-vaginal edge or vestibular edge) and an **upper** lateral **edge** directly connected to the crura of the clitoris are described. These edges are blunt and rounded. The posterolateral ends of the bulbs (which are sometimes referred to as the "bases" of the bulbs) are in contact with each great vestibular Bartholin's gland, which they cap in a frontal region (Fig. 13.2). When the bulbs are relatively symmetrical and short, the tangent at their base passes via the centre of the introïtus. The antero-medial ends (which are sometimes referred to as the "top" of the bulbs) are interconnected by the **commissure of the bulbs** (bulbar commissure). The bulbar commissure is, unlike the "pseudo-cavernous commissure", a true commissure (Figs. 5.5 and 5.13) each bulb merges with the bulb on the opposite side via a true intermediate spongious bridge: the "fer à cheval à extrémité postérieure" ("horseshoe with a posterior end") described by Grégoire.[10] This flattened bridge, with a concave end section, penetrates between the junction angle of the 2 crura clitoridis and the anterior surface of the distal urethra.[11] It thus forms a sort of Roman arch in the ogival region resulting from the convergence of the two clitoral crura (Fig. 5.5). The anterior end of the 2

[9]Thus, in the impressive treaty of Bourgery and Jacob, the bulb is not described in the "clitoris" paragraph but in the "structure of the vagina" paragraph (see the chapter on "female genital apparatus". Organes génitaux de la femme, in Tome V, Slanchnologie, p. 277–304 et planches 63à 75).

[10]The median septum of the pars intermedia may extend at the bulbar commissure and thus outline the formation of the bulbar septum. However, the latter is never complete and significant left-right direct communications persist between the 2 bulbs.

[11]See Chap. 13, the very close connections between the bulbar commissure and the urethral wall and Bibliography: M.S. Baggish et al.

bulbs and the bulbar commissure form, on the dorsal views, a nearly perfect half-circle (Fig. 5.8) whose diameter significantly varies (between 2 and 4 cm) according to the anatomical arrangement, the shape and the axis of the bulbs. The distal terminal part of the urethra passes under this spongy tunnel (Figs. 5.8 and 13.1). Such as the bulbar commissure curvature differs from the straightness of the clitoral crura and from their angular convergence, other differences appear between the bulbs and the crura of the clitoris: The crura and their junction, in a more cranial location, seem to overlap the bulbs and their commissure. The large axis of the bulbs and that of the crura cross each other and form an extended X. While the crura, positioned against the ischio-pubic rami, have an axis, which is close to the lower edge (the axis of this bone segment), the bulbs, which are less oblique and sometimes almost vertical, are practically perpendicular to the large vaginal axis (Fig. 5.5).

It should be recalled that from a histological viewpoint, the bulbs and their commissure consist of a spongy tissue, wrapped in a thin connective tissue sheath, the bulbar albuginea, which does not have the thickness or the mechanical characteristics of the albuginea of the corpora cavernosa. The bulbs are residues of the deep part of the foetal corpus spongiosum which, due to the female genital dehiscence, has positioned itself "where it remains in place", i.e. on either side of the urethral and vaginal ducts. Thus are explained the differences in size and shape existing between the right and left bulbs, the available volume not being automatically identical on either side.

The commissure of the bulbs is not, as could be believed according to the photos of dissected parts, dissociated from the clitoris beneath which it penetrates. As has been pointed out from the start, in this chapter, the spongy and cavernous formations belong to a single bulbo-clitoral organ. The anterior surface ("back") of the commissure of the bulbs is closely connected to the posterior surface of the body of the clitoris at the inflection angle between the ascending and descending segments of the body. The junction area, well known since Köbelt, is called, such as has already been noted, the pars intermedia.

Moreover, the commissure of the bulbs has an extension, which protrudes downwards and backwards, under the post-angular descending part of the clitoral body, in the direction of the glans. This extension is also a residue of the foetal spongy tissue. We propose to call it the **infra-corporeal residual spongy part**.

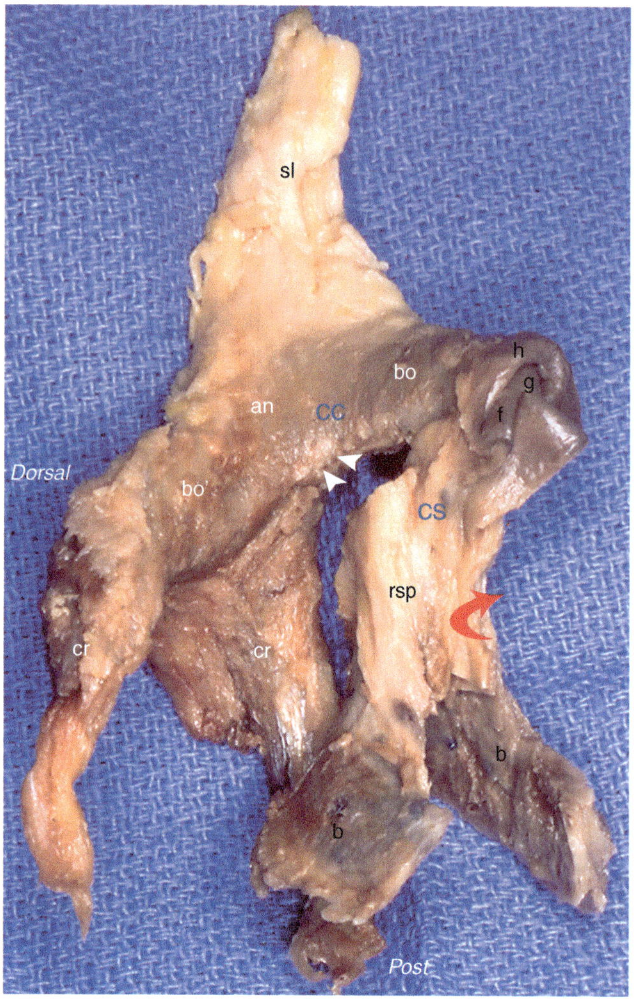

Fig. 5.14 Advanced dissection of a bulbo-clitoral organ (view of ¾ right) with an almost complete separation of the spongy and cavernosa parts. The spongy part is lightly moved forwards (*red curved arrow*). *CC* corpus cavernosum, *CS* corpus spongiosum, *an* angle (or elbow), *b* bulb (spongiosus), *bo* body of the clitoris (descending part), *bo'* body of the clitoris (ascending part), *cr* crus clitoridis, *f* frenum (frenulum clitoridis), *g* glans clitoridis, *h* hood, *rsp* residual spongy part, *sl* suspensory ligament, *white arrowheads* they show two vascular openings of the intermediate network of Kobelt

5.3.2 Infra-corporeal Residual Spongy Part

This portion (**icrsp** or by abbreviation, **rsp**), which is not well known and often forgotten, remains an important component of the bulbo-clitoral body. Recognising this part is essential as its existence facilitates the understanding of the organogenesis of the apparatus itself. This extension of the bulbar commissure (Fig. 5.14), which has an irregular aspect and a diameter nearly equal to that of the cavernous cylinder of the clitoral body, follows the posterior edge of this body, to which it is closely connected via the distal part of Köbelt's pars intermedia. It resembles a cord of spongious tissue, which is more or less involutive due to the anarchistic architecture including irregularly arranged gaps and a conjunctive interstitial tissue, which is more abundant than in the bulbar spongy tissue. This spongy portion, which is "infra-corporeal" on the parts studied ex situ, deserves to be qualified as "residual" because it is what remains from the foetal corpus spongiosum, whose initial purpose (surrounding the phallic urethra and the urethra of the glans) could not be fulfilled due to the involution of the above-mentioned urethral

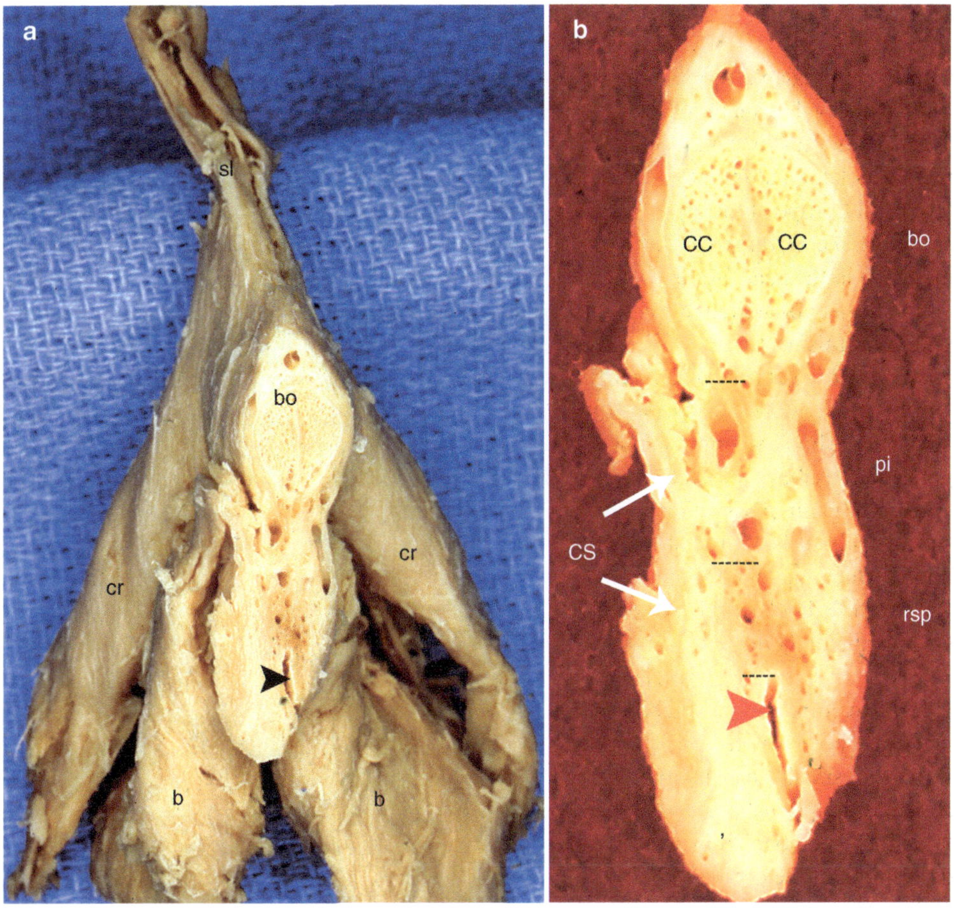

Fig. 5.15 Position of the pars intermedia of Köbelt on a frontal section of a bulbo-clitoral organ. (**a**) Level of the section (ventral view of a bulbo-clitoral organ); (**b**) enlargement of the section's surface. The section, perpendicular to the axis of the clitoral body, interests the clitoral shaft (descending part of body). *b* bulb, *bo* clitoral body (pars descendens), *CC* corpora cavernosa, *cr* crus clitoridis, *CS* corpus spongiosum (residue), *pi* pars intermedia (Köbelt), *rsp* residual spongy part, *sl* suspensory ligament, *black* and *red arrowheads* showing the end of the vestibule's roof

portions, but which has remained connected to the corpora cavernosa in their development in the clitoris, such as for men in relation to the formation of the penis. The similarity exists up to the glans, which will be formed by the spongy tissue. This is the case for men and women.

This extension was especially studied by A.A.W.M. van Turnhout et al. who were able to use many samples from plastic surgery operations related to "transgender" transformation (from woman to man). At the time of these operations, referred to as metoidioplasty (or metaoidioplasty) surgery, the clitoris, hypertrophied by a preliminary hormonal treatment, is transformed into a penis, and one of the surgical stages involves resecting the spongy tissue of the icrsp, which forms a sort of indurated cord under the body of the clitoris. For our part, we could, on each laboratory specimen, perfectly identify this infra-corporeal residual spongy extension of the bulbar commissure, ending at the glans of the clitoris. All of all the spongy structures also have a relatively surprising aspect (Figs. 5.2 and 5.14) recalling a sort of monstrosity and the anatomist can only be disconcerted by the contrast between this aspect and the delicate appearance of the free part of the glans, which luckily is the only part which is visible!

The icrsp communicates with the cavernous cylinder of the clitoral shaft, such as the commissure of the bulbs communicates with the corpora cavernosa at the corporeal inflection angle. However, for the icrsp, the connection vessels of the pars intermedia are shorter and more spaced out.

The vestibule, which is more posterior, is located immediately beneath the icrsp and the vessels of its lateral walls also communicate with the icrsp. The vessels of the spongy tissue diverge when they meet the ridge of the vestibular cavity (Fig. 5.15) and separate to penetrate the thickness of each wall of the vestibule and develop communications with the vestibular vessels which, in turn, are anastomosed with those of the labia minora. This explains the well-known contribution of the labia minora to the physiological modifications occurring during the sexual excitation phase in women. Overall, this represents continuity between the corpora cavernosa of the clitoral body, pars intermedia, bulbs and icrsp or between icrsp and vulvar vascular network (Fig. 5.15).

5.3.3 The Pars Intermedia

It was perfectly described in 1851 by Köbelt in his book entitled *De l'appareil du sens génital des deux sexes* ("About the genital sense apparatus of both sexes". Despite the limited technical means of the time, the drawings of this "pars intermedia" (see Fig. 5.16) carried out by this author

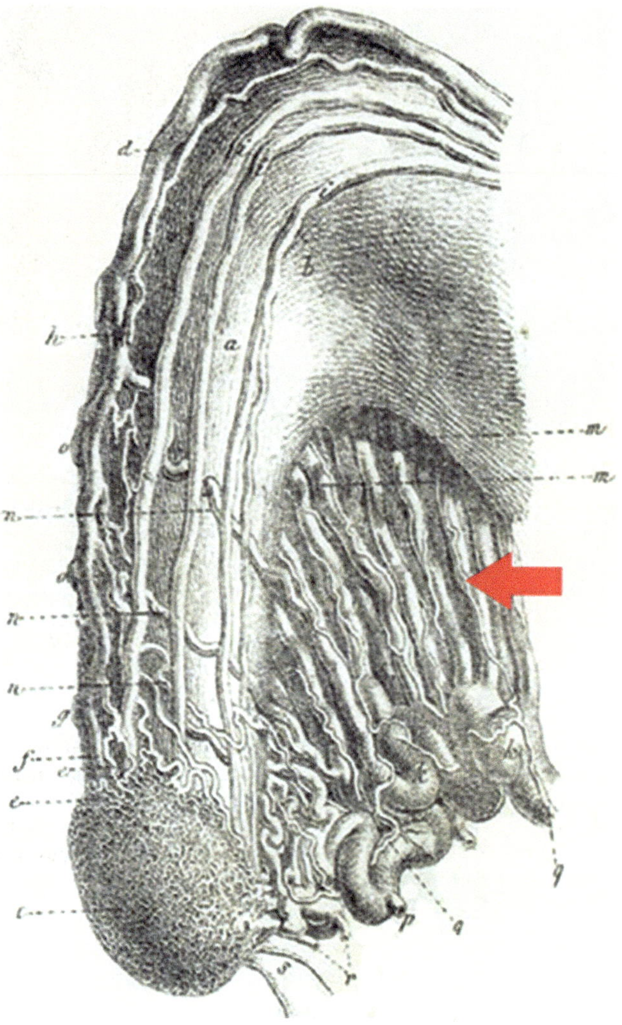

Fig. 5.16 The communicating vessels of the pars intermedia (*red arrow*) (drawing of Köbelt's book)

according to his dissections are astonishingly accurate and comply perfectly with the histological, anatomical and radiological documents, which we, happy researchers of the twenty-first century, have at our disposal! This network is particularly dense at the junction point between the commissure of the bulbs and the clitoral inflection (elbow of the clitoris). At the corner junction angle, the pars intermedia is dense and spread out and generally consists of 2 rows of more or less sinuous vertical veins, the majority of which are large, separated by a thick median fibrous septum and aligned on the cavernous septum, which it seems to extend (Figs. 5.17 and 5.18). The height of the venous rows and their median septum is variable, but in most cases, it is equivalent to that of the median septum of the clitoral body (cavernous septum). The veins of the pars intermedia connect the junction area of the crura clitoridis and 2 bulbs (Fig. 5.17). They are nearly always large and have a longitudinal axis (parallel to that of the septum). More rarely, these anastomotic veins form an actual venous maze (within which the septum is no longer visible), which provides a nearly direct communication between the spongy and cavernous tissues of the bulbo-clitoral organ. The pars intermedia is not limited to the inflection angle of the clitoral body. It extends up to the distal end of the body of the clitoris, thus developing a close connection between the corpus cavernosum of the clitoral body and the infra-clitoral residual spongy part. It is, however, less dense and more dispersed than at the inflection angle. At this level, it has an aspect, which has already been perfectly described by L. Köbelt: "double symmetrical row of venous communications, which are directed towards the lower surface of the body of the clitoris, where they penetrate". The vascular orifices corresponding to the cross section of these vessels, when separated by dissection, the cavernous part of the body and the icrsp, are perfectly visible, even to the naked eye (Fig. 5.14). The lateral veins (also described by Köbelt), which "surround the lateral surface of the body of the clitoris" to reach "the dorsal vein of which they form the lateral roots", also develop from this pars intermedia.

However, Köbelt's network is not the only means of communication and fusion of the cavernosa and spongiosa corpora. There also exist interpenetrations between these 2 erectile components and areas, which the spongy venous lakes penetrate directly within the small cavernous cavities, without any intermediate vascular interposition, which can be clearly seen by microscopic observation (Fig. 5.17). This refers once again to the existence of a single bulbo-clitoral organ where the 2 erectile tissue components are not only connected but closely related.

5.3.4 The Glans

Although its free part has already been studied in the "morphology" chapter (Chap. 4), the glans, a very particular formation, deserves a global structural study. The glans, which derives, as we have observed, from the genital tubercle, such as the male penis, consists of, such as the latter, a spongy tissue. However, the lack of any "hollowing" of a pre-meatic urethra has contributed, such as for the infra-corporeal residual spongy part, which did not find a phallic urethra below the corpus cavernosum of the clitoral body, to modifying this spongy tissue of the genital tubercle. This tissue has persisted but has an in fine, involutive aspect, such as the "icrsp", which it extends. This explains the numerous discussions on the erectile or non-erectile character of the glans, reported by R. Stefani et al.

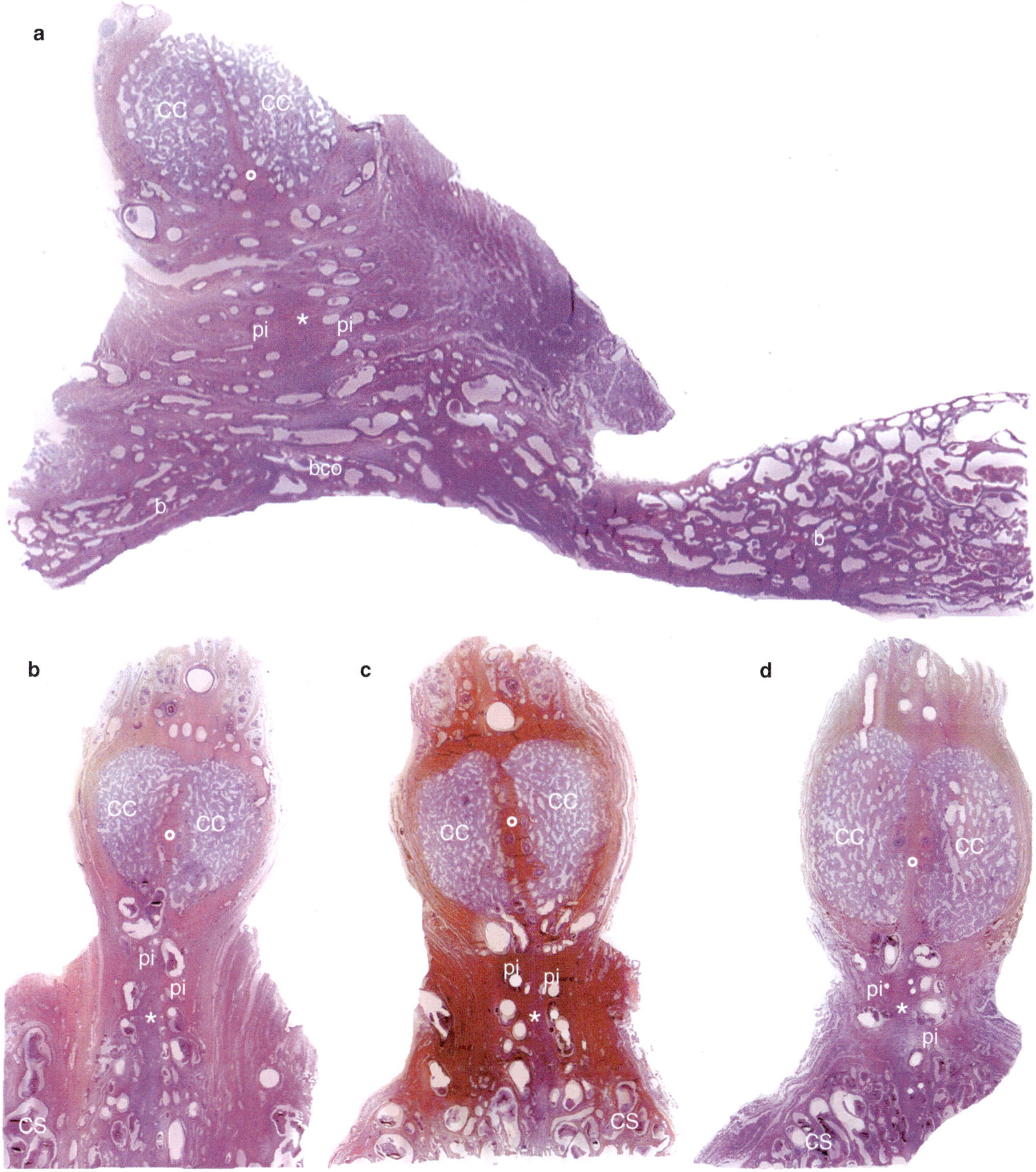

Fig. 5.17 Frontal sections of 4 bulbo-clitoral organs. Photomicrographs showing the aspects of the pars intermedia (the connections with the bulbs are well visible only on **a** specimen). *b* bulb, *bco* bulbar commissure, *CC* corpora cavernosa, *CS* corpus spongiosum, *pi* pars intermedia, ° median septum of CC, * septum of the pars intermedia. On (**a**–**c**), the septa of CC and of the pars intermedia are aligned. On (**d**), the aligning is less evident and the vessels more plexiform

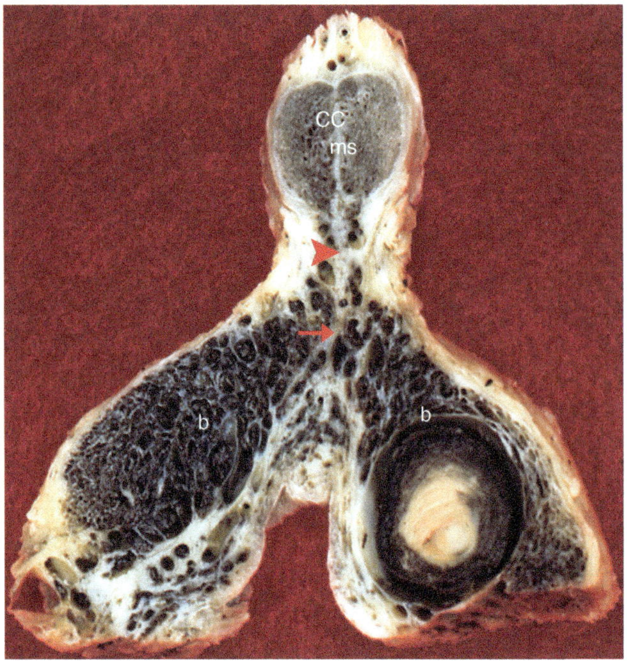

Fig. 5.18 The septa of the bulbo-clitoral organ's erectile bodies (frontal section of a bulbo-clitoral organ). *b* bulb, *CC* corpora cavernosa clitoridis in the clitoral body, *ms* median septum, *red arrowhead* septum of the pars intermedia, *red arrow* septum into the bulbar commissure. Note: The aligning of the septa. An old haematoma visible into the left bulb

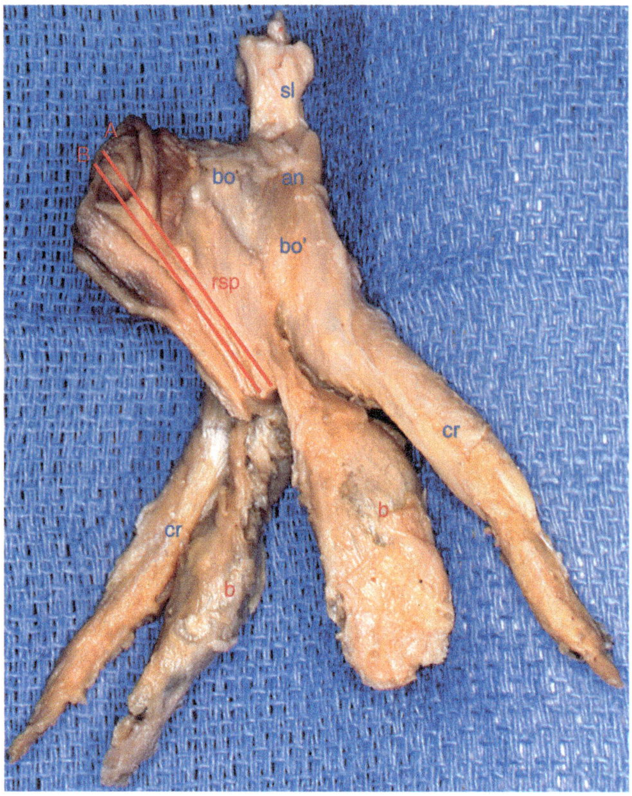

Fig. 5.19 Lateral view showing the two constituents of a bulbo-clitoral organ. *Red legends* for spongy tissue (Two transverse Sections A and B have been made in the axis of RSP, in order to observe its architecture (see photomicrographs of these transverse sections)). *Blue legends* for cavernous tissue. *an* angle, *b* bulb, *bo'* clitoral body, pre-angular part (ascending part), *bo* clitoral body, post angular part (descending part), *cr* crus clitoridis, *rsp* residual spongy part, *sl* suspensory ligament

In any event, from a structural viewpoint, two parts of the glans can be identified (Fig. 5.12):

- The distal part or free part, consisting only of a specific but involutive spongy tissue, covered with a non-keratinised squamous epithelium, and to which is attached the preputial hood and glans frenula.
- The proximal part, which is not visible and is a sort of hidden part of the glans, inside which terminate the tapered distal ends of the corpora cavernosa of the clitoral body, which are always separated by the median fibrous septum. However, the complexity of this proximal part of the glans does not end here.
  - Firstly, the cavernous septum (median septum) extends beyond the cavernous ends, thus forming the ligament described by Forster,[12] which opens up in the distal glans into several bundles (such as sort of palm tree) (Fig. 5.12), thus forming a true fibrous reinforcement of the clitoral glans (this reinforcement is even capable of reaching the end of the free glans).
  - Otherwise, this arrangement only occupies the "upper" half of the glans, at the level of the plane of the corpora cavernosa. The "lower" half (Figs. 5.19 and 5.20) is only occupied, such as the free part of the glans, by the involutive spongy tissue.

Overall, the clitoral glans benefits from tumescence possibilities, which allow both its vascular structure and its location (as it caps the end of an erectile clitoral body, to which it is attached by a fibrous reinforcement!).

[12]This ligament is often reinforced by a fibrous contingent from one or the two halves of the albuginea.

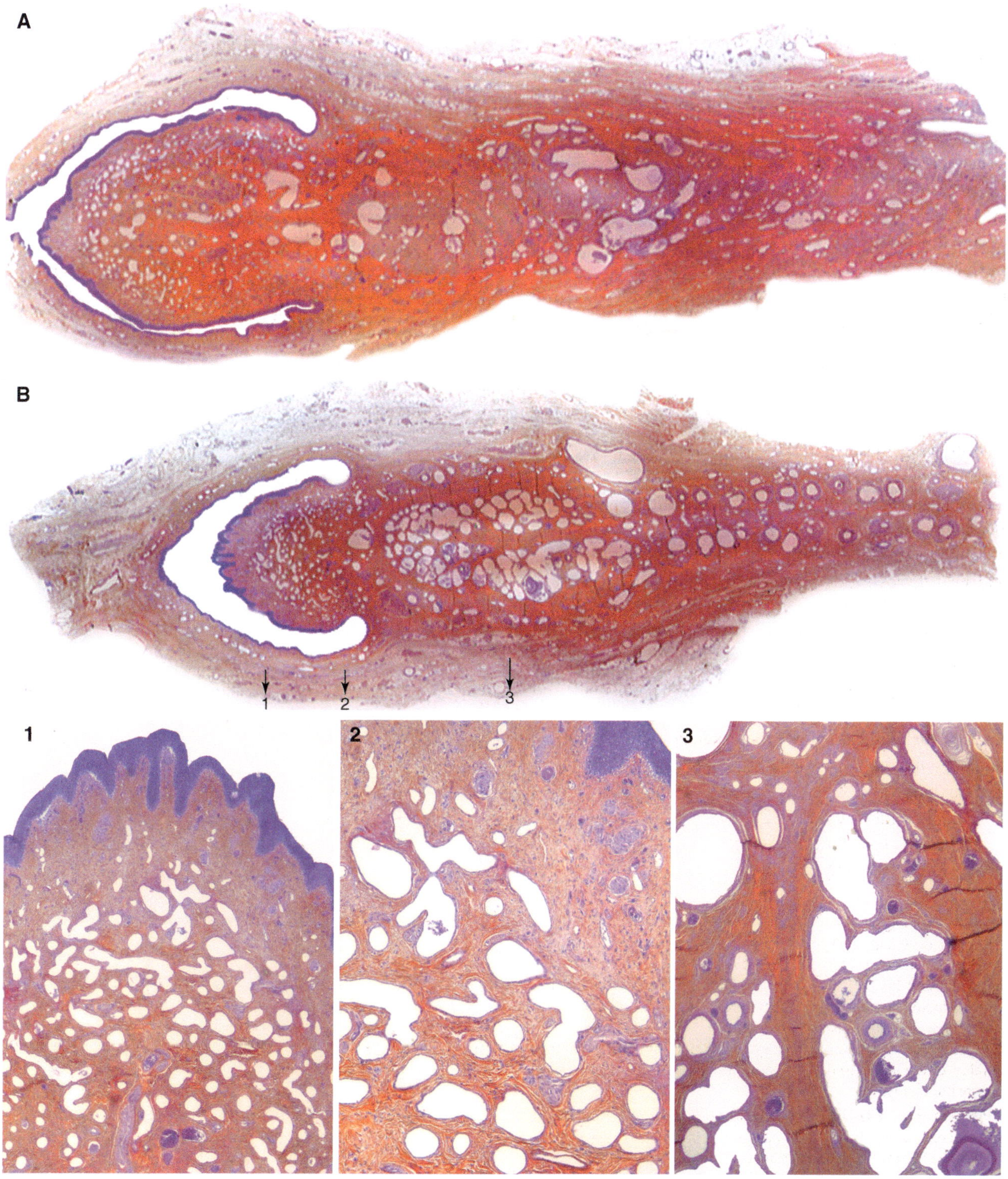

Fig. 5.20 Photomicrograph of the transverse section A through the RSP of the clitoral body's post-angularis part (see the level of section on the previous picture). Photomicrograph of the transverse section B through the RSP of the clitoral body's post-angularis part. Microscopic aspects of the spongy tissue of the RSP on frontal Sections 1, 2 and 3

6 Microscopic Study of the Bulbo-clitoral Organ

6.1 Cavernous and Spongy Tissue

6.1.1 Cavernous Tissue

The tissue of the corpora cavernosa clitoridis, or cavernous tissue, is erectile tissue, meaning its histological structure enables these bodies to fill with blood, increase in size and harden. This tissue is made up of a complex network of venous sinuses that appear as intercommunicating anfractuous cavities (cavernous labyrinth) and of a thick fibrous envelope, the albuginea, which was long believed, erroneously, to be inextensible. The sinusoid cavities, or sinuses of the corpora cavernosa, most often have wide-open lumina (though sometimes collapsed) that are very irregular. Their dimensions vary greatly. Their shapes are more often polyhedral or narrow and elongated rather than circular or oval. These sinuses present numerous recesses. The inter-sinus communications are clearly visible and the main characteristic of the sinuses is that they are richly anastomotic (Fig. 6.1). The wall of the sinuses is indeed a vessel wall, comprising an endothelial layer and a conjunctive axis made of collagenous bundles in which smooth muscle fibres may be seen. Specific stains (orcein) demonstrate the existence of a rich network of elastic fibres (Fig. 6.2). It should be noted that the walls of the sinuses are thin, especially if compared to the size of the vessel lumina. The endothelial cells, made apparent by the immunolabelling with an anti-factor VIII antibody, have junctions of varying tightness and rest on a basal lamina. Extensions originate from the sinus walls and float in the vessel lumina. Their role is probably to direct blood flow during filling. Most of these extensions appear as fingers that are curved to varying degrees (like a bent index finger). They may also take the shape of small mounts, pyramids, bull horns, points, hooks or clubs. There are also complete or incomplete bridges between two opposite sinus banks, the sinus septa, which contribute to making the sinus labyrinth even more complex. Numerous small vessels (arterioles or venules) are also housed in the sinus walls. These small vessels are very sinuous and their walls are often thick. The arterioles and their branches are called "helicine" due to their often helicoidal arrangement (see Chap. 10). The larger vessels are located at intersections in the inter-sinus connective tissue. Otherwise, a fair number of nerve endings are found in the sinus walls themselves and in the inter-sinus connective tissue (Fig. 7.1).

It is now proven that the work on corpora cavernosa of the penis can be transposed to the clitoral corpora cavernosa (A. Toesca et al.). Without negating the previous description, this work, notably that of A.M.B. Goldstein et al., demonstrates that instead of viewing the cavernous tissue as a labyrinth of sinuses with walls made of collagenous fibres and smooth muscle fibres, it should be considered as a fibrous collagenous tissue, rich in smooth muscle fibres, in which a network of intercommunicating vascular spaces are housed.

The inter-sinus connective tissue is made up of collagenous fibrous bundles which have the morphological characteristic of being wavy when in a state of rest (Fig. 6.3). According to E. Wespes, who for the most part, studied the penis corpora cavernosa, it is the presence of elastic fibres arranged in a bridge in contact with the collagenous bundles that enables the collagenous fibres to maintain this pleated appearance when in the flaccid phase and, most of all, to return to it as sexual arousal wanes.

The smooth muscle fibres of the clitoral corpora cavernosa were clearly identified using immunohistochemistry methods as the anti-desmin antibody affixes to them specifically. Alpha-smooth actin staining may also be used (Fig. 6.2). These muscle fibres have a longitudinal or transversal orientation within the inter-sinus connective tissue (A. Toesca). They are arranged as bundles or packets and end on collagenous fibres (Fig. 6.3) from which they will eliminate the folds to enable clitoral tumescence. These smooth muscle fibres are also present at the large arteriole and venule adventitia and then disappear on the small vessels. All of these smooth muscle fibres are interconnected. They therefore play a manifest contractile role in the regulation of sinus blood flow, as indicated by the author cited above. Their spontaneous myogenic activity corresponds to the flaccid

V. Di Marino, H. Lepidi, *Anatomic Study of the Clitoris and the Bulbo-Clitoral Organ*,
DOI 10.1007/978-3-319-04894-9_6, © Springer International Publishing Switzerland 2014

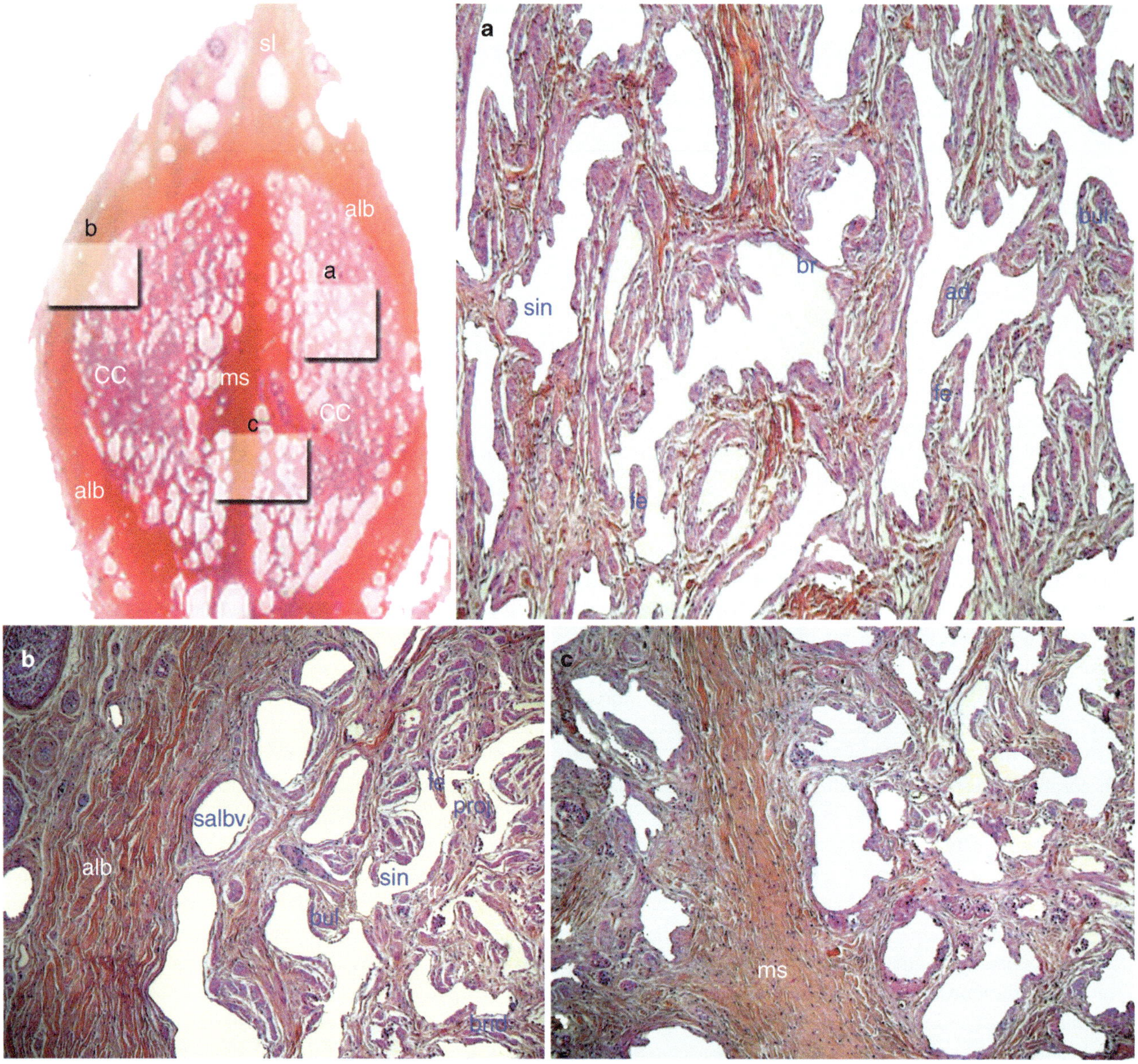

Fig. 6.1 Details of a frontal section through the descending part of the clitoral body showing the organisation of the corpora cavernosa. Part (**a**) of the section: the cavernous labyrinth; part (**b**) of the section: tunica albuginea and corpus cavernosum; part (**c**) of the section: the median septum and the corpora cavernosa. *ad* advance, *alb* tunica albuginea, *brid* bridge, *bul* bulge, *cc* corpus cavernosum, *fe* finger-like extension, *ms* median septum, *proj* projection, *sin* sinus, *sl* suspensory ligament, *salbv* sub-albugineous vein

state and their release only occurs during sexual arousal when the NO neurotransmitter is produced (see Chap. 7). Studied by electron microscopy (E. Wespes), these smooth muscle fibres demonstrate all characteristics of myogenic cells: uniform nucleus, fine basement membrane and above all, in the cytoplasm, mitochondria, contractile filaments and glycogen granules (energy nutrient).

Our microscopic observations demonstrated the existence of a sub-albugineal venous plexus in the clitoral corpora cavernosa, analogous to its well-known counterpart of the corpora cavernosa penis (Fig. 6.4). It is therefore surprising that certain authors negate the existence of this plexus (including A. Toesca). It is the compression of the veins of this plexus against the albuginea by the dilated cavernous sinusoids that prevents blood from leaving the corpora cavernosa and causes rigidity. Therefore, if this venous plexus did not exist, the clitoris would not have erectile capacity during sexual arousal.

At the body of the clitoris, the corpora cavernosa are surrounded by a single thick peripheral envelope of connective tissue, the **albuginea of the clitoral body's corpus cavernosum**. A fibrous sagittal partition, resulting from the

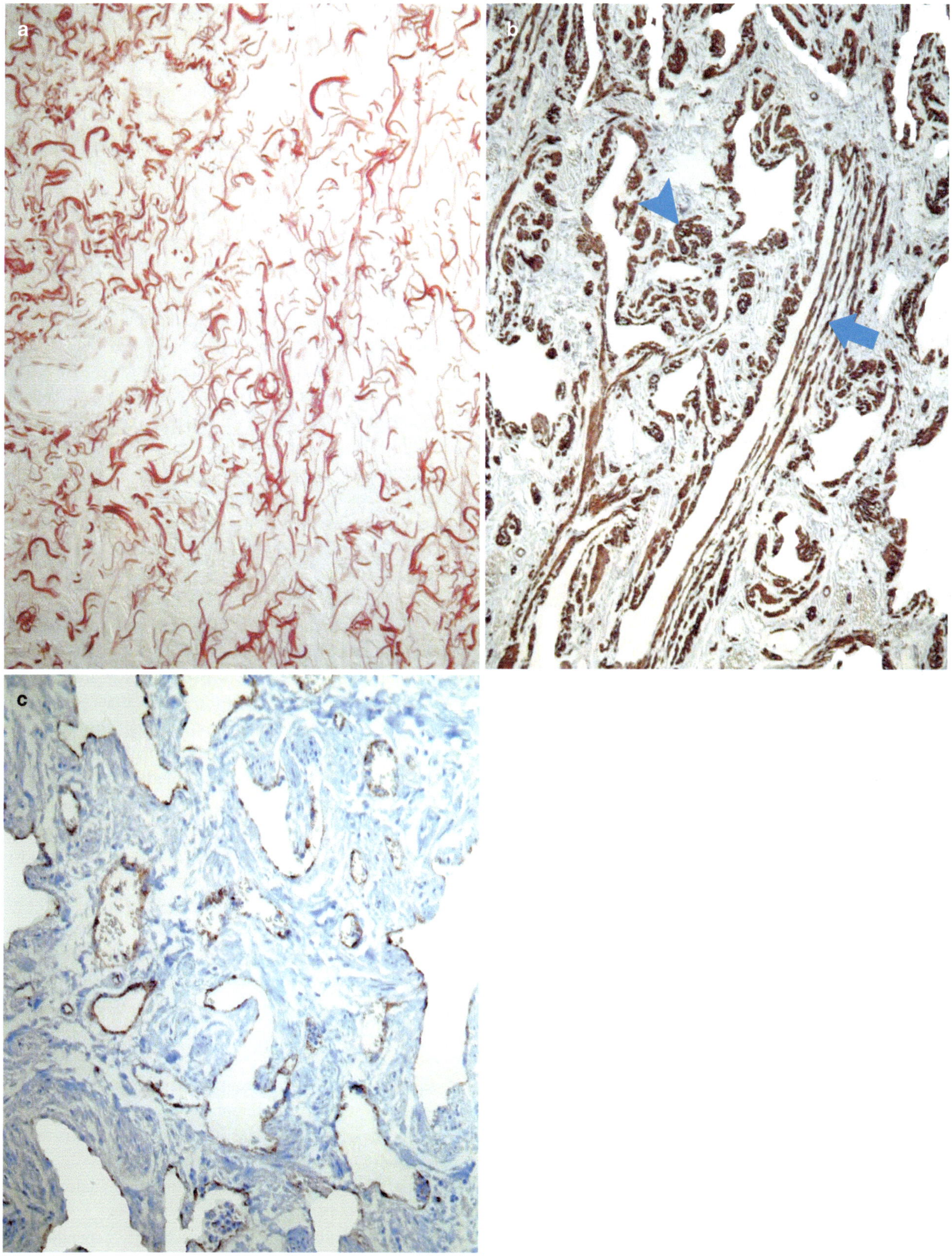

Fig. 6.2 Microarchitecture of the clitoral corpora cavernosa (histochemical study by specific staining of frontal sections of the clitoral body). (**a**) Staining by orceine: highlighting the elastic fibres. (**b**) Staining by smooth actin: highlighting the bundles of smooth muscle fibres (*wine coloured*) oriented in all directions, transversely (*blue arrow*) or longitudinally (*blue arrowhead*). (**c**) Staining by factor VIII: highlighting the endothelial cells: *brown colour* of the endothelia of the cavernosus maze and of the vascular endothelia. These endothelial cells secrete NO!

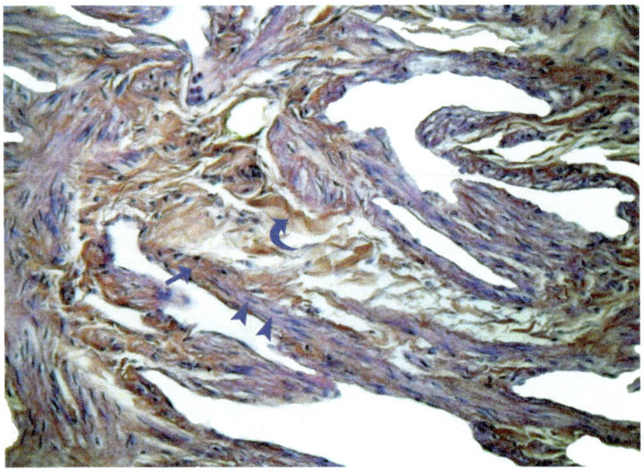

Fig. 6.3 Microarchitecture of the clitoral corpora cavernosa (Details of the cavernous network on a section of the clitoral body). The collagen fibres and the fibroblasts have an *orange colour*. The smooth muscular cells have a mauve colour. Note: At rest, the collagen fibres have an undulating shape (*curved blue arrow*). The smooth muscular cells, with flat and elongated nuclei (*blue arrowheads*), settle on the bundles of collagen fibres (*right blue arrow*)

median junction area of the crura's corpora cavernosa (the clitoral septum or median septum) originates from this albuginea.

The existence of the albuginea is essential because it conditions the phenomena of intumescence and rigidity by providing resistance to the sudden filling of the clitoral vascular labyrinth. Studied by numerous authors, the clitoral albuginea is now better known. The measurements we took on sections perpendicular to the clitoral axis (in the middle part of the descending portion of the body) demonstrate an average thickness of 300 μm. Several characteristics are now known:

- The albuginea is not a rigid, nondeformable structure: Its thickness increases and decreases during clitoral tumescence (even if these variations are very limited). However, the albuginea cannot expand beyond a certain increase in its dimensions. It therefore opposes the pressure of blood flow and the increase of the sinus diameter, thus limiting tumescence, i.e. the swelling of the clitoral body.
- The albuginea includes two layers (Fig. 6.4), like that of the penis corpora cavernosa: an external layer of longitudinal collagenous fibre bundles and an internal layer of circularly oriented bundles. However, only the internal circular layer is clearly visible on the clitoris. The external longitudinal bundles are sometimes sparse in the external most part of the circular layer or even absent in certain quadrants. This is perfectly understandable given that the elongation of the clitoris remains very moderate and certainly does not have the same purpose as penile elongation for coitus.
- The collagen bundles of the albuginea are made up of fibres that are indeed inextensible, but they are pleated, wavy and therefore have a capacity for elongation. The significant size of the circular layer of this albuginea therefore demonstrates the relative expansion of the clitoral diameter in the sexual arousal phase. It is the smoothing of the waves in the collagenous fibres that enables this expansion; the albuginea also thins at the same time. During detumescence and return to flaccidity, the wavy morphology of the collagenous fibres is acquired once again (resting morphology) due to elastic fibres along with the collagen bundles.
- The fibrous columns of the sinus trabeculae originate from the deep surface of the albuginea. A.M.B. Goldstein et al. are to be credited for having demonstrated that a sort of fibrous frame exists within the cavernous tissue, creating a true "fibrous skeleton" of the corpora cavernosa and preventing them from rupturing. The columns of this frame elongate during tumescence, enabling an increase in the diameter of the sinus lumina. The fibrous frame also surrounds the vessels and nerves (Fig. 10.1), which are then completely protected in true fibrous tunnels when the intracavernous pressure becomes very high and erection occurs.
- The tunica albuginea makes a complete crown around the corpora cavernosa but it continues above the clitoral body, with collagenous bundles of the suspensory ligament ("noon" position on the crown) and below the body with the collagenous bundles of the intermediate part (6:00 on the crown). A fibrous partition originates from the albuginea at this last position: the **clitoral septum** (septum cavernosum of clitoral body) or **median septum**.

  This septum is a thick, sagittal, median partition that is part of the cavernous skeleton (the sinus partitions are attached to it, as well as to the albuginea). Its base has a width close to the thickness of the albuginea. It thins to a point at its anterior end (noon position!). More often than not it forms an incomplete partition such that the two corpora cavernosa meet well above the end or tip. The sinuses of the left and right corpora cavernosa have oblique axes converging at the communication zone which is therefore essential for intercavernous drainage. The arteries of the corpora cavernosa are located on either side of the middle part of this median septum and are attached to it by their fibrous sheaths.

  Other septa (paramedian or lateral) may exist. They always originate from the albuginea but are thus arranged obliquely and are often very short.

Some variants exist in the microscopic anatomy of the cavernous formations, depending on the levels:

- The size and number of sinus lacunae vary, depending on the cavernous segment considered (Figs. 5.6 and 5.10): These elements vary in an inversely proportional manner. The largest and most sparse sinuses, with equal surface area or

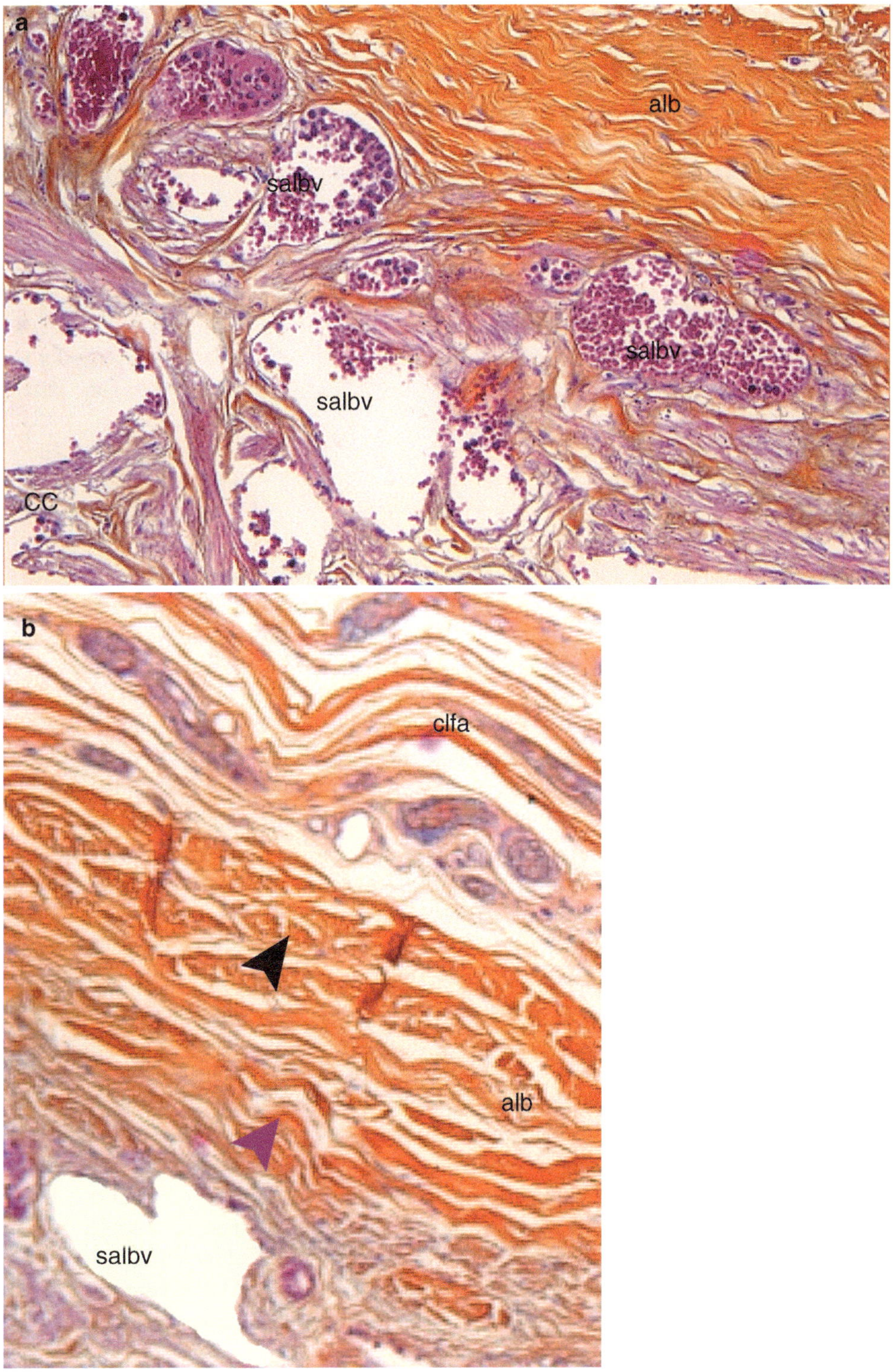

Fig. 6.4 Microscopic anatomy of the clitoral tunica albuginea. (**a**) Low magnification (x50) showing the layer of sub-albugineal veins (*when these veins will be compressed by the dilated sinuses expanding against the tunica albuginea, the evacuation of the blood emptying the corpora cavernosa will stop, making the clitoris rigid*). (**b**) High magnification (x100), showing the two layers of collagen fibres in the tunica albuginea: *black arrowhead* locating the longitudinal bundles of the outer layer and *mauve arrowhead* locating the circulary oriented fibres of the inner layer. *alb* tunica albuginea, *clfa* clitoral fascia, *CC* erectile tissue of corpora cavernosa, *salbv* sub-albugineal veins

volume of cavernous tissue, are located at the crura or ascending portion of the body. The smallest and most abundant sinuses are located at the descending portion where they are very tightly and densely packed. At the crura, the sinus lumina are larger towards the centre than at the periphery.

- The albuginea evidently surrounds the crura and body of the clitoris. However, the albuginea of the crura is thinner and cannot be seen without dissection. Each crus is completely covered by the ischiocavernosus muscle (see the Chap. 11) which truly sheathes it.

- The form of the corpora cavernosa varies depending on the levels considered; more or less cylindrical at the crura and the ascending portion of the body, the corpora cavernosa take on a semilunar shape at the descending part (Fig. 6.1). At this level, the two corpora cavernosa face one another by their medial plane surface, and the former centre parts of the albugineae merge in a median septum, whereas the remaining parts of the two primitive albugineae merge as one single albuginea: the clitoral body albuginea (Fig. 6.1).

It should be recalled that the corpora cavernosa are richly innervated, which will be covered in the "Innervation" chapter (Chap. 8).

6.1.2 Spongy Tissue

While it belongs to the same bulbo-clitoral organ and extensively communicates with the cavernous tissue, the spongy tissue is quite different from the latter. It does not have an erectile property in the true sense of the word (as is true for spongy tissue in men), even if it does contribute to achievement of tumescence and rigidity from a vascular point of view.

Observation of the slides demonstrates these differences immediately:

- The spongy tissue, to be filled with blood upon sexual arousal, like the cavernous tissue, is made up of intercommunicating cavities; but here the cavities or lacunae are of a less aberrant shape, rounded or oval for the most part and of irregular dimensions, but always much bigger than those of the cavernous tissue sinuses (Fig. 6.5). The tissue really does resemble a sponge, hence its name. The trabeculae that separate the lacunae are finer than those in the cavernous tissue, especially when compared to the dimensions of the lacunae themselves. The lumen of the lacunae is bordered by an endothelium that at first appears to be identical to that of the cavernous lacunae but for which the anti-factor VIII antibody immunolabelling is very weak (Fig. 6.6). The functions of these endothelial cells are most definitely different than those of the cavernous endothelium cells (different antithrombotic properties?[1]).
- Observation under greater magnification shows that the intermediate connective tissue is less abundant than in the corpora cavernosa. Here it is also made up of wavy, collagenous fibre bundles, demonstrating potential capacity for elongation. In the spongy tissue, near the collagenous fibres, we can see smooth muscle fibres which also go to prove the possibility of filling or emptying the lacunae by action on the trabecula collagenous fibres. Specific staining of the elastic fibres using orcein shows, however, that the spongy tissue is richer in elastic fibres than the cavernous tissue (A. W.M. Van Turnhout); once again recalling the comparison to the sponge which returns to its initial shape after being pressed to empty it of the liquid absorbed.
- The peripheral fibrous layer that surrounds the spongy tissue is much thinner than the albuginea of the corpora cavernosa, such that we cannot really call it a spongy albuginea but rather a fibrous peripheral envelope or tunic. However, a fibrous architecture exists which is similar to that of the cavernous frame, and the lacunae trabeculae are in continuity with the fibrous peripheral envelope. The thickness of this envelope exceeds that of the trabeculae by very little.
- If we consider the large size of the spongy lacunae, the relative thinness of the trabeculae separating these megalacunae and the fineness of the peripheral envelope of the bulbs, it is understood that the spongy tissue, especially at the bulbs (where it is particularly exposed) is very fragile. This explains the large perineal haematomas that may occur during labour when the foetus is expelled or the development of centro-bulbar haemorrhagic cysts (see Fig. 5.18).

The spongy tissue presents specific traits depending on its topography:

- **At each of the two bulbs**, the fibrous peripheral envelope is covered on its outer surface by bundles of striated muscles of the homolateral bulbospongiosus muscle (Figs. 6.7 and 11.4) (see the Chap. 11). To visualise the fibrous envelope, the muscle must be resected. We can then see the blue-grey appearance of this envelope, semitransparent due to its thinness. Upon sectioning, the bulbar lacunae demonstrate a variable general direction: parallel to the longitudinal axis of the bulb (obliqueness in front and inside) for elongated bulbs with a leech-like appearance. The direction is perpendicular to the longitudinal axis for bulbs in the form of small pouches (Fig. 6.5). It should also be noted that the vascular lacunae with the largest diameters are observed at the large posterior end of the bulbs.
- **At the commissure of the bulbs (commissura bulborum)** (Fig. 5.17), the lacunae have a tendency to become horizontal and follow the general arc form of a supraurethral circle. In addition, these lacunae are transversally very flat and elongated. The lacunae of the two bulbs extend into these commissural lacunae and the septum of the overlying pars intermedia does not always descend to the bulbar commissure (see Fig. 5.13). The fibrous envelope now only exists on the posterior surface (supraurethral concavity) of the bulbar commissure. The anterior surface (back) of the commissure continues with the vascular lacunae of the pars intermedia.

[1]Note that on histological sections, the spongy lacunae are always filled with erythrocytes (Fig. 6.7)!

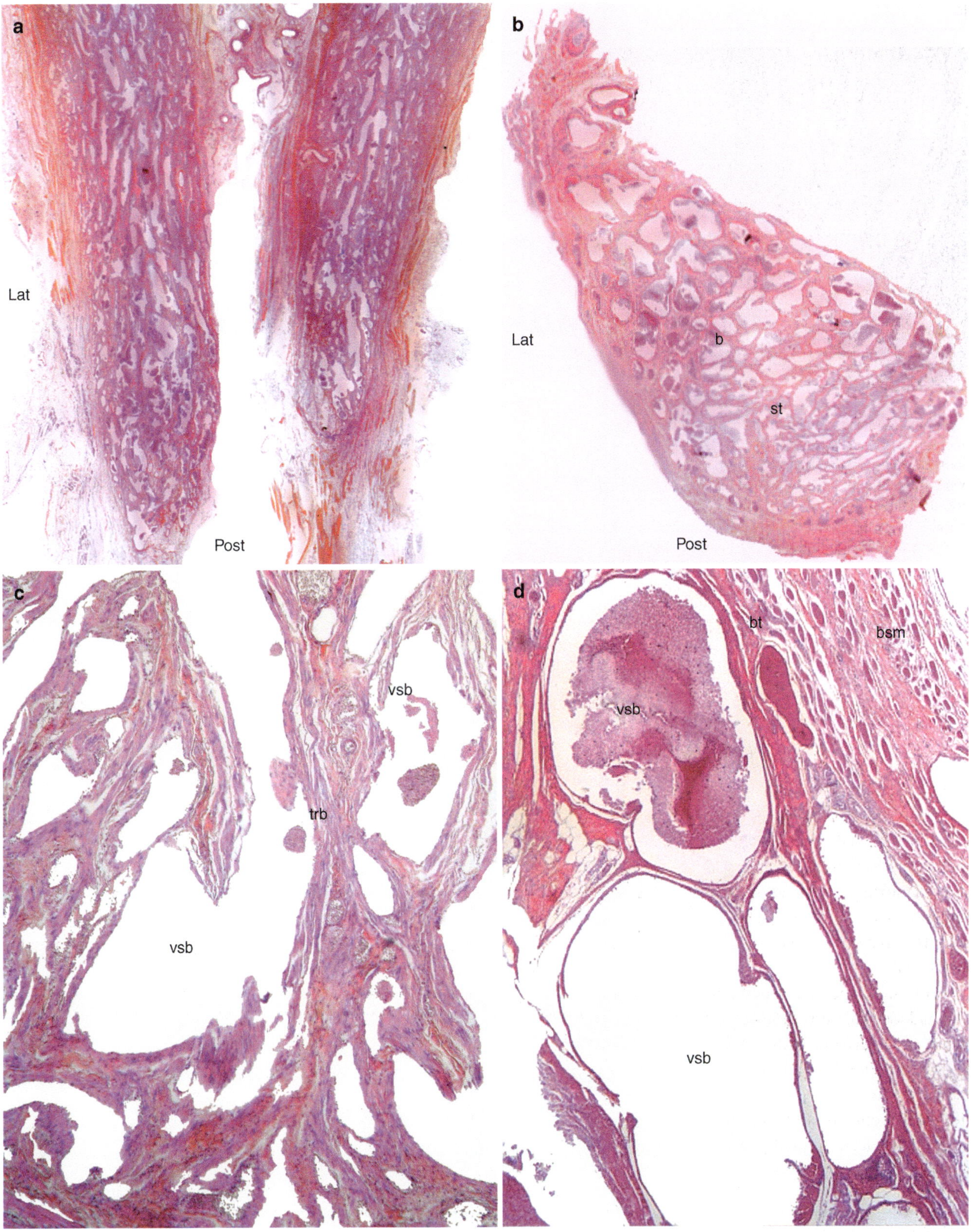

Fig. 6.5 The spongy tissue of the bulbs (bulbo-clitoral organ). (**a**) Photomicrograph of two bulbs in collapsed state; (**b**) photomicrograph of an isolated fig-shaped bulb; (**c**) microscopic view of the spongy tissue (low magnification); (**d**) microscopic view (strong magnification) showing the rounded vascular spaces. *b* bulb, *bsm* bulbo-spongiosus muscle, *bt* (fibrous) bulb tunica, *st* spongy tissue, *trb* trabeculae of the bulb, *vsb* vascular spaces of the bulb (spongy sinus). Note: The frequent presence of residual blood in the vascular spaces of the bulbs

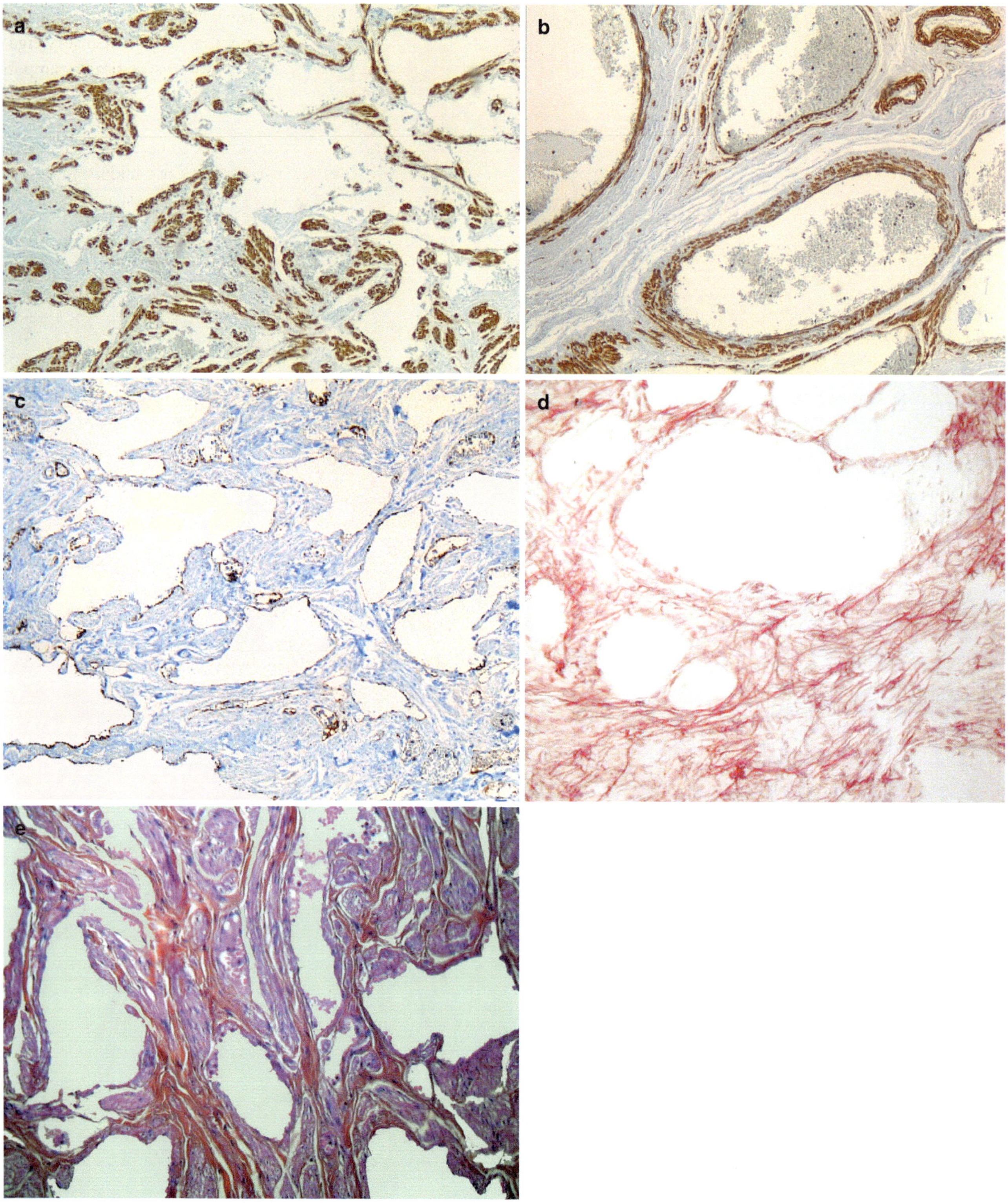

Fig. 6.6 Microarchitecture of the bulbo-clitoral organ's corpus spongiosum (histochemical study by specific staining of the bulb's sections). (**a**, **b**) Immuno histochemical staining by alpha-smooth actin: highlighting the bundles of smooth muscular fibers (*wine-coloured*). (**c**) Immuno histochemical staining by Factor VIII: highlighting the endothelial cells. (**d**) Staining by orcein: highlighting the elastic fibers. (**e**) Classical histological staining (HEAS). (**a**, **b**): The smooth muscular fibers are abundant into the spongy stroma (**a**), and around the vessels (**b**). (**c**) The impregnation of the spongious tissue's endothelial cells is poor. (**d**) The red elastic fibers draw a sort of reticulum around the rounded spongy cavities. (**e**) The collagen fibers and the fibroblasts have an orange colour. The smooth muscular fibers have a purple colour and their elongated nuclei are dark blue

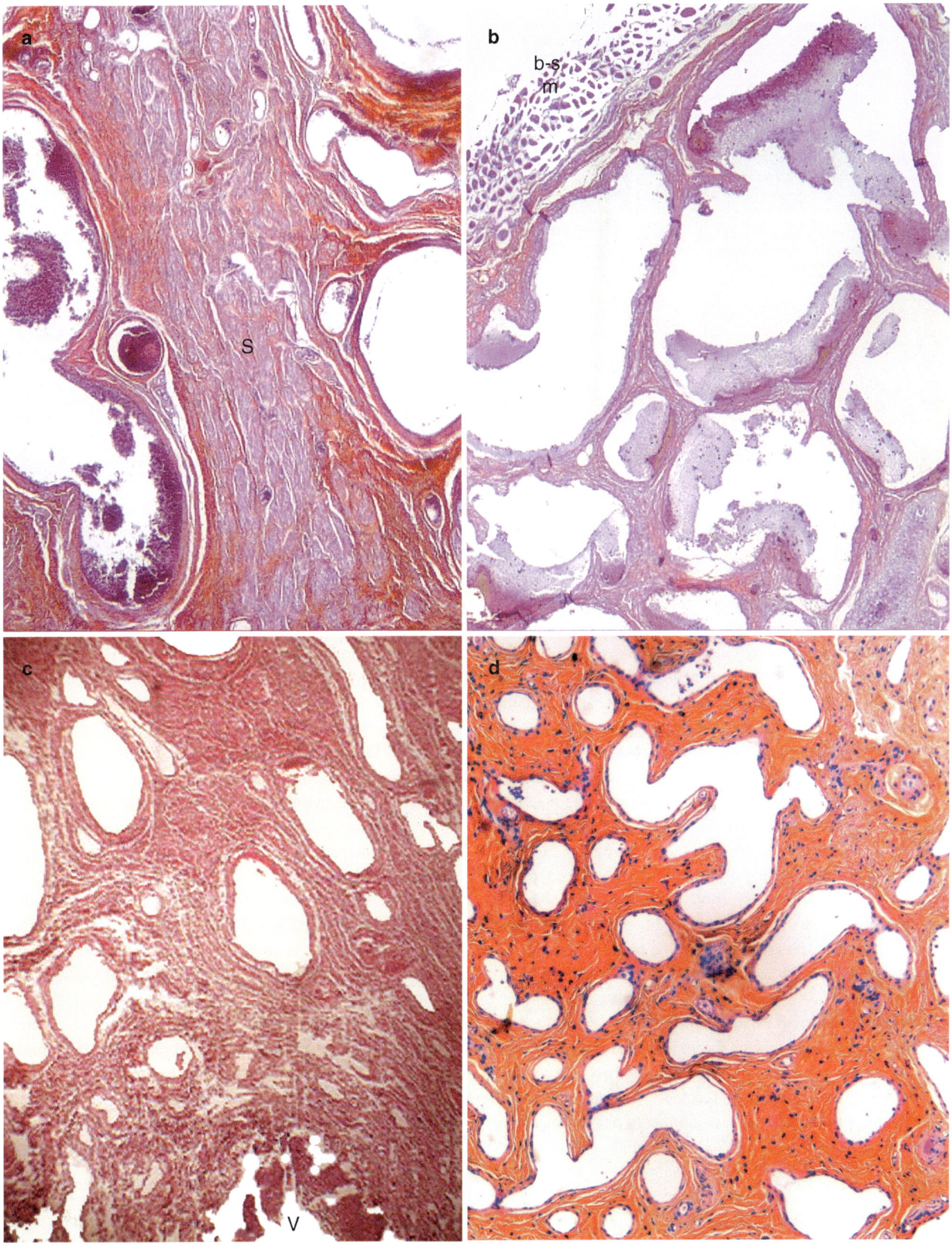

Fig. 6.7 The four different aspects of spongy tissue according to its location. (**a**) Spongy tissue of the pars intermedia; double row of closely set venous canals separated by a large septum (*S*) rich in smooth muscle fibres (*purple coloured*). (**b**) Spongy tissue of the bulbs, large rounded or polyhedric vascular spaces containing many red cells; striated muscle fibres are those of the bulbospongiosus muscle (*b-s m*). (**c**) Spongy tissue of the residual spongy part (*rsp*), above the vestibular roof (*V*). (**d**) Spongy tissue of the glans clitoridis (lower half)

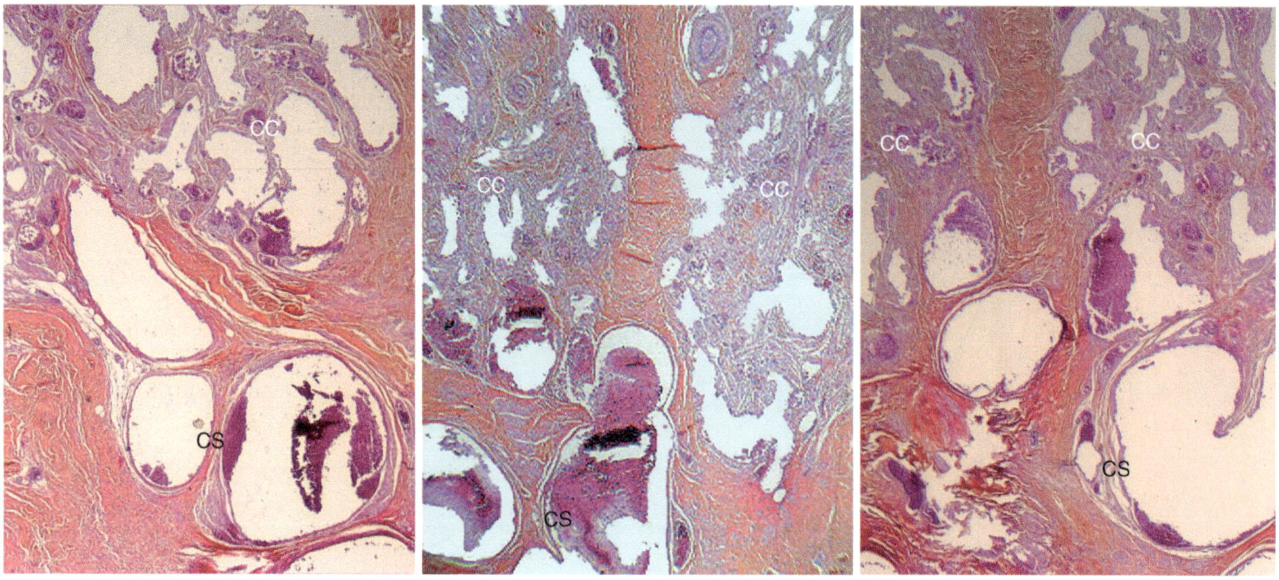

Fig. 6.8 Three examples of interpenetration of spongy and cavernous tissues. *CC* corpus cavernosum, *CS* corpus spongiosum

- **At the pars intermedia**, the vessels are sinuous with a large parallel axis in a comb-tooth pattern. Upon sectioning, they appear as lacunae with a greater diameter than the bulbar lacunae, arranged one on top of the other in a "stack of plates" (see Figs. 6.7 and 5.17). They often have a tendency to form two vertical columns, left and right, parallel to the median septum that separates them. This septum of the intermediate part is a thick, fibrous septum located in line with the median septum of the clitoral body which it follows, becoming progressively enlarged (Fig. 5.17). This septum is usually interrupted at the commissure of the bulbs. The pars intermedia lacunae communicate with the sinusoids of the posterior part of the elbow and of the clitoral body, and we can even see the spongy lacunae going through the albuginea, erupting into the homolateral cavernous tissue (Fig. 6.8)! The vascular rows of the pars intermedia are well protected because they are located within the thick network of collagenous fibres in connection with the overlying clitoral albuginea.
- **The infra-clitoral residual spongy part** (**icrsp**) extends from the commissure of the bulbs, as shown previously, and is positioned below the descending part of the clitoral body. Obscured or even refuted by some authors, the "icrsp" is on the contrary, a significant part of the spongy tissue of the bulbo-clitoral organ in which we can observe the presence of large vascular lacunae with large vertical axes. The diameters of these lacunae tend to decrease as we approach the glans. This icrsp is connected to the clitoral body by the pars intermedia previously studied. Below the icrsp, the spongy lacunae are divided into two contingents, right and left, which reach the lateral walls of the vestibule to anastomose with the veins present in these walls (Figs. 6.7 and 5.15). The visible turgescence of these veins during sexual arousal can be explained by the fact that these vestibular veins are in continuity with those of the labia minora (turgescence transmitted by blood engorgement of the erectile bodies).
- As we saw previously, the spongy tissue continues from the icrsp up to the free part of the glans. But this spongy tissue becomes very involutive as it approaches the glans (spongy tissue in which a terminal urethra did not form, as it does in the male glans penis), such that at the glans (especially in the "lower half", we only find scattered, rounded vascular lacunae of varying diameters, arranged more towards the centre than at the periphery, within interlacunar connective tissue that is more abundant and more dense (slightly wavy, collagenous fibres, arranged with a nice microscopic architecture (Fig. 5.20)). In this stroma, we also find elastic fibres and some smooth muscle fibres. In the "upper half" of the glans, the presence of Forster's ligament and its tree structure causes the spongy residues to be dispersed within the connective frame thus constructed (Fig. 5.12).

7 Physiology of Cavernous and Spongy Tissues

7.1 Their Role in the Erection of the Clitoris

7.1.1 Can These Tissues Be Referred to as "Erectile Tissues"?

The adjective does not apply to the two types of tissue in the same manner.

The **spongy tissue** has no actual erection capacity (this is all the more obvious compared to the male penis in men). Such as we have just observed, this tissue can be saturated with blood and become tumescent. However, due to the absence of any true albuginea, the thinness and the actual structure of the fibrous envelope, which is not constant (it is only present at the level of the bulbs), and especially to the absence of any sub-albugineal venous network, a blocked blood evacuation or a backpressure, such as can be observed at the level of the corpora cavernosa, is not possible.

This does not apply to the **tissue of the corpora cavernosa** where all parts are organised such as to make an erection possible: presence of a thick, extremely resistant and double-layered albuginea, presence of a considerably developed sub-albugineal venous network (Fig. 6.4), vascular network with significantly anastomosed sinus. **The tissue of the clitoral corpora cavernosa is precisely therefore an erectile tissue** and, as a result, there is no doubt concerning the fact that the clitoris has an erection capacity.

This capacity has, however, often been denied, even by eminent specialists, who have referred to the absence of any sub-albugineal venous network (which is completely inaccurate), to the fixity of the clitoral curve related to the presence of the suspensory ligament (but this is forgetting that this connection only really concerns the area of the angle), to the reduced dimensions of the corpora cavernosa and to the fact that these corpora cavernosa do not reach the end of the glans (which, however, corresponds to the same configuration as in the penis). Others, such as Pernkopf in 1943, acknowledged the erectile capacity, while adding that it was an "ineffective erection". Kobelt, whose remarkable work concerning the clitoris has already been mentioned in this study, strongly supported the idea (as from 1844) that the clitoris could enter into erection during the phase of sexual excitation. It is quite certain that, considering the dimensions of the clitoris, this erection remains discrete and limited, resulting however in a slight straightening of the descending portion, an increase in the size of the crura and clitoral body, a rigidity of the descending portion of the body and an extrusion of the glans outside the hood.[1]

7.1.2 Physiology of the Erection of the Clitoris

7.1.2.1 The Biochemical Cascade

This biochemical cascade is well known thanks to the remarkable work completed during recent years by pharmacologists and biochemists, who were searching for new drugs capable of treating cases of male impotence (Bennett AH.).

The process only starts when the phase of sexual excitation occurs, whether this phase is caused by mechanical stimulation (masturbation or other), erogenic stimulation (visual, auditive or other) or during sexual intercourse. This process involves the autonomic nerve endings of the trabecular structures and their capillaries (Fig. 7.1).

Specific parasympathetic nerve endings (nitrergic neurons), under the action of nNOS (neuronal form of the NOS enzyme, nitric oxide synthase), and the endothelial cells of the cavernous sinusoids, under the action of eNOS (endothelial form of the NOS enzyme) (Burnett Al et al.) will then play an eminent part by secreting NO, i.e. nitric oxide (non-adrenergic and non-cholinergic and yet the first neurotransmitter of an erection). At the same time, the sympathetic nerve system is inhibited. The NO will then activate an enzyme, the soluble guanylyl-cyclase (sGC), which will,

[1] However, during the orgasmic phase, the glans may retract under the hood (See chapter 15, The bulbo-clitoral Organ in the Sexual Act), page 129.

V. Di Marino, H. Lepidi, *Anatomic Study of the Clitoris and the Bulbo-Clitoral Organ*,
DOI 10.1007/978-3-319-04894-9_7, © Springer International Publishing Switzerland 2014

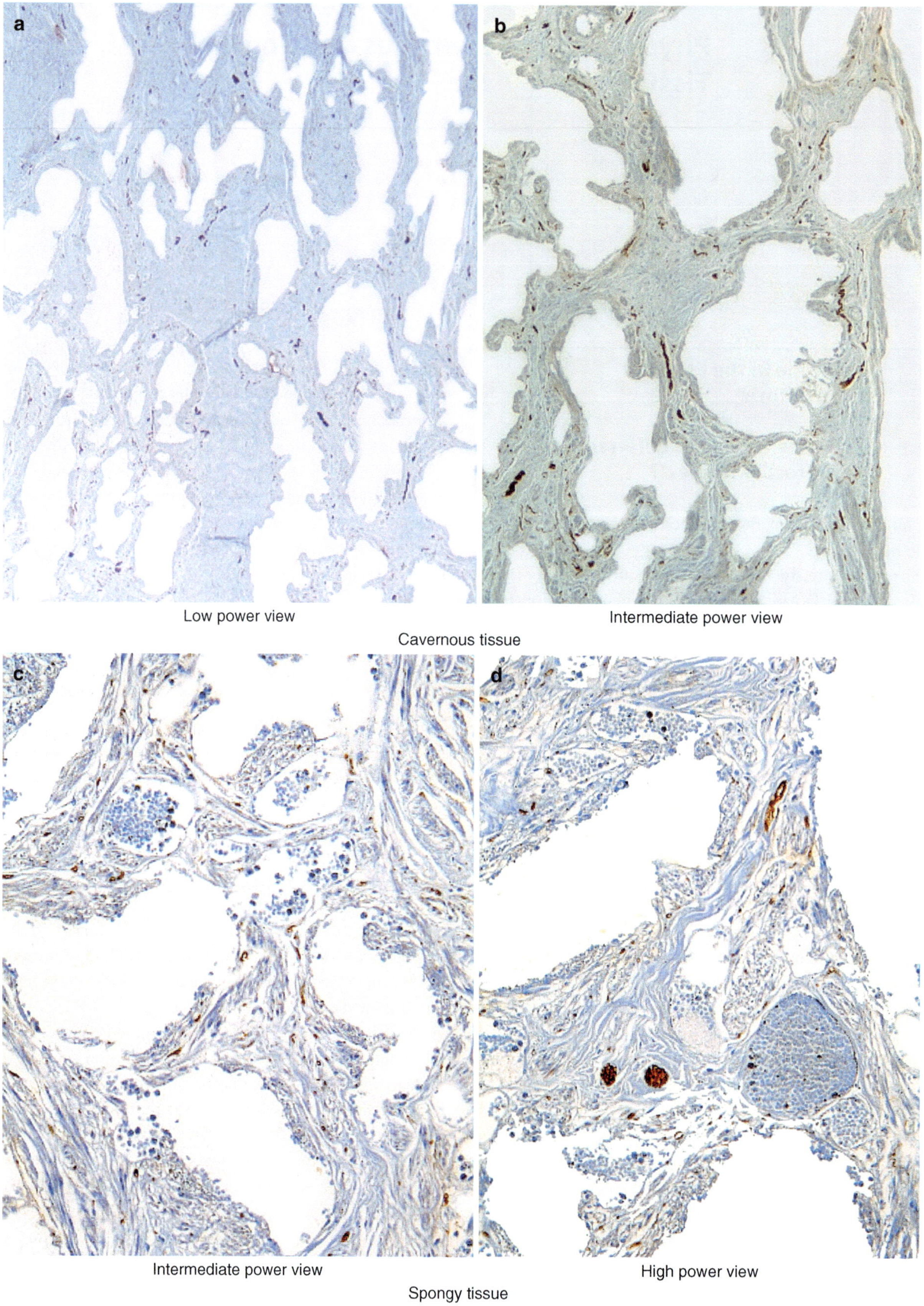

Fig. 7.1 Innervation of the cavernous and spongy tissues (PS100 immuno-staining) (dark-brown-stained nerves). (**a**) Low power view. (**b**) Intermediate power view. (**c**) Intermediate power view. (**d**) High power view

in turn, increase the intracellular concentration of cGMP, cyclic guanosine monophosphate (second neurotransmitter of erection). A hyperpolarisation through the activation of PKG (cGMP-dependent protein kinase) then occurs in the cell. This process activates the calcium-activated potassium channels (BKCa), thus causing a calcium leak and generating in fine, a relaxation of the smooth muscle fibres (Fig. 6.3), which is the starting point of the vascular and tissue phenomena initiating erection.

When sexual excitation decreases, the cyclic guanosine monophosphate is broken down by the type-5 phosphodiesterase (PDE-5), which will gradually cause the return of calcium inside the cell, the retensioning of the smooth muscle fibres and, consequently, the return of the undulations of the collagen fibres (this state being also favoured by the action of elastic fibres) through the restoration of adrenergic sympathetic tonicity stimulating the beta 2 adrenoreceptors of the cavernous tissue. The flaccid state is thus correlated with a state of spontaneous contraction of the smooth muscle fibres (recordable basal myogenic activity). However, there exists, within the clitoral smooth muscle fibres, receptors capable of either causing the relaxation of these fibres or restoring their initial myogenic state. Vasoconstrictive adrenergic alpha 1 receptors and several vasoconstrictive substances (neuropeptide Y, endothelin, PGF-2 alpha prostaglandins) have therefore been identified. Similarly, substances favouring the relaxation of the smooth muscle cells, such as the vasointestinal polypeptide (VIP), PGE 1 prostaglandins, etc., have also been identified.

7.1.2.2 Vascular and Tissue Phenomena

The relaxation of the muscle fibres occurs at two levels:

- Relaxation of the smooth muscle fibres of the superficial clitoral arteries (dorsal arteries) and deep arteries (paraseptal arteries of the corpora cavernosa and their intra-trabecular helicin arteries) by direct action of the parasympathetic fibres on the muscarinic receptors and VIP receptors of these vessels, thus causing their vasodilatation. As a result, a considerable increase of the arterial flow in these vessels occurs, thus allowing the cavernous lacuna to be filled rapidly.
- Relaxation of the smooth muscle fibres of the sinus trabecular walls. These sinuses then open (they become spherical), which is favoured by the relaxation of the collagen fibres of their wall but also by the relaxation of the collagen fibres of the cavernous albuginea. While the corpora cavernosa are being filled, the sub-albugineal veins are pushed back against the internal surface of the albuginea. The corpora cavernosa extend and expand, while their albuginea becomes thin. At the same time, identical phenomena occur in the spongy tissue, and in particular, in the bulbs. They are filled with blood under the effect of the blood flow increase and of the reception of the excess blood from the corpora cavernosa (for which they are used as a spillway), via the pars intermedia network and the direct bulbo-clitoral anastomoses (Fig. 6.8). The cavernous and spongy sinuses are saturated very rapidly, the sub-albugineal veins collapse and the evacuation of intra-sinus blood can no longer occur normally. When the albuginea is at its maximum tension and the filling process is no longer possible (no venous evacuation pathway), rigidity occurs (accentuated by the contraction of the ischiocavernosus muscles) and the clitoris enters into erection.[2]

7.1.2.3 Detumescence

When sexual excitation decreases, the cyclic guanosine monophosphate is broken down by the type-5 phosphodiesterase, which will gradually cause the retensioning of the smooth muscle fibres through the restoration of the noradrenergic sympathetic nerve tonicity and, consequently, the stimulation of the alpha 1 receptors of the clitoral arteries; the vasoconstriction and reduction of their flow; the restoration of the function of the sub-albugineal veins, which can once again combine with the cavernous veins and the emissary veins of the deep dorsal vein to evacuate sinus blood; the return of the undulations of collagen fibres thanks to the action of the elastic fibres arranged as a bridge; and the restoration of the albuginea's usual thickness and diameter. Such as we have just observed, the stimulation of the parasympathetic nerve system inhibits the stimulation of the sympathetic nerve system and, vice versa. Current neuroanatomical work tends to show that it is the hypothalamus which ensures the coordination of the two types of activity of the autonomic nervous system.

[2]The sub-albugineal network merges with the deep dorsal vein of the clitoris via emissary veins, which cross through the albuginea (Fig. 7.2). These veins provide a reduced drainage, which is residual when the sub-albugineal network collapses. This minimum drainage is essential to ensure that the clitoris is not threatened of asphyxiation and that the oxygenation of the tissues can be renewed. If this mechanism does not operate correctly, a pathological phenomenon may appear: clitoral priapism, of which several cases have been reported in medical literature.

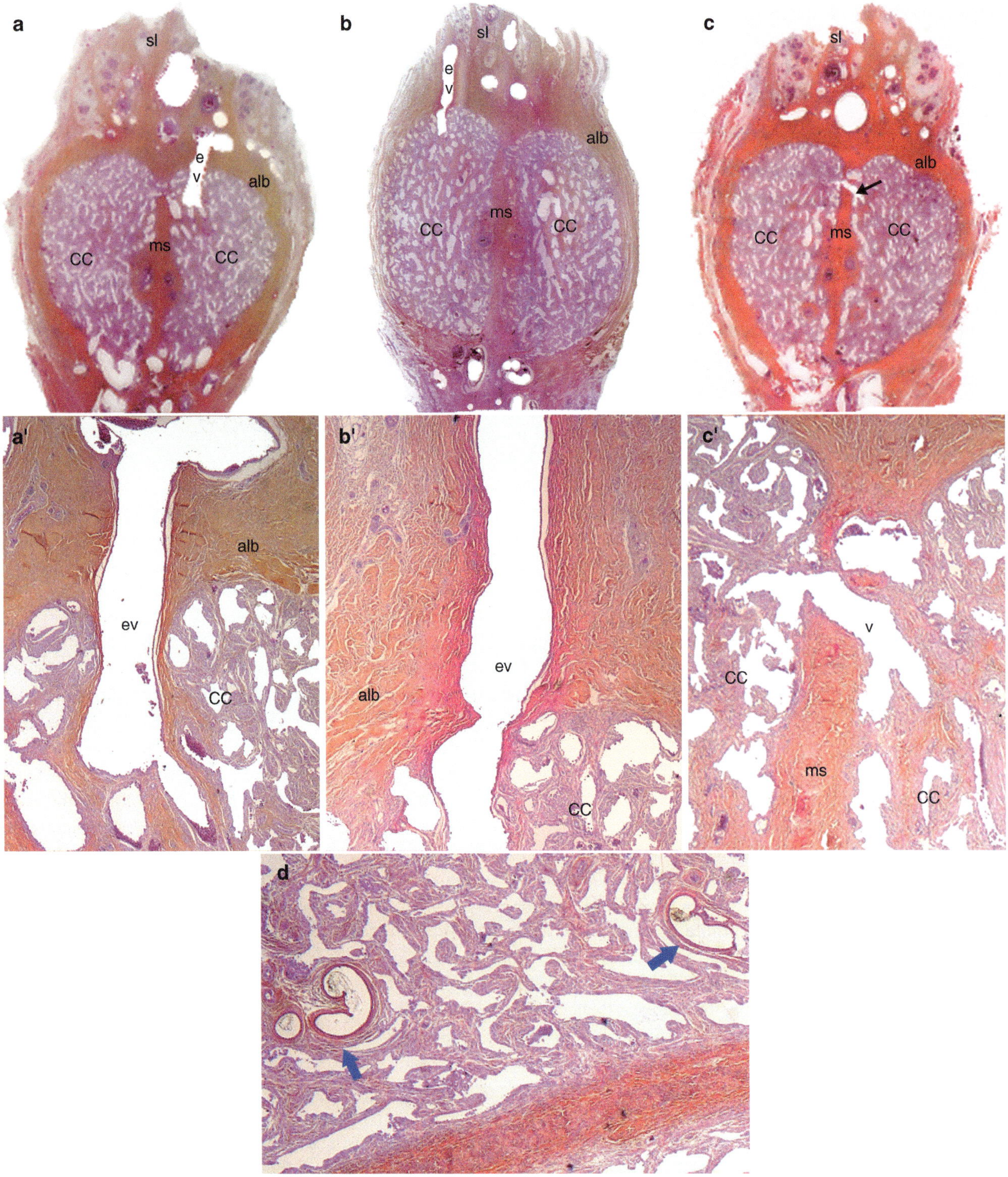

Fig. 7.2 Examples of emissary and draining veins of the corpora cavernosa. (**a**, **b**) Drainage to the suspensory ligament's veins (**a′**, **b′**: microscopic magnifications of **a** and **b**). (**c**) Communicating vein between the 2 corpora cavernosa (**c′**: microscopic magnification of **c**). (**d**) An example of intra-cavernous vein (usual drainage to the deep vein of clitoris). *alb* tunica albuginea, *cc* corpus cavernosum, *ev* emissary vein, *ms* median septum, *sl* suspensory ligament, *v* draining vein, *black arrow* transeptal vein, *blue arrows* they show intra-cavernous veins

8 Innervation of the Bulbo-clitoral Organ

8.1 General

This innervation is extremely important, not only due to its density but also due to the major role it plays in sexual life. It consists of a double innervation, which is, on the one hand, cerebrospinal and, on the other hand, vegetative.[1] The pudendal nerve (shameful nerve), cerebrospinal contingent, is the key component and has therefore been referred to as "king of the perineum" (R. Robert et al.). However, it is not the only nerve structure to be involved. The other key nerve contingent is the vegetative contingent, originating from the inferior hypogastric plexus: the cavernous nerve of the clitoris. Such as will be noted in the following sections, the two components converge and meet to ensure the extraordinary innervation of the bulbo-clitoral organ.

8.1.1 Pudendal Nerve

The anatomy of the pudendal nerve (*shameful nerve* according to former authors) is better known since a number of specialists, and a team from Nantes[2] (already mentioned above) in particular, have focused their attention on the investigation of the causes of positional perineal pains and the discovery of neuralgia by "pudendal nerve entrapment syndrome".

The pudendal nerve is the terminal branch of the pudendal nerve plexus dedicated to external genitals and the perineum. The pudendal plexus is most generally plexiform and formed by the fusion of the fibres from the ventral branches of spinal nerves S2, S3 and S4. This plexus originates from the true pelvis and ends as a single trunk: the pudendal nerve.

As soon as it has formed, the nerve leaves the true pelvis, crosses the greater ischiatic incisure (greater sciatic foramen) in the infrapiriformis canal (sub-pyramidal canal) and reaches the gluteal region (buttocks region). It rapidly leaves the latter and passes through the true pelvis again by circumventing the ischial spine (sciatic spine) and by crossing the lesser ischiatic incisure (lesser sciatic foramen). It passes through the true pelvis, with the artery and the pudendal veins (shameful internal vessels), against the internal surface of the ischial tuberosity, and follows the lateral wall of the ischio-rectal fossa by engaging through a fibrous tunnel, Alcock's pudendal canal (shameful canal). This canal consists of the fascia of the obturator internus muscle and arched fibres stretched between this fascia and the falciform process of the sacrotuberous ligament (greater sacro-sciatic ligament). It is in the first centimetres of this fibre canal that the pudendal nerve forms an important collateral branch: the inferior rectal nerve. It then divides into its 2 terminal branches: the perineal nerve and the dorsal nerve of the clitoris.[3]

The **perineal nerve** will, in turn, divide at the posterior edge of the anterior perineum (urogenital perineum), into a superficial perineal nerve and a deep perineal nerve:

- The superficial perineal nerve is a sensory nerve. It is dedicated to the teguments of the perineum (posterior labial nerves dedicated to the posterior part of the labia majora) and to the external orifice of the urethra.
- The deep perineal nerve, once it has circumvented the posterior edge of the deep transverse muscle, is successively routed in the deep space and then in the superficial space of the anterior perineum. Its motor contingent supplies the ischiocavernosus and bulbospongiosus muscles, muscles of the erectile bodies of the bulbo-clitoral organ. Its sensory fibres are dedicated to the major vestibular gland, the spongy bulb and partly to the vestibule.

[1] The central nerve system (whose centres are cortical) is dependent on the conscience and will. The vegetative or autonomous system, which is independent of the conscience and will, ensures the operation of the organs.

[2] The potential compression areas are located in the space between the sacrotuberous and sacro-spinous ligaments, at the level of the falciform process and, lastly, in the pudendal canal. (Alcock's canal). See R. Robert et al, Neurosurgery, 55, 453–469.

[3] The termination of the pudendal nerve can also be located before the penetration in the pudendal canal, above the ischial tuberosity.

V. Di Marino, H. Lepidi, *Anatomic Study of the Clitoris and the Bulbo-Clitoral Organ*,
DOI 10.1007/978-3-319-04894-9_8, © Springer International Publishing Switzerland 2014

The **dorsal nerve of the clitoris** extends in the direction of the pudendal nerve and remains in the pudendal canal, along the ischio-pubic branch, which it will follow to join the clitoris.

Such as we have just observed, all is clear for the anatomy of the pudendal nerve. So much cannot be said for the anatomy of the dorsal nerve of the clitoris. Not many studies have been conducted on this nerve until recently.

Considering the importance of this nerve and the observations that we have been able to make during our dissections, its specific study proves to be essential and will be conducted later.

8.1.2 Cavernous Nerve of the Clitoris

It is this nerve which leads the contingent of fibres of the autonomic nervous system (vegetative nerve fibres) to the erectile bodies. Its fibres originate from the uretro-vaginal part of the **inferior hypogastric plexus** (Fig. 8.7), plexus pair located in the sub-peritoneal cellular tissue, which coats the lateral surfaces of the female pelvis.

Each plexus is formed by the plexiform convergence of ortho- and parasympathetic fibres.[4] The following fibres are included in its composition:

- Orthosympathetic nerve fibres: They are the sacral splanchnic nerves—they come from the pelvic sympathetic ganglia (sacral sympathetic ganglia).
- Parasympathetic nerve fibres, an important contingent from the pelvic splanchnic nerves (Eckard's erectile nerves), collateral of the pudendal plexus, originating from the ventral roots of sacral nerves S2, S3 and S4.
- Mixed nerve rami from the superior hypogastric plexus (inter-iliac plexus or presacral nerve of Latarjet), which conveys both the orthosympathetic and parasympathetic fibres.

The efferent branches of the inferior hypogastric plexus are dedicated to the rectum (medium rectal nerves), the uterus, the vagina, the bladder, the clitoris and the bulbs.

Thus, **the vegetative nerve fibres dedicated to the clitoris are routed thereto via the 2 cavernous nerves**, which originate from this inferior hypogastric plexus.

8.1.3 Descriptive and Topographic Anatomy of the Clitoral Innervation

8.1.3.1 Dorsal Nerve of the Clitoris

The dorsal nerve of the clitoris has, up until recent times, rarely been studied and therefore is badly known, especially since few authors have had the curiosity to reread the remarkable work conducted by Kobelt and, even more exceptionally, the will to study the said nerve by dissection. However, respecting the integrity of this nerve guarantees that the functionality of the clitoris will not be compromised and, therefore, that an essential and key part of female sexuality will not be affected.

This nerve is at the centre of many surgical acts: treatment of fractures of the pelvis, reduction of symphysis disjunctions, surgery of the urethra, repair surgery of certain lower urogenital congenital anomalies (Baskin L.S.), clitoridoplasties, surgery of tumours of the pubis, installation of transobturator tape for treating urinary incontinence (Achtari C. et al.), "transgender" plastic surgery, etc. Major iatrogenic consequences, which are generally final, can only be avoided through perfect knowledge of the anatomy. It is necessary to identify, for each of the 2 dorsal nerves of the clitoris, several segments, which each have a specific significance: the initial ischio-pubic segment, the ascending clitoral segment, the pre-symphysis (prepubic) segment, the corporeal clitoral dorsal segment and the termination at the glans. These various segments are separated by direction changes and actual kinking, which also deserve to be analysed in a specific study, especially the first during which we will observe how the cavernous nerve meets the dorsal nerve of the clitoris.

- The **initial or ischio-pubic segment** of the dorsal nerve of the clitoris is the superior division branche of the pudendal nerve of which it extends the direction, thus remaining in the fibrous pudendal canal with the artery and pudendal veins. It should be noted that this dorsal nerve of the clitoris is difficult to release in this initial segment, where it is routed against the ischio-pubic branch, below the fascia of the obturator internus, above the insertions of the deep fascia (superior fascia) of the urogenital diaphragm. It provides, at this level, the nerve of the external sphincter of the urethra, which will cross the above-mentioned fascia to join the sphincter fibres in the deep space of the perineum. Beyond the pudendal canal, the dorsal nerve of the clitoris will cross the pudendal artery, which will be in the more medial position. It surfaces by crossing through the superior fascia of the urogenital diaphragm and penetrates the deep space of the perineum, where it becomes accessible for perineal dissection. It may then, depending on the situation, remain above the inferior fascia of the urogenital diaphragm or, on the other hand, resurface by crossing through this fascia (Fig. 8.1). This penetration has been the subject of recent studies (A. Vaze et al.): according to these authors, it is located, on average, at 2 cm, 7 from the external orifice of the urethra. In both cases, the nerve will continue its forward and inward diagonal route towards the inferior branch of the pubis, which shelters and protects it and to which it is closely connected. At the inferior edge of this branch, a more or less deep groove can be observed. This groove has been described by Sedy J et al. as follows: the groove of the

[4]The ortho- and parasympathetic fibres (whose actions are antagonistic) especially control the motricity of the vessels, the smooth muscles, the viscera and the glands.

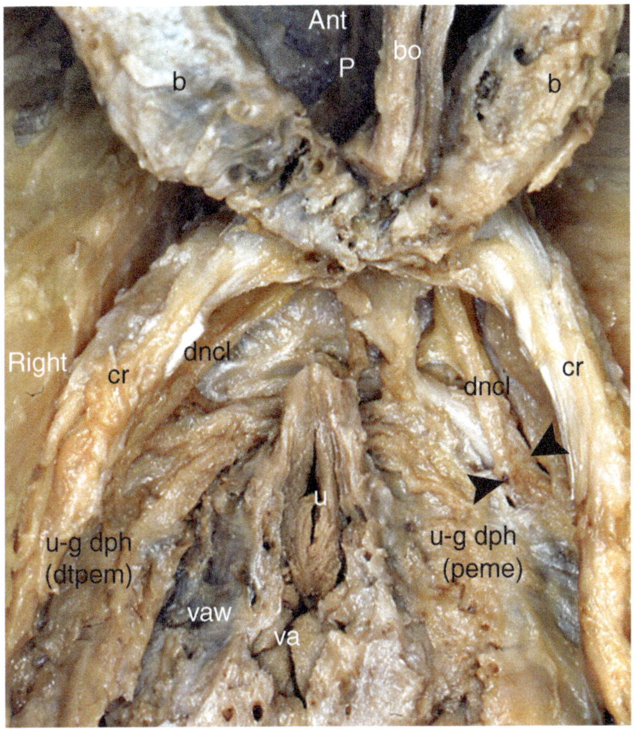

Fig. 8.1 The dorsal nerves of the clitoris in their ischio-pubic initial segment (dissection of the female anterior perineum, gynaecological position). The bulbs and the clitoris were folded against the pubis. The crura were released from their fibrosis attachments and pushed slightly outside. *b* bulb, *bo* clitoral body (descending part), *cr* crus of clitoris, *dncl* dorsal nerve of clitoris, *p* pubic symphysis, *u* external urethral orifice, *u-g dph* uro-genital diaphragm, *dtpem* deep transverse perineal muscle (visible on the right, after resection of the perineal membrane), *u-g dph (peme)* perineal membrane, *va* vaginal orifice, *vaw* vaginal wall, *black arrowheads* they line the banks of the incision of the perineal membrane making the nerve, which still remains in the deep perineal space, visible. Note the oblique course of the dorsal nerve of the clitoris, going forwards and within

dorsal nerve of the clitoris (nearly 70 % of cases) is more or less marked, 15.8 mm long on average and even extends, according to these authors, at the inferior part of the ventral surface of the pubis. According to Sedy, this protective groove, which contains the nerve (which is often covered with a fibrous tissue), is as important as the pudendal canal because, in some cases, such as the latter, it can be too narrow and become a site of nerve compression (for cyclists in particular!), the consequence being a "syndrome of the dorsal nerve" (slightly different from the "pudendal nerve entrapment syndrome" and involving evocative symptoms such as an erectile malfunction). In order to meet the clitoris, the nerve finally reaches the inferior edge of the body of the pubis and its very dense prepubic fibrous coating. More precisely, it appears in the lateral part of the infra-pubic region, in front of the transverse perineal ligament (Henlé's or Krause's ligament), below the caudal symphysis edge (inferior edge), coated with the significant fibrous layers of the arcuate pubic ligament (Lauth's ligament). Therefore, the nerve is in a very dense fibrous atmosphere, on which no author has insisted (except for Hruby S. et al., who described it in men) and which, however, plays a major protection role. Each of the 2 nerves will be wrapped at this level, in an actual protective sheath, a fibrous sleeve (with smooth internal wall), formed by the numerous fibrous components which surround the infrapubic region. In some cases, certain nerves are wrapped in a fibrous canal formed by the duplication of the arcuate pubic ligament or even in a tunnel duplicated from the presymphysis tissue. The incision of the most superficial lamina of this envelope releases the nerve, which then seems to sprout out from an actual fibrous cone (Fig. 8.2). It should be noted that the homologous pudendal artery is also well protected, inside an underlying fibrous sleeve. The above details can be perfectly well observed by slightly tilting the clitoris, released from its pubic attachments, ventrally and backwards (Fig. 8.2). The thin subpubic urethral plate, which spreads out laterally up to the 2 nerves (which can be observed by transparency) must also be resected (Fig. 13.4).

- **The first elbow or pubic elbow** (Fig. 8.3): Each nerve will then modify its trajectory to rise up in front of the pubic symphysis, against the dorsal surface of the clitoris. However, beforehand, it will be considerably densified by the arrival of the cavernous nerve, which converges towards it while supplying the vegetative contingent. Due to this contribution, the dorsal nerve of the clitoris (beyond this anastomosis) will become a particularly exceptional nerve, a true "mixed nerve", belonging both to the cerebrospinal nervous system and to the vegetative nervous system, as a motor component for the deep transverse muscles and external sphincter of the urethra, a sensory component for the corpora cavernosa, the glans and prepuce, and a vegetative component for the cavernous and spongious bodies. It is thus our aim to evoke, with the description of this first elbow of the dorsal nerve of the clitoris, the descriptive anatomy of the cavernous nerve.
- **The ascending clitoral segment** (Fig. 8.2): After having received and incorporated the autonomous fibres of the cavernous nerve, the dorsal nerve of the clitoris only remains visible for a short moment. It is routed medially and then becomes vertical as soon as it meets the inflection of the terminal portion of the crura, i.e. at the level where the crura will be supported. It then penetrates below the retro-crural fascia (Fig. 8.4) (fibrous structure that we have studied previously), which will, once again, provide an efficient protection.[5] The nerve is then in the

[5] When the bulbo-clitoral organ is unstuck from the pubis, the 2 dorsal nerves of the clitoris remain covered by their "fibrous cape" and there is no risk of lesion. On the other hand, the cavernous nerves, which benefit from a reduced protection at this level, are likely to be injured due to stretching if surgical operations are not carried out with precaution and care.

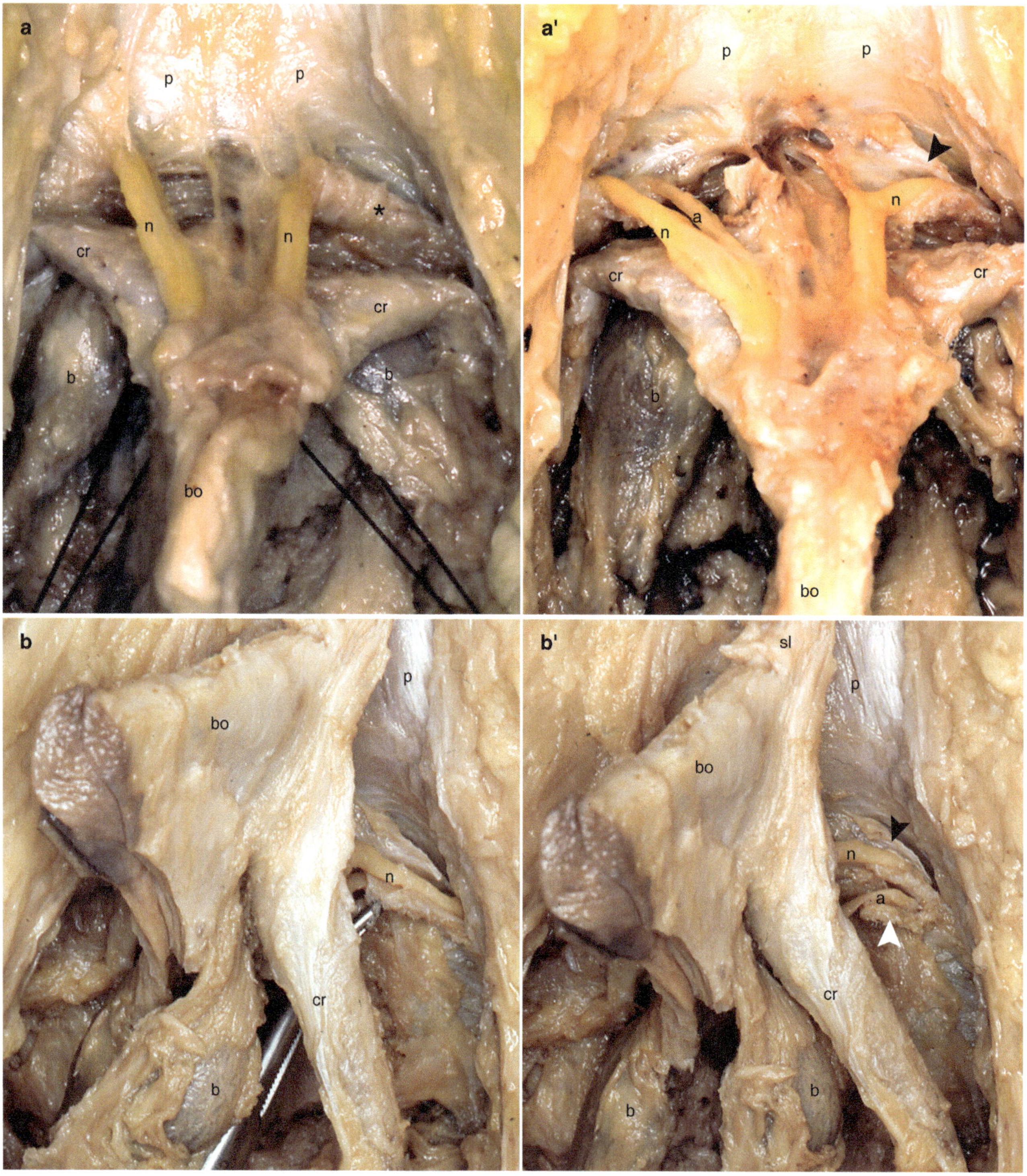

Fig. 8.2 The infra-pubic protective fibrous sheaths of the dorsal nerve and dorsal artery of clitoris. (**a**, **a′**) Views of the prepubic area (clitoral body retracted backwards and downwards). (**b**, **b′**) Side views of the bulbo-clitoral organ (detached from the pubic symphysis). (**a**) Note the left fibrous cone-shaped sheath (*), protecting the left dorsal nerve of clitoris. (**a′**) Note the opened fibrous sheaths. The opening of the left fibrous sheath (*black arrowhead*) shows the dorsal nerve of clitoris before it receives the cavernous nerve. (**b**) Note the opened fibrous sheath for the left dorsal nerve of clitoris (tracted by a forceps). (**b′**) Note the left opened fibrous sheaths of the nerve (*black arrowhead*) and of the artery (*white arrowhead*). *a* dorsal artery of clitoris, *b* bulb, *bo* body of clitoris, *cr* crus clitoridis, *n* dorsal nerve of clitoris, *p* pubis, *sl* suspensory ligament

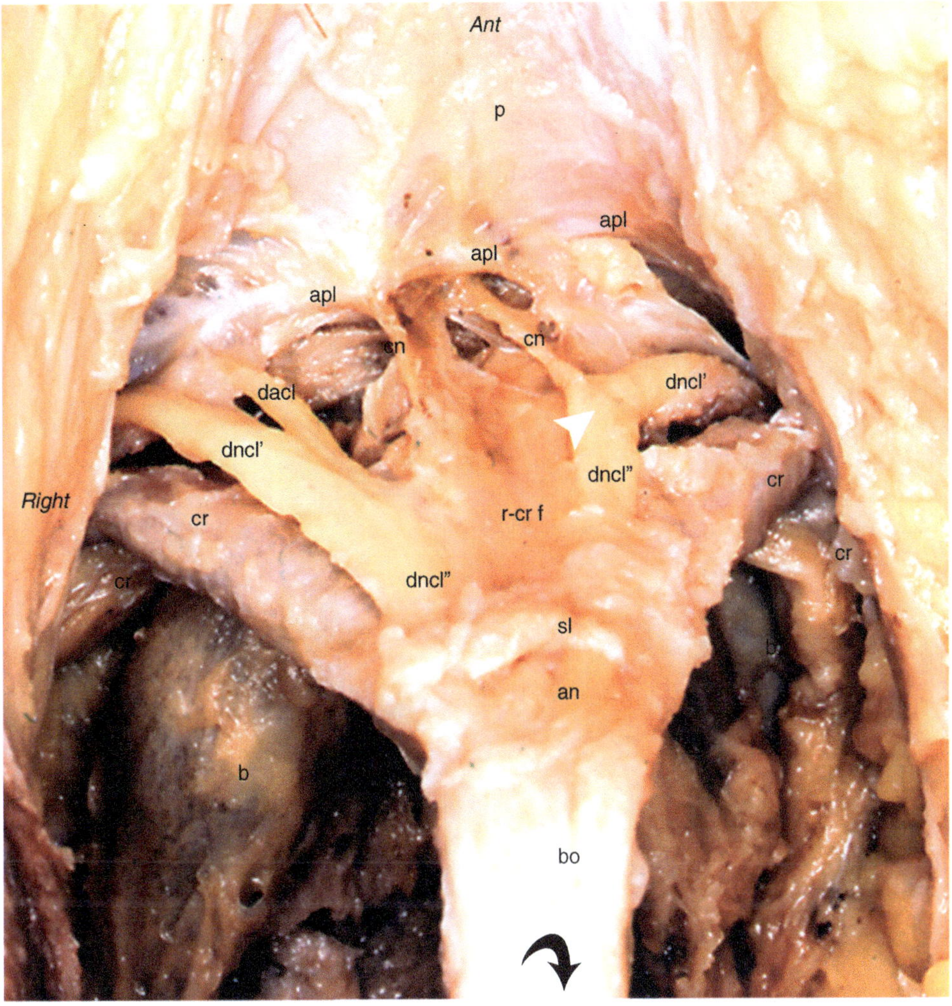

Fig. 8.3 Dissection of the prepubic area. It shows the constitution of the dorsal nerve of clitoris just before entering clitoris (the clitoral body has been retracted back and down: *black curved arrow*). *an* angle of the clitoral body, *apl* arcuate pubic ligament, *b* bulb, *bo* clitoral body (descending part), *cn* cavernous nerve, *cr* crus clitoridis, *dacl* dorsal artery of clitoris, *dncl′* dorsal nerve of clitoris (before receiving cavernous nerve), *dncl″* dorsal nerve of clitoris (after incorporation of the cavernous nerve), *p* pubic symphysis, *r-cr f* retro-crural fascia (partially resected), *sl* suspensory ligament, *white arrowhead* it shows the change of direction of the dorsal nerve of clitoris, successively oblique forward and inward, then vertical and ascending, after receiving the cavernous nerve

retro-crural space (Fig. 5.9c) . It is routed with the dorsal artery of the clitoris, whose pathway is more medial. The veins essentially emerge from the centre line. It should be recalled that the retro-crural fascia is placed against the fibrous coating, which covers the pubic symphysis. It can be detached therefrom without much difficulty. The vertical pathway will make it possible for the nerve to pass behind the ascending portion of the clitoral body and rise up to the angular top of the body to penetrate the base of the suspensory ligament of the clitoris.

- **The second elbow or clitoral elbow**: The elbow of the nerve will follow the change of direction of the clitoral body by adopting an angulation identical to that made by the clitoral body to pass from its ascending portion to its descending portion. In order to adapt to the elbow of the clitoris, each dorsal nerve passes in the basal part of the suspensory ligament, which, in turn, plays its role of protective component. The nerve is routed very deeply, against the external wall of the albuginea, at a lower level than that of the large dorsal vein of the clitoris (which is on the centre line), on either side of the small dorsal veins (which are almost median and also against the albuginea). The 2 nerves are located at 11 a.m. and 1 p.m., respectively. The microscopic examination perfectly shows the pluri-fascicular structure of each nerve and the relations of these bundles with the numerous components, which pass via the base of the suspensory ligament of the clitoris (Fig. 12.3).
- **The clitoral descending segment** (Figs. 4.5 and 8.6): It is corporeal. It successively concerns the descending segment of the body, then the glans (to which it is distributed). While progressing along the descending segment and emerging beyond the suspensory ligament, the nerve slips between the 2 superficial and deep lamina of the clitoridis fascia, which are still outside the thick corporeal albuginea. At this level, the nerve is no longer really dorsal. It becomes lateral and tends to go deeper towards the end of the body. The protection of the nerve persists in the corporeal trajectory: It is no longer fibrous and now consists of a thick preputial coat, such as demonstrated by dissections (Fig. 8.5). Along this corporeal trajectory, the nerve provides very thin branches, perpendicular to its

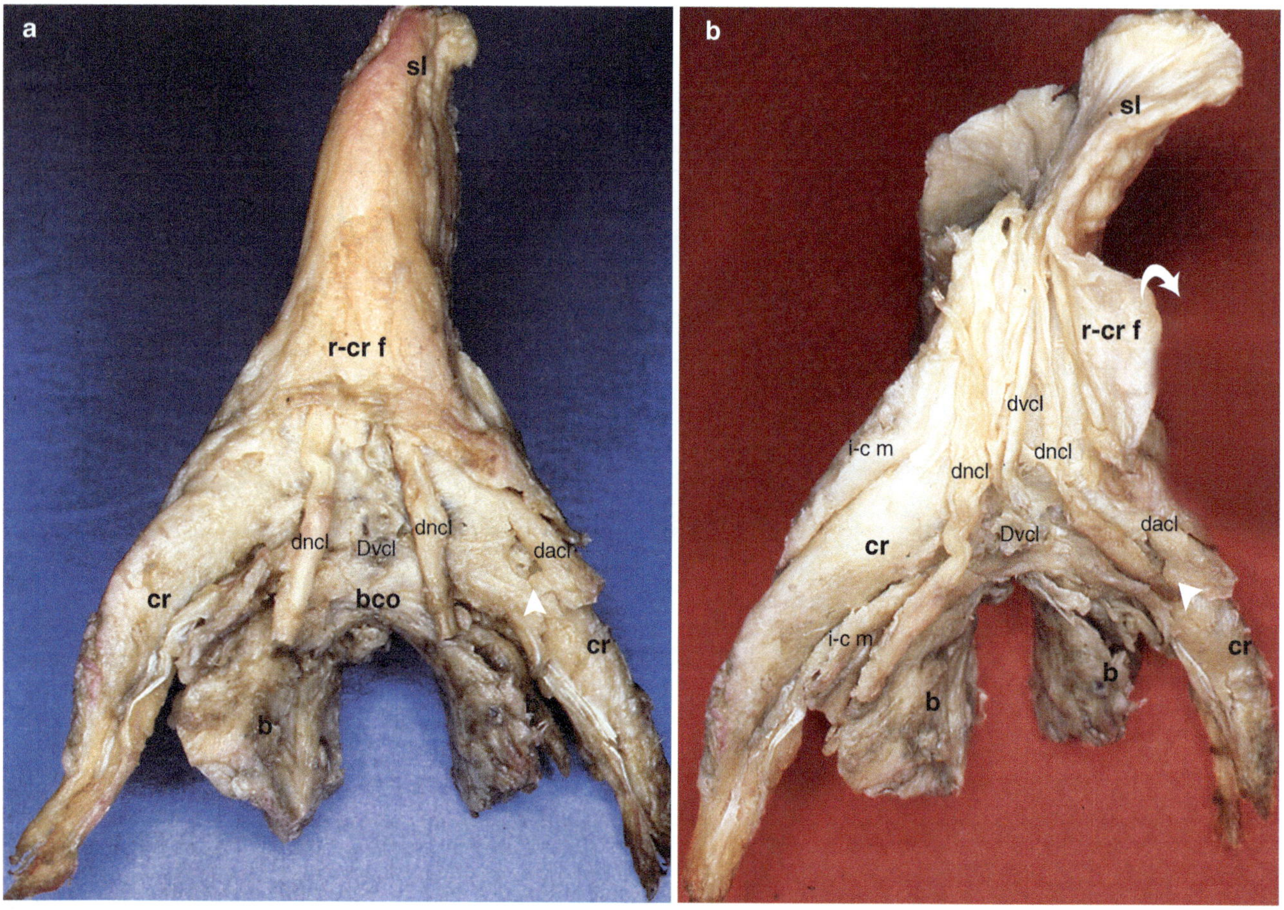

Fig. 8.4 Dorsal aspects of the bulbo-clitoral organ showing the second part (ascending clitoral segment) of the dorsal nerves of clitoris. (**a**) Both dorsal nerves of clitoris go into the retro-crural space (dorsally limited by the retro-crural fascia). (**b**) The retro-crural fascia has been cut medially and retracted outside (*white curved arrow*), showing the ascending course of the nerves. *b* bulb, *bco* bulbar commissure, *cr* crus clitoridis, *dacl* dorsal artery of clitoris, *dncl* dorsal nerve of clitoris, *dvcl* dorsal vein of clitoris, *Dvcl* deep vein of clitoris, *i-c m* ischio-cavernosus muscle, *r-cr f* retro-crural fascia, *sl* suspensory ligament, *white arrowhead* deep artery of clitoris

longitudinal axis (including several branches dedicated to the corpora cavernosa and perforating the albuginea). We can only be surprised by the considerable diameter of the nerve respect to the diameter of the clitoral body. In order to have an idea of the size ratio, the comparison with the diameter of the collateral digital nerves and that of a woman's little finger can be referred to! This underlines, in both cases, the significance of innervated components! The numerous undulations of the nerve trunk should also be noted in this corporeal trajectory (Fig. 8.5) as they correspond to as many extension reserves for an erection. When the nerve arrives near the glans, it develops an important collateral branch (Fig. 8.6) dedicated to the latter and then goes deeper. It initially passes under the superficial lamina of the clitoral fascia and then crosses the deep lamina to penetrate the glans.

- **The nerve ending**: The final development of the nerve is not immediate. Large subdivisions of the nerve are observed. They are routed relatively superficially in the direction of the apex of the glans. It is even possible to observe terminal branches of remarkable dimensions arriving in contact with the external layers of the glans, in the area of the corpuscles (Fig. 8.6). It is only with specific colouring, which shows innumerable "sensory terminations" and a great amount of corpuscular receptors and demonstrate the significant innervation of the glans, that it is possible to understand why the terminal branches of the nerve have such considerable and almost unusual dimensions (Fig. 8.6), resolving in innumerable microscopic rami only in the immediate vicinity of the Krause-Finger corpuscles. It should also be noted that the glans is not the only "nerve terminal". The microscopic observation of cross sections within the axis of the clitoral body shows that branches of the dorsal nerve of the clitoris are directed laterally to supply the clitoral hood (Fig. 9.10b) and contribute to the very dense innervation of the labia minora (Fig. 9.8). In addition, the observation of frontal sections of the bulbo-clitoral organ, on which the nerves have been identified by the S-100 protein, has shown branches of the dorsal nerve of the clitoris, which descend against the lateral walls of the

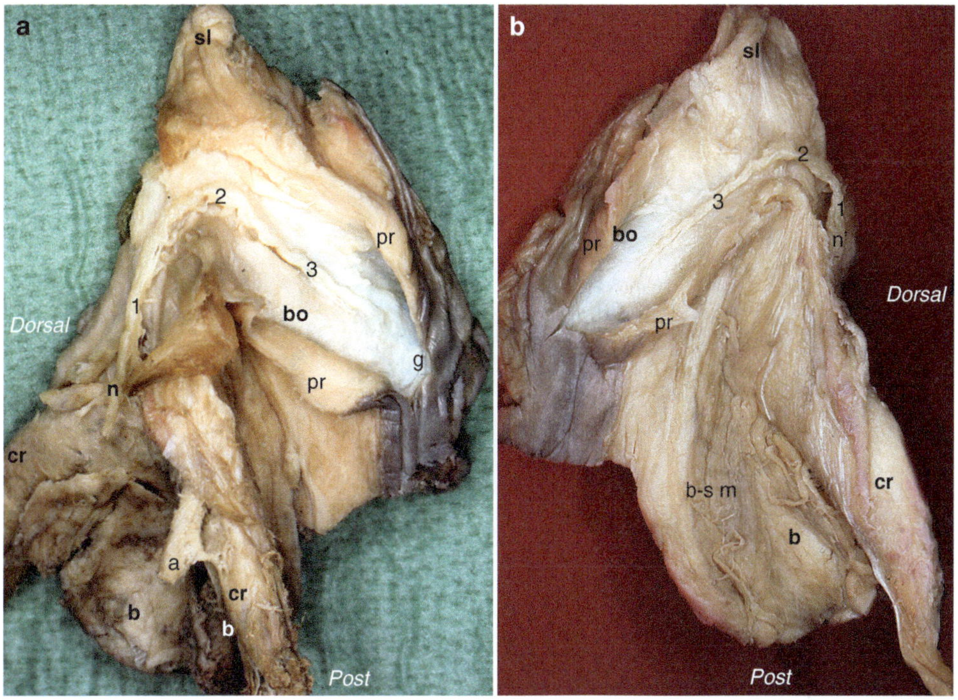

Fig. 8.5 Courses of the two dorsal nerves of clitoris. (**a**) Right nerve; (**b**) left nerve. *a* right dorsal artery of the clitoris, *b* bulb, *b-s m* bulbo-spongiosus muscle, *bo* clitoral body (descending part), *cr* crus clitoridis, *g* glans clitoridis, *n* right dorsal nerve of clitoris, *n'* left dorsal nerve of clitoris, (*1*) prepubic part, (*2*) angle, (*3*) descending part of the nerve, *pr* prepuce (opened on midline and pulled outside), *sl* suspensory ligament. Note that the nerve starts superficial, then goes deeper when it reaches the glans

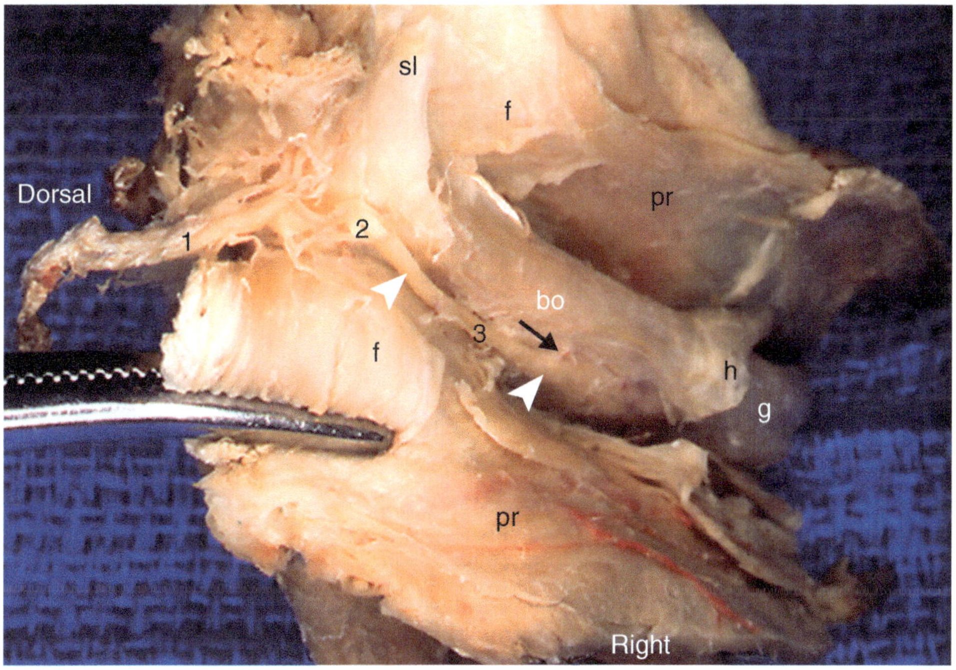

Fig. 8.6 Bird's-eye view of a dissection of the right dorsal nerve of the clitoris. The right half of the preputial sheath (sectioned medium) is retracted outside by a pulling clamp. *bo* body of clitoris, *f* clitoral fascia, *g* glans clitoridis, *pr* prepuce (or foreskin), *sl* suspensory ligament, *black numbers* distal parts of the dorsal nerve of clitoris, *1* pre-pubic part, *2* angle, *3* descending part, *black arrow* it indicates a nerve branch to the corpus cavernosum, *white arrowheads* they show the path of the nerve. Note that the nerve becomes less superficial when approaching the glans in which it will give its terminal branches. Note also that the path of the nerve is more lateral than dorsal

cavernous albuginea to laterally reach the spongy structures and contribute, from the centre line, to the remarkable innervation of each lateral wall of the vestibule.

8.1.3.2 Cavernous Nerve

The cavernous nerve that, as we have observed, joins the 1st elbow of the dorsal nerve of the clitoris, deserves to be studied due to its considerable significance. This nerve, such as we have recalled previously, therefore exits the inferior hypogastric plexus (Fig. 8.7) by passing under the middle part of the pubic symphysis, or more laterally, under the para-symphysis part of the body of the homolateral pubis. The nerve can be free but, in most cases, it penetrates the fibres of the arcuate pubic ligament. It is relatively short and does not exceed 2 cm as it is also small (approximately 1/6 of the diameter of the pudendal nerve). It is shortly routed under the thin subpubic urethral plate, on either side

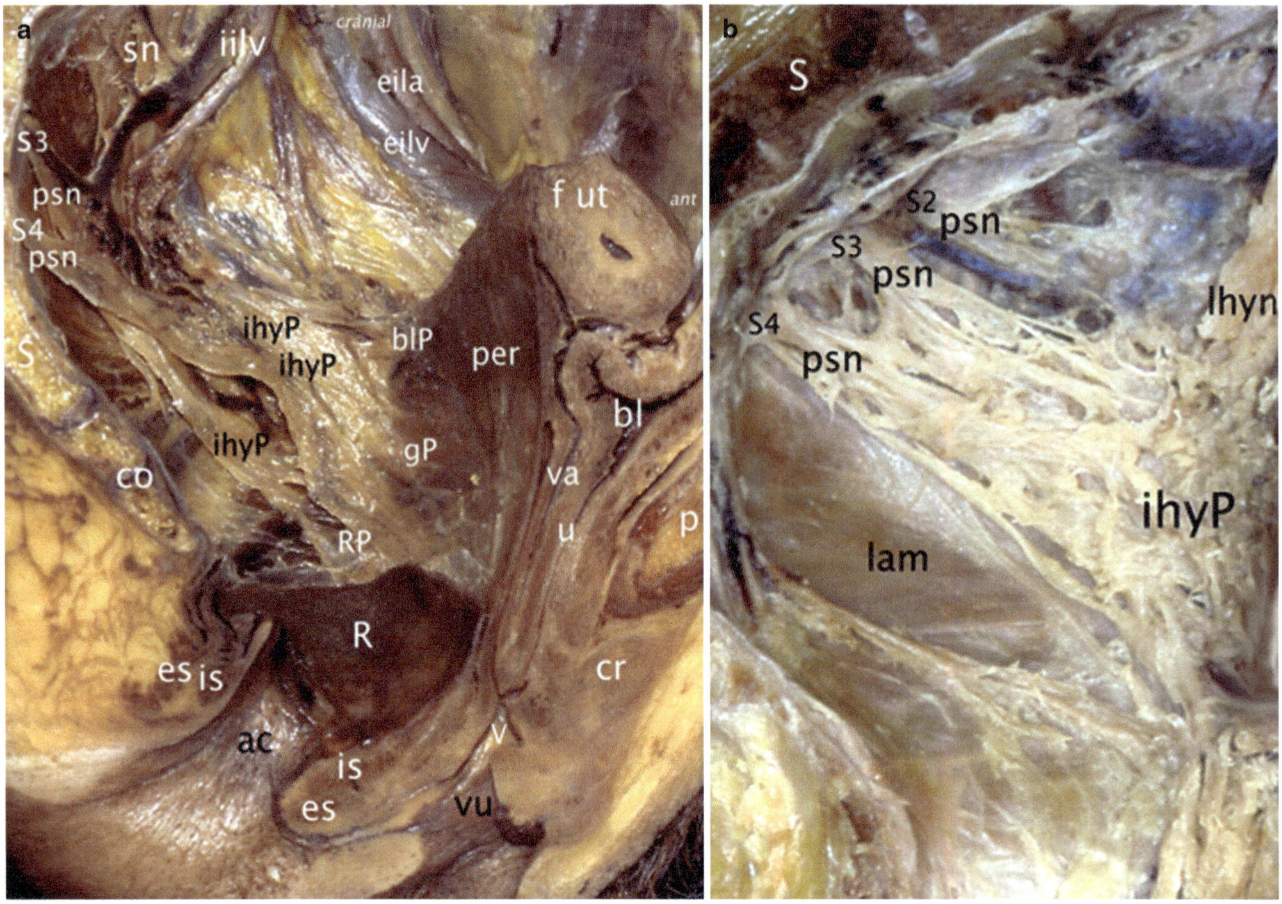

Fig. 8.7 The inferior hypogastric plexus (from which the autonomic nerves of the bulbo-clitoral organ originate) viewed on two dissections of the lateral wall of female left hemi-pelvis (after resection of the upper 2/3 of the rectum). (**a**) Overview; (**b**) details after dissection of the interlacing nerves. *ac* anal canal, *bl* bladder, *blP* bladder plexus, *co* coccyx, *cr* crus clitoridis, *eila* external iliac artery, *eilv* external iliac vein, *es* external sphincter (sphincter ani externus), *f ut* fundus of uterus, *gP* genital plexus (utero-vaginal plexus), *ihyP* inferior hypogastric plexus, *iilv* internal iliac vein, *is* internal sphincter, *lam* levator ani muscle, *lhyn* left hypogastric nerve (from superior hypogastric plexus), *per* peritoneum (anterior peritoneum of the utero-rectal recess), *psn* pelvic splanchnic nerve (nervi erigentes); they originate in the ventral rami of the pudendal plexus (S2, S3, S4), *R* rectum (remaining part after resection of the upper 2/3), *RP* rectal plexus, *S* sacrum, *sn* one sacral nerve, *S2*, *S3*, *S4* ventral rami of nerves, *u* urethra, *V* vestibule, *va* vagina, *vu* vulva

of the urethra and clitoral veins (which will join the Santorini's venous plexus), and then becomes straighter to slide in front of the inferior edge of the pubis and meet the dorsal nerve of the clitoris. Our dissections have perfectly demonstrated this convergence (Fig. 8.3). It is interesting to note that certain authors have reported similar conclusions on foetuses of 14–34 weeks old, by using layered and serial sections, histochemical staining and 3D reconstructions (S. Yucel et al.). The layered sections of the dorsal nerve of the clitoris are nNOS negative throughout the entire ischio-pubic trajectory. However, they become the site of intense nNOS activity from the pubic arch, which is a sign confirming that a positive nNOS contingent (in fact that of the cavernous nerve) has joined the dorsal nerve of the clitoris. On the other hand, other authors seem to ignore the existence of the cavernous contingent and believe that the function of the dorsal nerve of the clitoris is not clear!

8.1.4 Descriptive and Topographic Anatomy of the Innervation of the Spongy Part

We want to devote a paragraph to the corpus spongiosum due to its specificity and to the fact that it is generally unknown. Such as for the "clitoral" parts, this innervation is a dual-type innervation as it is both somatic and autonomic. It has a few alternatives according to the parts being considered:

8.1.4.1 Innervation of the Spongious Bulbs

- The sensory innervation passes via the bulbar nerve, collateral of the deep perineal branch of the perineal nerve, which itself originates from the pudendal nerve (Fig. 13.2). This bulbar nerve approaches the bulb at the level of the lateral edge of its posterior half, providing a thin collateral for the greater vestibular gland (Bartholin's gland). The bulbar nerve also contributes to the sensory innervation of the vestibule. As for its motor branches, they mainly ensure the motor function of the homolateral ischiocavernosus and bulbospongiosus muscles.

- Over the last years, vegetative innervation has been the subject of many work projects employing 3D reconstruction methods with serial sections of coloured embryos, specifically according to the method of Yucel, taken up by the French authors G. Benoit, F. Peschaud and D. Moszkowicz. This is therefore how it was possible to describe a group of nerve fibres, which are mainly parasympathetic and NOS positive. They originate from the posterior part of the vaginal plexus (itself efferent from the inferior hypogastric plexus) and reach the internal surface of each bulb, "improperly"[6] called bulbar nerves. These nerves directly reach the spongy bulb such that it is possible to say that, at the level of the bulb, somatic and vegetative terminal innervations are separate. There probably exist distal anastomoses between the 2 innervations. The bulbar branches of the deep perineal nerve undoubtedly receive, which is impossible for us to affirm, vegetative nerves from the bulbar autonomic contingent (similar to what has been observed for the dorsal nerve of the clitoris)! In addition, our own research, using standard dissection techniques combined with the histological sections stained with the S100 anti-protein, showed us that the cords of the dorsal nerve of the clitoris also contribute to the bulbar innervation: These rami are routed, such as we have observed previously, against the lateral surfaces of the corpora cavernosa of the clitoris and reach, along the "pars intermedia", the tapered anterior ends of the bulbs, thus contributing to their innervation. It is highly probable that a communicating network develops between the anterior and posterior vegetative fibres, i.e. between the vegetative fibres provided by the cavernous nerve and the "bulbar vegetative fibres".

8.1.4.2 Innervation of the Infra-corporeal Residual Spongy Part

The innervation of the infra-corporeal residual spongy part is ensured by the lateral collaterals of the dorsal nerve of the clitoris, in its descending segment.

8.1.4.3 Innervation of the Glans

The innervation of the glans is ensured, such as has been shown during the study of the dorsal nerve of the clitoris, by the terminal branches of this nerve, which cross through the clitoral fascia and penetrate the conjunctivo-vascular parenchyma, where they ramify and develop to ensure the transmission of information received by the multiple highly specialised receptors of the glans and its cutaneous environment. The nerves of the glans essentially include sensory somatic cords. They also carry vegetative nerve rami, dedicated to the smooth muscle fibres (of the vessels in particular) of this formation. This innervation is extremely dense, such as that of the preputial hood and labia minora.

8.1.5 Functional Anatomy of Female Sexual Arousal

It is better known since the current practice of MRI. The stimuli are multiple, whether they are tactile stimuli (especially in erogenous zones), or stimuli involving other sensory apparatus (especially visual apparatus but also hearing, olfactory and gustatory apparatus), even direct conceptual stimuli (erotic reading) or indirect conceptual stimuli belonging to erotic imagination (dreams, souvenirs, fantasies or even erotic conversations). The various types of abovementioned stimuli can also cumulate, interfere and be reinforced. Tactile stimuli will be conveyed via the pudendal nerve and pass via the spinal relay before reaching the specific sensory brain areas, via the spino-thalamic and thalamo-cortical pathways. Stimuli directly acting on specific cortical areas and the brain in general (amygdaloid body, nucleus accumbens, hypothalamus, limbic system, etc.) will, in turn, act on the spinal centres to trigger somatic and vegetative reactions representing sexual arousal in women. It should be noted that, at any moment, the brain (frontal areas in particular) can exert its control on reactions induced in this manner by playing a reinforcing or inhibiting role. It should however be recalled that, in women, as well in men, the spinal effector centres are significantly independent as, during periods of cortical rest (during sleep periods), sexual arousal with erection is possible!

It is therefore at the level of the spinal cord that the vegetative centres are located. The protoneurons (preganglionic neurons) of the ortho- or parasympathetic vegetative pathways are developed from these vegetative centres. The parasympathetic centres are located at the level of the sacral cord (from S2 to S4), while the sympathetic centres are at the level of the thoraco-lumbar cord (from T11 to L2).[7]

[6]The "bulbar nerve", term used by many authors, is an inadequate term because it leads to confusion: It is normally used to refer to the somatic branches of the deep perineal nerve, dedicated to the spongious bulb. Moreover, the term of nerve evokes a well-structured nerve structure, such as is the case for the cavernous nerve, whereas it actually concerns, for these bulbar vegetative fibres, superimposed thin rami.

[7]The pathways of the vegetative nervous system, whether they belong to the ortho- or parasympathetic contingent, are all bi-neuronal, with a first neuron, the protoneuron, which departs from the spinal centre and is articulated with a second neuron, the deutoneuron, which ends at the effector organ. The synapse between the 2 neurons occurs at a different site for each of the 2 contingents.

The parasympathetic pathways form the splanchnic pelvic nerves (Eckard's erector nerves), which detach from the spinal ventral roots from S2 to S4. These nerves reach the inferior hypogastric plexus. They contain the axons of the parasympathetic protoneurons, which cross through this plexus without acting as a relay. They then pass via the cavernous nerves or bulbar vegetative nerves to reach, such as we have observed previously, the erectile bodies, where the synapse with the effector deutoneuron (postganglionic neuron) occurs.

The orthosympathetic pathways develop from the intermedio-lateral zone of the spinal cord and from its lateral horn, from T11 to L2. The axons of these protoneurons depart from the ventral ramus of the corresponding spinal nerves, pass via the latero-vertebral sympathetic trunk and are then routed in the sacral splanchnic nerves to reach the inferior hypogastric plexus, where they act as a relay. The effector deutoneuron will then pass via the cavernous nerves and the bulbar vegetative nerves to reach the erectile bodies.

The physiology of sexual arousal can be thus represented. The various stimuli will involve local responses (responses at the level of the bulbo-clitoral organ):

1. Motor somatic responses: contraction of the pelvic striated muscles, ischiocavernosus and bulbospongiosus (innervation by the pudendal nerve)
2. Vegetative answers:
   - Inhibition of the sympathetic fibres, which control the contraction of the smooth muscle fibres during periods of non-excitation
   - Stimulation of the parasympathetic fibres with relaxation of the smooth muscle fibres of the vessels (which explains the generalised vasodilatation of all the vessels of the bulbo-clitoral organ) and erectile tissues (with the opening of the sinusoids and length and size modifications of the corpora cavernosa), all of which leads to tumescence and clitoral erection phenomena

Other phenomena will occur at the same time as sexual arousal and may precede an orgasm: genital secretions (glandular in particular), cutaneous vasomotor phenomena, various mammary reactions (nipple erection, areolar turgescence) and contractions of the levator ani muscles. We mention them in this section despite the fact that they do not directly involve the subject of our study, the bulbo-clitoral organ. When the sexual excitation ceases, the local and general phenomena will be reversed: the action of the parasympathetic nerve will be inhibited, the tonicity of the sympathetic nerve will be resumed and clitoral flaccidity will be restored.

9 Sensory Corpuscles

9.1 General

The bulbo-clitoral organ is, such as we have seen previously, provided with a significant innervation. The sensory somatic fibres form a very important contingent in this innervation. They also make it possible to permanently inform the overlying nerve centres of any stimulus occurring at the level of this organ. Very specialised and specific receptors ("corpuscles"), positioned at the ends of some of these sensory fibres, receive the stimuli corresponding to "sensuous sensations" and transmit them to the nerve fibres, which convey them to the sensitive and specialised cerebral areas. Nerve endings are extremely numerous at the level of the bulbo-clitoral organ, especially the glans, which explains the extreme sensitivity of the latter. It should be noted from the start that although the clitoral glans has the same number of sensory "terminations" and genital corpuscles as the penile glans, assessed at around 8,000, the density of these receptors, with respect to the size of each of these organs, is 50 times higher for the female glans. This means that the sensitivity of the clitoral glans is extreme[1] compared to that of the male glans, which is already very high!

These terminations and corpuscles make the clitoris, and especially the glans, an extraordinary organ, which is very specialised, and exclusively dedicated to female pleasure, whereas the penis and the penile glans are "multifunctional" as they are also used for urinating purposes, penetration and spermatic emission during the coitus act.

It should be noted that clitoral sensory corpuscles have been observed on animals (studies conducted by J.F. Tello on female mice, rats and ewes and by D. Ohmori on female rabbits, bitches and cats). However, these authors acknowledge that the genital corpuscles of the females under study are infinitely simpler than those of the human clitoris!

Several types of nerve endings (sensory nerve "terminations") can be identified in the human clitoris:

- Some are not specific to the clitoris and are similar to tactile terminations, the same nerve endings as those identified over all of our skin and more particularly in areas dedicated exteroception: palms of the hands, soles of the feet and labia majora.

  Among these terminations, some do not have any corpuscles: These are free terminations, "free nerve endings" and there exist others, which end at a receptor corpuscle or a tactile corpuscle; these terminations are referred to as encapsulated terminations, "corpuscular nerve endings". In this case, it is the unit consisting of the nerve fibre termination and the corpuscle, in which this termination has developed, which forms the receptor. Such as over the entire skin surface, the corpuscles observed at the clitoris are standard tactile corpuscles: Meissner's corpuscles, Ruffini's corpuscles and Pacini's corpuscles.
- Other hyper-specialised corpuscles, specific to the clitoris, Krause-Finger's corpuscles or pleasure corpuscles.

All these nerve endings have a certain number of similar characteristics:

- They are part of the general context referred to by J.F. Tello as "sensory terminations of external genitals", which are the "mucocutaneous specialised sensory end organs" of the English authors (R.K. Winkelmann).
- They only appear at late stage and develop at puberty (Winkelmann), which suggests a hormonal influence. The free nerve endings and Pacini's corpuscles are an exception to this rule as they can be observed and are already perfectly formed (J.F. Tello) in newborn children (R.K. Winkelmann) and even in foetuses (K.E. Krantz).
- Their degree of organisation and their complexity tend to increase with age (R.K. Winkelmann) and the following issue can arise (such as mentioned by H. Jaeger) "certain repercussions of the intensity of sexual life on the development of these formations".

[1] This hypersensitivity is an element, which must be taken into account by a male partner during the prelude to a sexual act, as certain women support direct clitoral stimulation very badly as it generates a truly painful sensation when it is intended to provide pleasure!

V. Di Marino, H. Lepidi, *Anatomic Study of the Clitoris and the Bulbo-Clitoral Organ*,
DOI 10.1007/978-3-319-04894-9_9, © Springer International Publishing Switzerland 2014

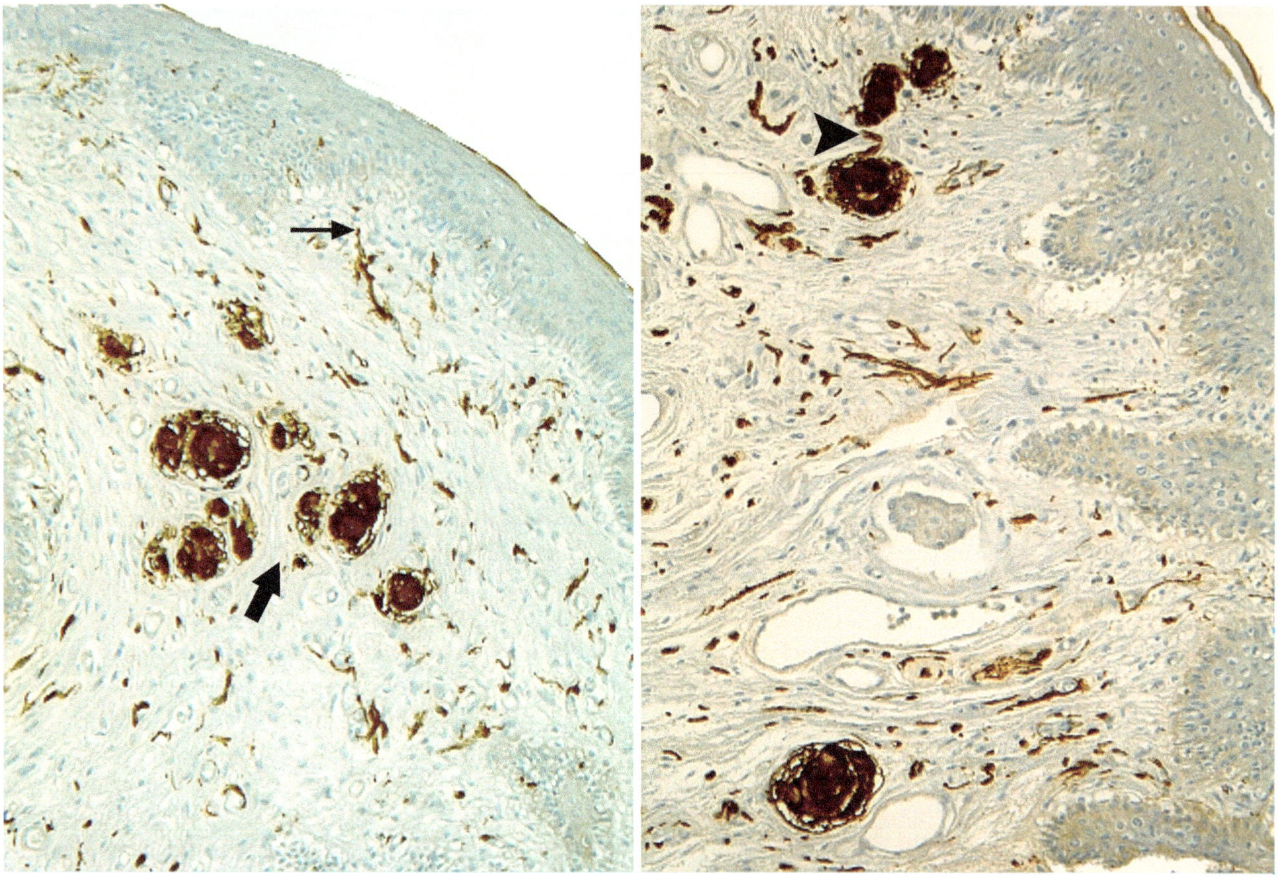

Fig. 9.1 Free nerve endings and bulbous corpuscles (PS100 staining). Note: A free nerve ending at the top of a dermal papilla (*thin black arrow*). A group of corpuscles (*thick black arrow*). The axon penetrating into the corpuscle (*black arrowhead*)

- They are more or less numerous according to the part of the external genitalia being considered, the clitoris being incontestably the organ which has the greatest number.
- Their number and their distribution vary according to individuals.

We will successively study non-specific nerve endings and specific nerve endings.

9.1.1 Non-specific Nerve Endings

Among the non-specific nerve endings, terminations in direct connection with the skin's epidermis and terminations located in the dermis can be mentioned.

- **Nerve endings connected to the epidermis**: These include, on the one hand, free nerve endings and, on the other hand, terminations with Merkel discs.

The free nerve endings correspond to axons, which exit their myelin sheath (Schwann's cell envelope), ramify and end in the dermis, in contact with the epidermis (Fig. 9.1), or in the epidermis itself, if they have crossed the lamina basalis, which remains exceptional (H. Jaeger). Small terminal swelling or bulbs may be present at the end of the ramifications (K.E. Krantz). They can anastomose to form, at the most superficial part of the dermis, a subepidermic nerve reticulum (H. Jaeger). They are distributed as thermoreceptors (sensitive to thermal variations, especially if they are sudden) and nociceptors (sensitive to pain). The extreme density of the distal axons and their arborisation and ramifications, which can be observed on histological preparations, account for the abundance of these free nerve endings, not only at the level of the clitoris but also of the labia minora.

The meniscoid corpuscles (discs of Merkel or "tactile menisci") are very superficial, in contact with the basal epithelium of the glabrous skin's epidermic ridges. They consist of axon ramifications, whose dilated ends come in contact with a particular meniscoid cell: the Merkel cell. C.J. Cold and J.R. Taylor observed them on the glabrous part of the external epithelium of the prepuce, i.e. at the level of the clitoral hood. Some are also present at the level of the labia minora. These discs detect localised light pressures and remain sensitive throughout the duration of this pressure. They therefore are rapidly adapting mechanoreceptors with an extended action.

- **Dermic nerve endings**: They are Meissner's corpuscles and Ruffini's corpuscles.

The encapsulated tactile corpuscles (Meissner's corpuscles) are also rapidly adapting mechanoreceptors. They

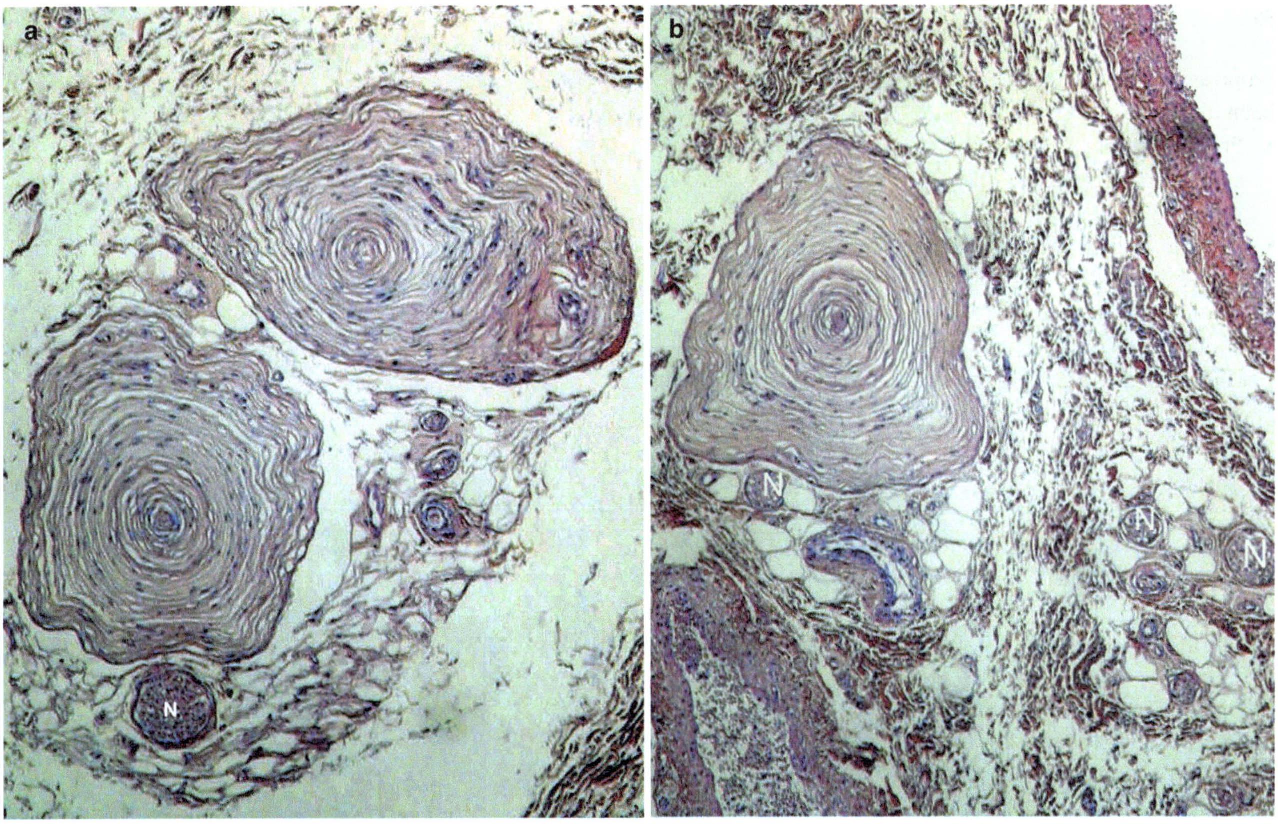

Fig. 9.2 Pacinian corpuscles in the hypodermis of the female prepuce (HPS staining). Observe: The nerve trunk (*N*) adjoining one of the two lamellated corpuscles (*left picture*). The numerous concentric in shape successive lamellae of each corpuscle. The central naked axon in each core. The adipose tissue in which you can find the lamellated corpuscles

have an alveolar aspect and are located in the top part of the dermis of the glabrous skin, in the middle of the dermis papillae. As they are sensitive to the slightest touch (they are capable of appreciating the speed at which a pressure is applied!) and therefore to brushing, they are also capable of detecting the slightest inequalities. They consist of spiral nerve ramifications, within a stack of lamellar cells, the lemnocytes, which are Schwann's cells and which are flattened and superimposed like piles of plates. They have a fibrous capsule (collagen fibres). They are especially observed on the medial surface of the labia minora and, in a smaller number, at the level of the clitoral hood.

Ruffini's corpuscles are located at a deeper level, in the dermis of the prepuce and of the labia minora. They are at both mechanoreceptors and proprioceptors. They are ovoid formations consisting of collagen fibres, between which axon ramifications having penetrated the corpuscle have developed. A nerve fibre exiting the corpuscle to reach the epithelium, as a free termination, or to end in a nearby Meissner's corpuscle (K.E. Krantz) is sometimes observed at their level. They are positioned parallel to the skin. They are slow adapting corpuscles, sensitive to cutaneous stretching (e.g. during shaving) or to an extended pressure (as they appreciate the related intensity and duration).

- **Dermo-hypodermic nerve endings** (present in the deep dermis and the hypodermis): These are the corpuscles discovered by Vater and then studied by Pacini,[2] a few years later (Vater-Pacini's corpuscle), which are now referred to as **lamellar corpuscles**. Their presence in significant numbers at the level of the clitoris deserves an in-depth description. These corpuscles are often large (transverse diameter of 5–6.5 μm) and are both mechanoreceptors and proprioceptors. They have an ellipsoidal shape, and their section is generally oval or circular even if we have observed irregular aspects (Fig. 9.2), polygonal on certain sections. They can be isolated and positioned at a certain distance from each other. They are often routed in groups of two, in similar sizes, or on the contrary, in very different sizes. They can also be grouped in clusters, in a half circle or even in a single line (Fig. 9.3). On a structural level (Fig. 9.4) the corpuscle consists of several cellular rings "the lamellae", concentric as an onion bulb and surrounding a central axon, which has lost its myelin sheath by penetrating into the axial space. The peripheral conjunctive envelope, as well

[2] It is Filippo Pacini (1812–1883), an Italian anatomist, who first described the lamellar corpuscles. However, they had already been observed by Abraham Vater (1684–1751) (a German anatomist)!

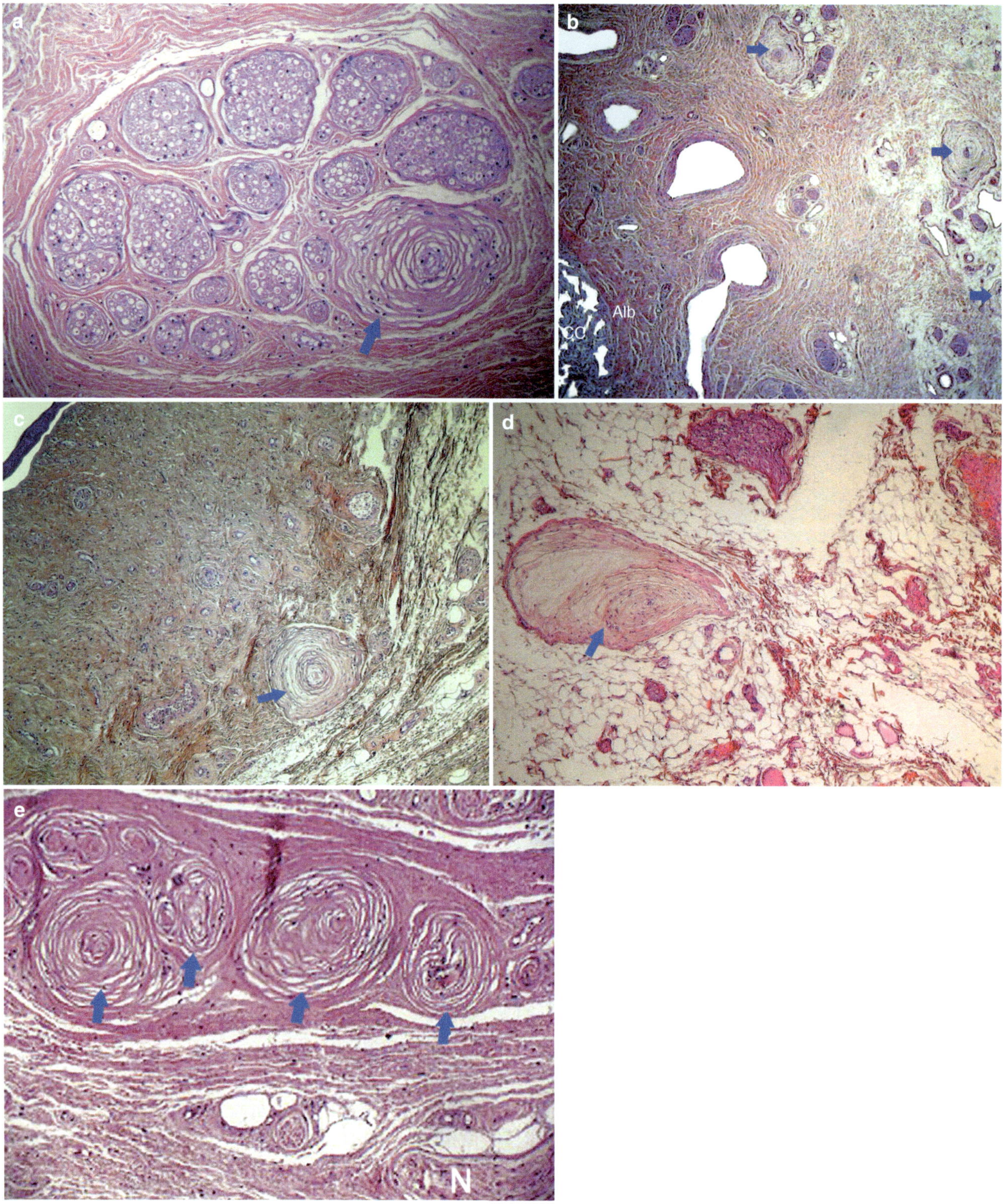

Fig. 9.3 Usual places of the clitoral lamellated corpuscles (The *blue arrows* show the corpuscles). (**a**) A corpuscle inside the epineurium (connective sheath of the nerve). The corpuscle occupies the place of a bundle of axons and is surrounded, just like it, by a sort of perineurium confounded with its peripheral connective layer. (**b**) Three lamellated corpuscles inside the cellular tissue around the tunica albuginea (*Alb*) of the corpora cavernosa (*CC*). (**c**) Lamellated corpuscles at the level of the junction of hypodermis with dermis. (**d**) A lamellated corpuscle inside the adipose tissue of the preputial hypodermis. (**e**) Several corpuscles in single file, inside the basis of the glans clitoridis. Note the different shapes of the corpuscles: Classical, oval-shaped (**a**), rounded (**b**, **e**), triangular (**c**), bludgeon-shaped or "mussel-shaped" (**d**), (evocative forms of seafood are frequent)

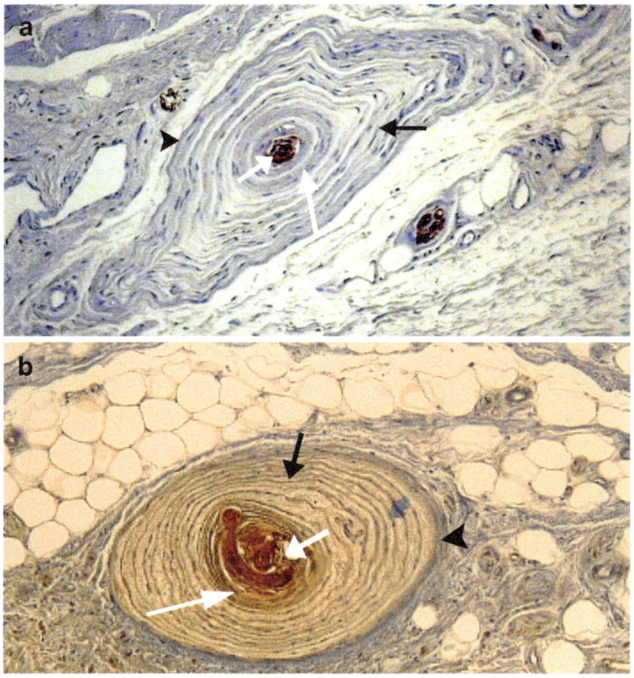

Fig. 9.4 Structure of the lamellated corpuscles of clitoris. (**a**, **b** Two examples of corpuscles located behind the angle (elbow) of the clitoral body) PS100 Immuno-staining for (**b**). *black arrow* external lamellae (capsule), *long white arrow* internal lamellae (intermediate growth zone), *short white arrow* core of the corpuscle (granular substance + naked axon, wrapped by joined lamellae), *black arrowhead* layer of connective tissue wrapping the corpuscle

as the external lamellae (capsule), is within the extension of the perineurium of the axon. The internal lamellae (teloglial lamellae), forming the intermediate growth area, consist of flattened cells, resembling fibroblasts but originating from Schwann's cells, which are more closely applied than the external laminae. Even more internal and more closely arranged lamellae form the "central club" of the corpuscle ("core" according to the Anglo-Saxon authors), in which is located the receptor nerve fibre, ending with a bulge. In electronic microscopy, these internal lamellae are only half-lamellae facing each other and separated by a radial slit. The total number of lamellae in a corpuscle is variable (20–30 on average in an adult). However, around 30 or more is frequent (Figs. 9.2 and 9.4). There are 5–10 lamellae in a foetus; this number increases with time (growth in the area of the teloglial lamellae) and, in particular, during puberty. As of birth, the lamellar corpuscles are already formed (J.F. Tello): It is already possible to identify the external lamella/strips separated by wide spaces and the internal lamellae, which are very close to each other, especially when approaching the centre. For this author, at puberty, the corpuscles will increase in volume, the number of lamellae will increase, the lamellae will tend to become closer to each other and the central area will generally tend "to be reduced to its simplest expression". However, corpuscles with a larger core can be observed in adults. It should be recalled that the inter-lamellar spaces of a mature corpuscle are not empty but filled with a viscous liquid film, in which collagen fibrils can be observed by electronic microscopy (C. Cavallotti et al.). The central axon, which has penetrated the corpuscle, is routed up to the core end and often ends in by one or more bulges. Along its route in this core, irregularities, varicosities and thickness inequalities are also observed on this axon (H. Jaeger). For J.F. Tello, the presence of corpuscles as of birth and the fact that they are formed in foetuses as of the 6th month, in contact with the nerve components, should be explained by their role in the early acquisition of protopathic sensitivity.

In the clitoris, lamellar corpuscles can be found in various locations: Firstly, in the vicinity of nerve cords (nerve ramus developed from the branches of the pudendal nerve), which they follow more or less closely, sometimes even inside the epineurium of the nerve (Fig. 9.5), especially to transmit the slightest vibratory excitation. They may even be positioned alongside the nerve endings to form a sort of sensitive pack. It is in the retro-crural small space (in the cell tissue limited by the retro-crural fascia) and in the latero-cavernous areas of the descending part of the body (Fig. 9.3) that their presence is the most obvious. They are also found in the dermis (especially the deep dermis) but the favoured sites are the subcutaneous cell tissue and the hypodermis, in contact with fatty tissue (especially at the level of the prepuce), and the suspensory ligament of the clitoris (see Chap. 12). A small number also exists in the albuginea of the cavernous bodies and more exceptionally in the septula of the cavernous bodies (K. Yamada). Such as we have observed, it is primarily the clitoral part of the bulbo-clitoral organ, which is equipped with lamellar corpuscles. However, it is not the only part to be provided as such. Lamellar corpuscles are identified at the end of the residual spongy part and glans, which form the distal part thereof (E. Lastly); the presence of lamellar corpuscles in the epithelium of the prepuce and at the labia minora (K.E. Krantz) is also to be recalled. It should be noted that all the above-mentioned formations have a greater amount of lamellar corpuscles than the penile glans or the male prepuce (C.J. Cold and J.R. Taylor).

It is well known that lamellar corpuscles are rapidly adapting mechanoreceptors and that they are not only very sensitive to pressures (they perceive deformations of a few microns!) but also particularly sensitive to vibrations![3]

[3] The lamellar corpuscles have an extreme sensitivity: they are capable of perceiving slight skin deformations (of around a few microns). They are also very sensitive to high-frequency vibrations (around 300 Hz) but have a "vibratory range" from 30 to 1,500 Hz, which is well known by "sex-toys" designers!

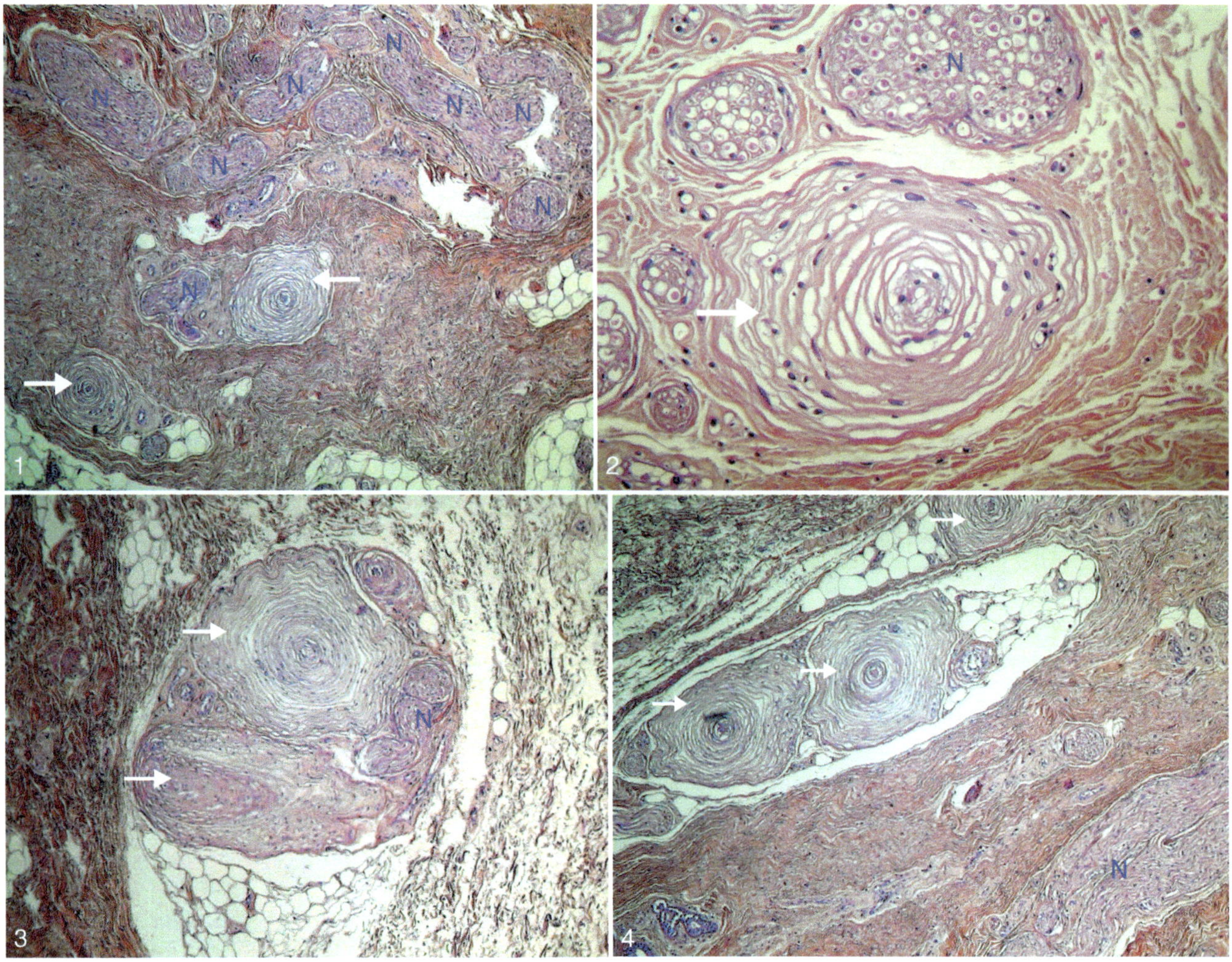

Fig. 9.5 Relationships between the lamellated corpuscles and the nervous ramifications inside the clitoris. *1* Large intradermal nervous ramifications (*N*) accompagnied by 2 lamellated corpuscles (*white arrows*). One of the corpuscles is stuck on a nervous branch. *2* A lamellated corpuscle (*white arrow*) located in the immediate neighbourhood of several nervous ramifications. *3* Assembling of nervous ramifications with 2 lamellated corpuscles in the hypodermis. *4* Three lamellated corpuscles (*white arrows*) whose 2 in single file, parallel to close nervous ramifications. Note on the photomicrographs *3* and *4*, the presence of adipose tissue close to the lamellated corpuscles

They also detect, with the free nerve endings, the sensory information related to tickling.

Their activity is biphasic: production of an impulse when the compression occurs and production of a new impulse when the compression stops (however, they are indifferent to a constant pressure). Another particularly important concept: Certain lamellar corpuscles could, according to K.E. Krantz, be integrated into the group of the corpuscles involving sexual activity; they are the corpuscles, which are positioned at the site of a nerve fascicle inside the epineurium (Figs. 9.3 and 9.5). According to this author, when sexual excitation occurs in the genital sphere, the blood supply to the vasa nervorum (arteries of the nerves) increases, which involves a pressure increase inside the entire surface circumscribed by the epineurium. "This pressure increase acts on the nerve endings of the lamellar corpuscles, which immediately lower the threshold necessary for a nerve discharge".

9.1.2 Specific Nerve Endings

They are all specialised nerve endings, characteristic of areas dedicated to generating sexual excitation in women. Many are observed in the clitoral part of the bulbo-clitoral organ. They are nerve endings provided with particular corpuscles. It is Krause and then Dogiel, who described them first as "Genitalkörperchen" or "endkolben". Other authors have also studied these formations, which are described as Golgi-Mazzoni corpuscles. Finger specified, by means of a graphic representation (in terms of pear, pea, heart shapes, etc.), the morphology of certain Krause corpuscles, which

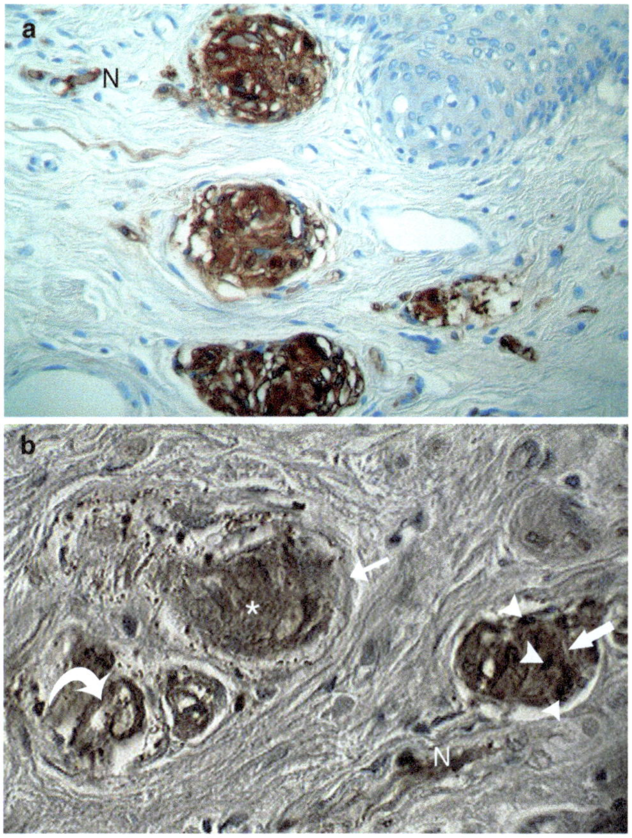

Fig. 9.6 Structure of the genital corpuscles ("corpuscles of pleasure") of the glans clitoridis (Immuno-staining by PS100). (**a**) Three large bulbous corpuscles superimposed, under epithelium. (**b**) Three corpuscules examined under high-magnification factor. *N* afferent nerve fibre, *asterisk* granular contents, *white thin arrow* capsule, *white large arrow* corpuscular syncitium, *white curved arrow* it shows a large and coiled intra-corpuscular nerve fibre, *white arrowheads* syncitial nuclei

seem to be more specifically dedicated to the excitation of specific sexual erogenous areas,[4] which explains their name, the Krause-Finger corpuscles.

These "wollustkörperchen", due to their key function in women, deserve this terminology, which is translated **into French** by the term "**corpuscules de la volupté**" ("corpuscles of pleasure"). The term of **genital corpuscles**, by which they are often referred to, indicates that they are also observed in extreme genitalia (male as well as female). Their name is related to their globular shape: **bulbous corpuscles**. Such as we have observed during our study, the distribution of these corpuscles is not identical in all women but their numbers and density seem to be extremely high in all specimens, which have been examined. Nevertheless, for K.E. Krantz, there are significant variations in terms of the quantity, quality and localisation of these structures that could explain the differences in sensitivity and excitability of the clitoris. In addition, certain authors have mentioned the possibility of an involution of these formations after the menopause (J.F. Tello). For Tello, signs of involution should be visible, at the level of the free terminations: broken or segmented fibres, terminal bulges with highly increased volumes, as well as at the level of the corpuscles—lower content in fibres and more stroma. In our research, no histological evidence supported the idea of a possible involution and we found similar corpuscles in young and old women. As opposed to lamellar corpuscles, bulboid corpuscles are not present in newborn babies or foetuses.

These corpuscles have especially been studied by many authors over the last years. They all agree to describe a general common structure (Fig. 9.6):

- Spherical or ovoid aspect.
- Large size (which can reach 150 μm in length on sections, in our studies).
- Lamellar conjunctive peripheral capsule (frequent but "not mandatory").
- Important central cavity filled with a fine granular substance, in which bathe the cores forming a syncytium, without cellular limits. This cavity is occupied by a nerve fibre (or by several nerve fibres), which has lost its myelin sheath and which ramifies or is wound up and whose end includes club-type bulges.
- Dermic location (surface or deep).

This explains the excellent definition suggested by F. Sfameni for this type of corpuscle: "nerve system organ, covered or deprived of a connective 'involucrum' and consisting of one or more nerve fibres, which, once they have been stripped of their myelin sheath, develop in the centre region and around a granular and nucleated substance".

- Primary location at the glans clitoridis.
- Multiplicity (great number of corpuscles in the same area).

Particularities are to be noted concerning this general diagram:

(a) For corpuscles organised in packs, significantly long nerve fibres are observed. These fibres generally are relatively thin and their route is "tortuous, undulating, full of elbows and undulations". Therefore, they are described as a true "neurofibrillary tangle" (H. Jaeger).

[4] Two types of erogenous areas are observed in women (R.K. Winkelmann):

The specific erogenous areas, which are areas whose stimulation produces the strongest sensations and which are necessary to generate an orgasmic reflex: prepuce, clitoris, vulva, perianal skin, labia, oral cavity and breasts (areolas and nipples, which also contain lamellar corpuscles and bulboid corpuscles).

The non-specific erogenous areas include cutaneous areas with many non-specific receptors and whose simple stimulation can generate sexual excitation with intumescence but which is not sufficient to generate an orgasmic reflex: sides of the neck and nape of the neck, armpits, internal surface of the arms, lateral walls of the thorax, internal surface of the thighs, infra-umbilical area, inguinal region, perinea, fingers, toes, scalp, etc.

(b) For corpuscles ramified, from a central fibre, significantly small branches and an abundant ramification are observed. The thickness of the central fibre varies along its pathway. "Thickness inequalities, constriction and widening phenomena, varicosities, spots, spherical particles, ampullar dilatation" are easily visible at the highest magnification of a microscope (H. Jaeger).

(c) Corpuscles organised in packs are generally formed by several fibres.

(d) The syncytium, in which bathes the nerve fibres, consists of a fine granular substance (K. Yamada) within which are arranged Schwann cell-type nuclei (H. Jaeger). This syncytium is more abundant and visible in ramified corpuscles.

With K. Yamada, who used the previous research work carried out by Ikui, 3 types of genital corpuscles can be identified[5]:

1. Type-1 corpuscles, in which the nerve fibre(s), which have penetrated the corpuscle, often change their pathway or thickness and wind into a ball to form a glomerular termination. The internal nuclei (round or oval) are numerous and have a very irregular arrangement. These corpuscles are generally provided with an external capsule. They have a more or less spherical shape.
2. Type-2 corpuscles have variable shapes. They can be round but generally are "long and narrow". In most cases, they have a capsule. The main characteristic lies in the fact that the fibre(s) (as well as their possible ramifications), which penetrate(s) the corpuscle, remain(s) in the central part while the peripheral part is only filled with syncytial nuclei. Moreover, the nerve fibre is generally thick and this thickness is not uniform over the entire pathway. These type-2 corpuscles can be isolated or join together as "aggregate corpuscles". In the last case, the same nerve fibre can be divided into several branches and each branch ends in the various nearby corpuscles. Moreover, these aggregate corpuscles can be surrounded by a single conjunctive sheath: the common sheath.
3. Type-3 corpuscles do not have a capsule. They are terminal nerve fibres, which lose their myelin sheath before ramifying or even winding to form a pack. Nuclei scattered along these terminal ramifications and the actual aspect of the nerve fibres show that they are corpuscular formations and only their external capsule is missing. All these corpuscles are distributed in the entire clitoris, but a great number of them can be found in "the portion enveloped by the epithelium (prepuce, hood), whether they are free or not" (J.F. Tello).

According to K. Yamada: If type-1 corpuscles are primarily located in the dermis of the glans clitoridis, they can also be found in the external and internal layers of the prepuce and even in the cavernous bodies. Type-2 corpuscles are located, in order of frequency, especially in the dermis of the glans, in the internal layer of the prepuce, in the cavernous body of the glans and its surrounding area, in the external layer of the prepuce, at the periphery of the cavernous bodies and in the tunica albuginea itself. Still according to K. Yamada, type-3 corpuscles prevail in the thick conjunctive tissue and therefore in the tunica albuginea, in the cavernous septula and under the glans.

For our part, we have essentially highlighted type-1 and type-2 corpuscles, not only in the clitoris, in the prepuce and its hood but also in the dermis of the vestibule (Fig. 9.7) and in that of the labia minora (Fig. 9.8). However, in these two last locations, these formations are scattered and never have the density, which can be observed in the clitoris. As for type-3 corpuscles, we have observed them not only in the clitoral conjunctive tissue but also in the dermal papillae of the external epithelium of the labia minora (Fig. 9.8).

In the dermis of the glans clitoridis, the location of the corpuscles is as variable as their frequency (Figs. 9.9 and 9.10). Similarly to H. Jaeger, the glans clitoridis can be divided into 5 areas: a central median area, the deepest, 2 medium intermediate areas and 2 more superficial lateral areas, which are subdermic (the point is part of the superficial areas but has a particular specificity). In most cases, the corpuscles are observed in the medium and lateral areas. A central location is relatively exceptional. As for the point, it may be provided with corpuscles or not. When it contains some corpuscles, the site of these corpuscles can obviously favour sexual excitation in a woman (Fig. 9.11). Otherwise, the frequency of these tubercles, along and at the rear of the preputial plate (epithelial preputial lamina), on both sides of the hidden part of glans should also be underlined (Fig. 9.9). It is interesting also to observe the corpuscles in the end of the residual spongy part, at the lower half of the glans (Fig. 9.12).

It should also be noted that some authors have classified bulboid tubercles according to their dimensions as very large, large and small tubercles. According to our observations, it appears that the very large tubercles, including some which can reach significant dimensions (Fig. 9.10), are more readily located in the medium intermediate part. The large tubercles occupy medium areas as well as surface areas. Lastly, the small tubercles are generally located in the surface area and many of them are in a subepidermic position (this could mean that it is not necessary for a corpuscle to

[5]A slightly different classification was established per H. Jaëger: This author describes 5 types but, on the one hand, his type 1 concerns corpuscles organised in a bundle, comparable to Meissner's corpuscles, and on the other hand, his types 2 and 3 are comparable to Yamada's type-2 corpuscles. Finally, his type 4 is identical to Yamada's type 3 (corpuscle placed in the middle of conjunctive fibres without their own capsule). Lastly, Jaeger isolates a type 5, which he calls the "interposed corpuscles" or intermediate corpuscles, thus named due to the fact that the nerve fibre does not end there and penetrates them to become the terminal fibre of another corpuscle.

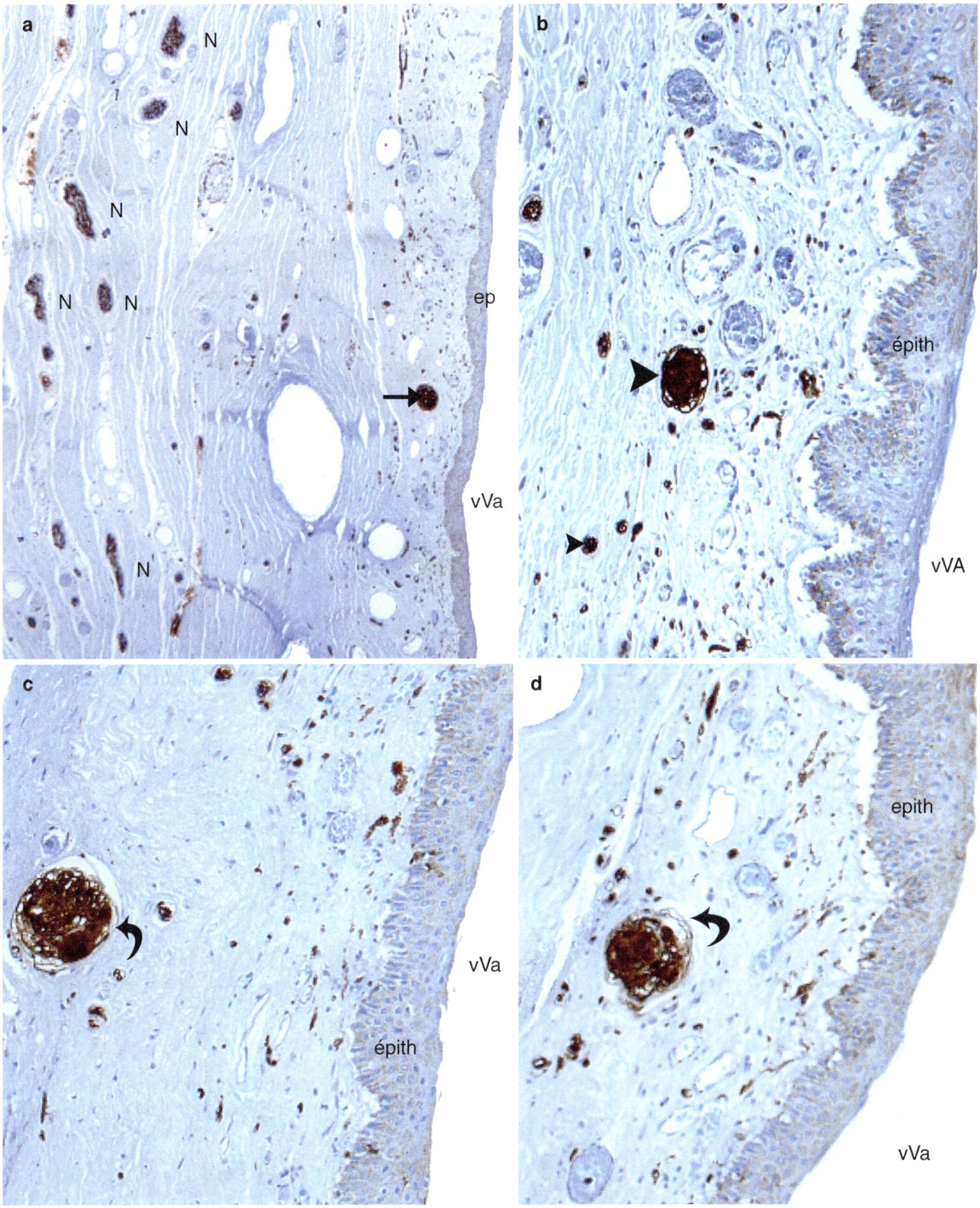

Fig. 9.7 Microscopic aspects of the vestibular corpuscles (PS100 immuno staining). (**a**) Right wall of the vestibule (low magnification); (**b–d**) morphology of the corpuscles (high magnification). *Epith* (*ep*) epithelium of the vestibule, *vVa* vestibule of the vagina, *N* nerve of the right wall of the vestibule, *black right arrow* (**a**) it shows a bulbous corpuscle (isolated, on the observed area), *black arrowheads* (**b**) they show two bulbous corpuscles, *curved black arrow* (**c**, **d**) they show the capsules of two large bulbous corpuscles. Note: scarcity of the vestibular bulbous corpuscles and their typical morphology. Note also the nerve wealth of the vestibular walls with many nerve endings under the epithelium, in the dermal papillae (visible under high magnification)

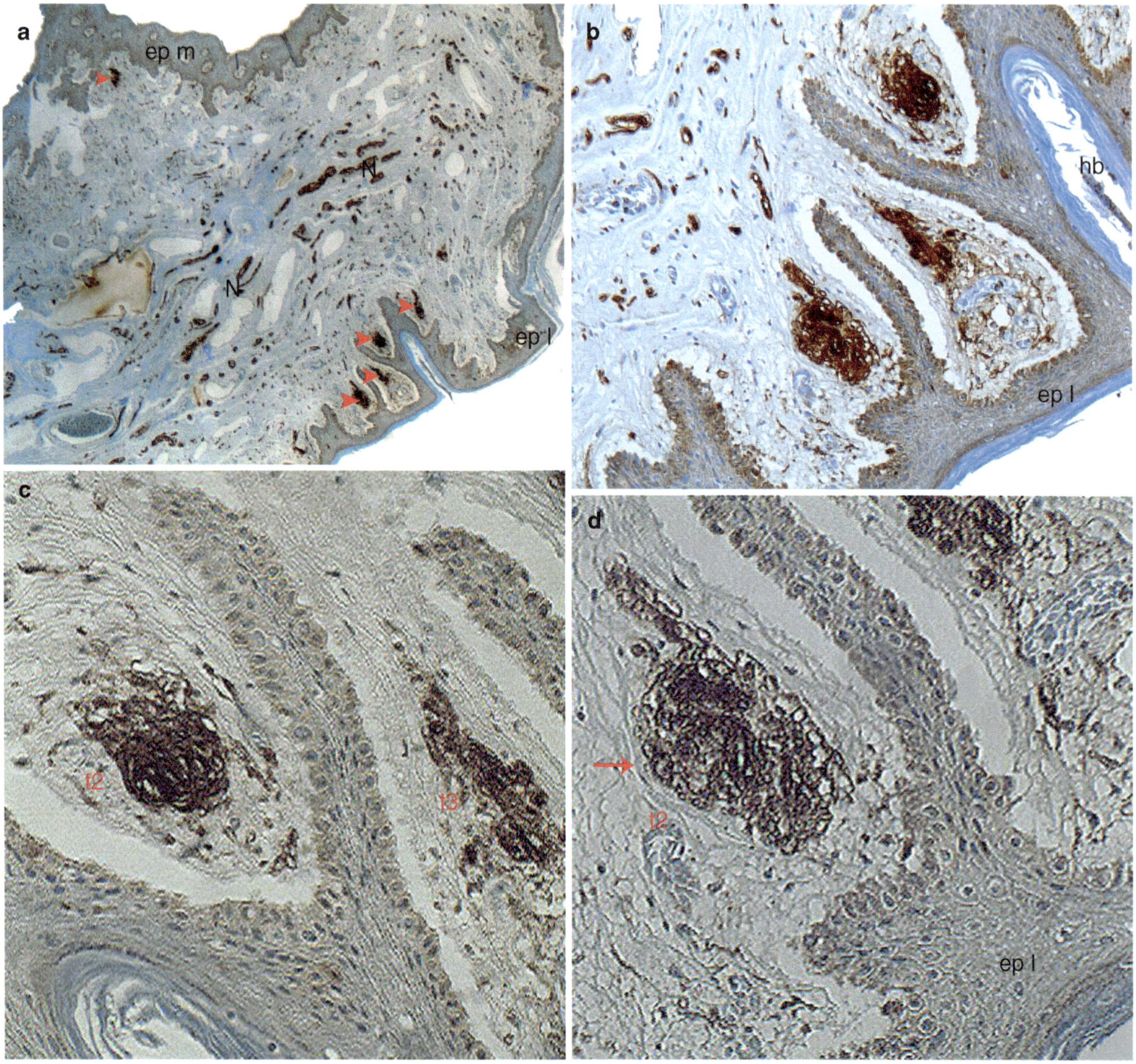

Fig. 9.8 Microscopic aspects of the genital corpuscles of the labia minora (PS100 immuno staining). (**a**) Transversal section of a labium minus (low magnification); (**b**) observation of the area in which several large corpuscles are visible; (**c, d**) high magnification of the genital corpuscles. *ep l* epithelium of the lateral wall of the labium minus, *ep m* epithelium of the medial wall, *hb* hair bulb, *t2* large rounded corpuscles, of type 2, *t3* elongated corpuscule, of type 3, *red arrowheads* they show the genital tubercles, *little red arrow* it identifies the draft capsule around the corpuscle, visible in (**d**). Note: Most corpuscles under the epithelium of the labium minus' lateral wall. The intradermal position of the corpuscles, housed at the bottom of the dermal papillae. On (**d**), the nerve afferent fibre of the corpuscle is clearly visible

have a large diameter to be effective, if it is located near to the surface of the receptor area!). The median central area, on the other hand, either has no bulbous corpuscles or a very low content thereof.

At the end of this study, it will be necessary for us to analyse the functional role of these bulbous corpuscles. To do so, we must gather the major observations which we could make:

1. There exists, at the end of the clitoris and therefore in a very reduced area, a great amount of nerve terminations and an extraordinary accumulation of nerve corpuscles, including a majority of bulbous corpuscles.
2. The density of the bulbous corpuscles at this level is more important than anywhere else in the external genitalia and in all the "erogenous areas".
3. Each clitoral bulbous corpuscle (such as any corpuscle of the male penis) is a multi-innervation site (penetrated by several fibres),[6] which causes, according to Ohmori, "an addition of sensations".
4. The significant size of the corpuscles seems to be explained by the need for containing the greatest length possible of

[6]This multi-innervation concept has been discussed by certain authors, including J.F. Tello "due to the difficulty of distinguishing if they are separate fibres or branches from the same fibre".

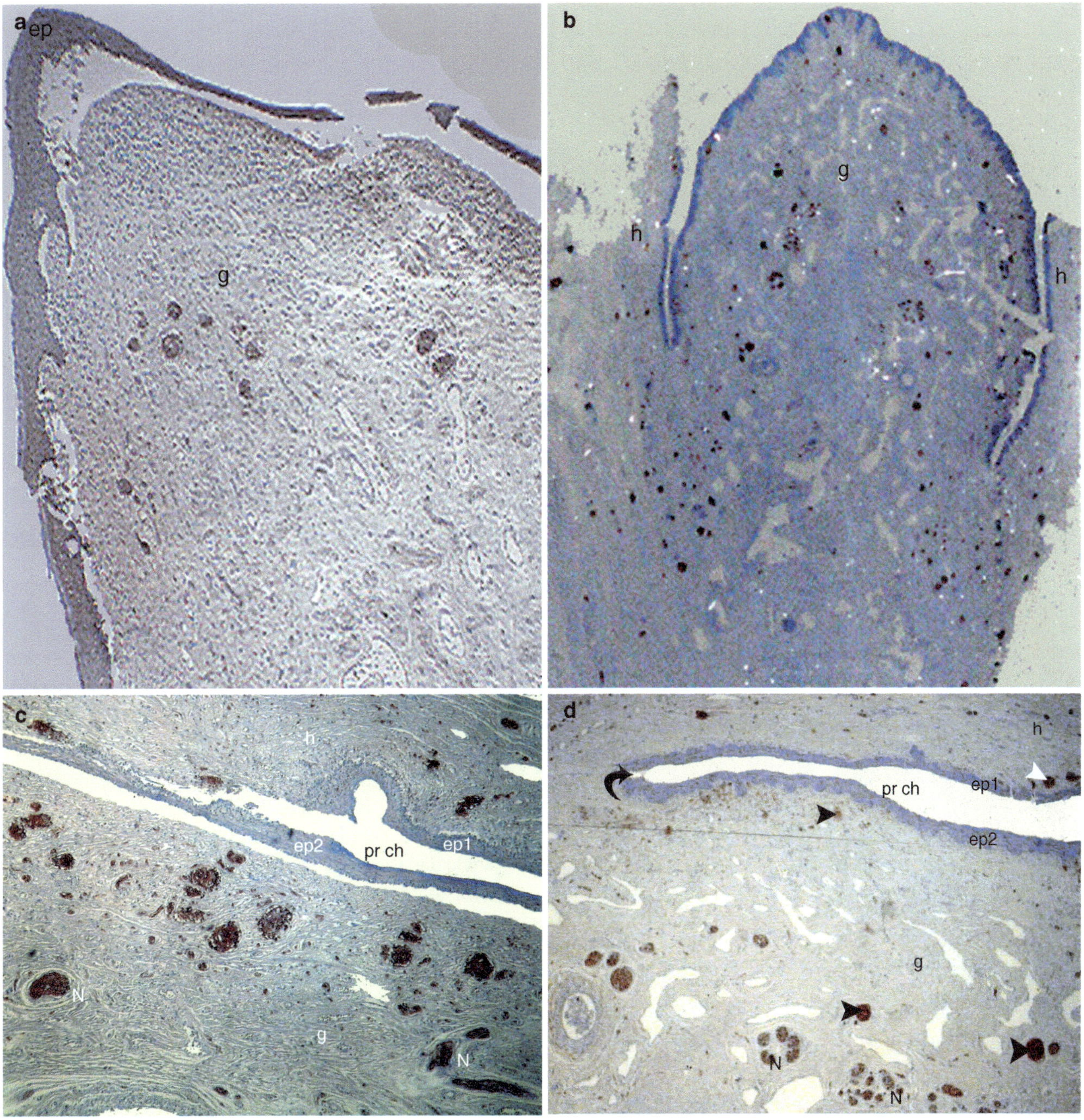

Fig. 9.9 Morphological aspects of the genital corpuscles (corpuscules of pleasure) of the glans clitoridis (PS100 immuno staining). (**a**) Corpuscles (*dark purple*) grouped at the end of the glans clitoridis. (**b**) Scattered corpuscles (*brown-black*) not only occupying the whole glans but also the distal part of the clitoral body; (**c**) many corpuscles of wide diameter, arranged under the epithelium of the glans, and along the preputial chamber too; (**d**) some big deep bulbous corpuscles and a multitude of punctated superficial corpuscles (*light brown* under the epithelium of the glans). *ep* epithelium, *ep1* epithelium of the medial surface of the hood, *ep2* epithelium of the glans, *g* glans, *h* hood, *N* nerve rami, *black curved arrow* neck of the glans, *pr ch* preputial chamber, *Arrowheads* (*black* for the glans and *white* for the hood) show some corpuscles

sensory nerve tissue, "the greatest mass of nerve tissue, the greatest spread" of neurofibres, so that there is a summation of the excitation and, as a result, the transmission of a very strong sensory message, which sometimes borders on pain![7]

5. Similarly, the arrangement of the aggregate corpuscles, supplied by the same fibre(s), also contributes to collecting sensory messages enriched by the passage through several successive corpuscles. Thus, H. Jaeger wonders

[7]In addition to this theory of nerve substance concentration, there is another particularly interesting theory (made by F. Sfameni), whereby the function of each bulbous corpuscle would be to modify the stimulus received by the peripheral free nerve endings, to which it is connected, and transform it into a sensation of pleasure. The granular substance of the corpuscles would contain, for this purpose, an endocrine secretion, produced by the nuclear syncytium, and would therefore play a major part in this modulation of the sensory message.

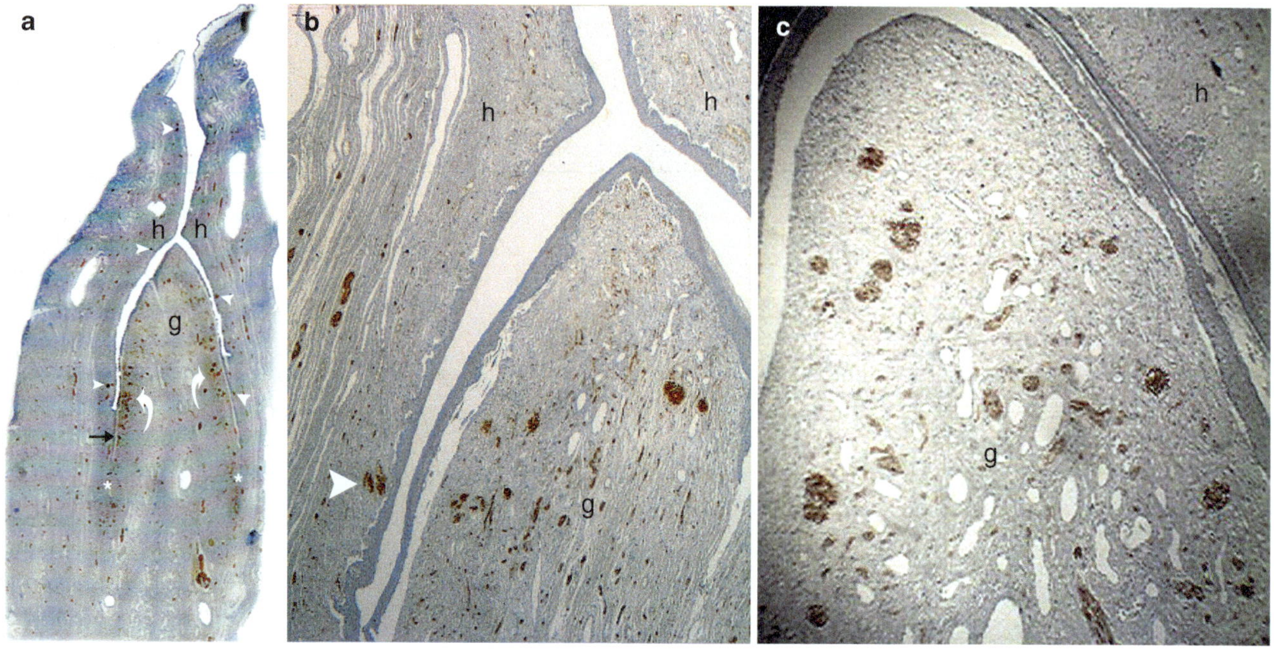

Fig. 9.10 Examples of distribution of the genital tubercles (corpuscles of pleasure) in the glans clitoridis. (**a**, **b**) Predominance of lateral distribution (curved arrows on **a**); (**c**) harmonious distribution without predominance. Three transverse sections of the glans clitoridis (PS100 immuno staining). The corpuscles of pleasure are stained *dark brown*. (**a**) Low-power view (note the presence of corpuscles* below the neck of the glans; actually, on both sides of the hidden part of glans). (**b**, **c**) High-power views. *g* glans clitoridis, *h* hood of the glans, *black arrow* epithelial preputial plate. Note on Sections (**a**) and (**b**), some corpuscles located on the hood (*white arrowheads*)

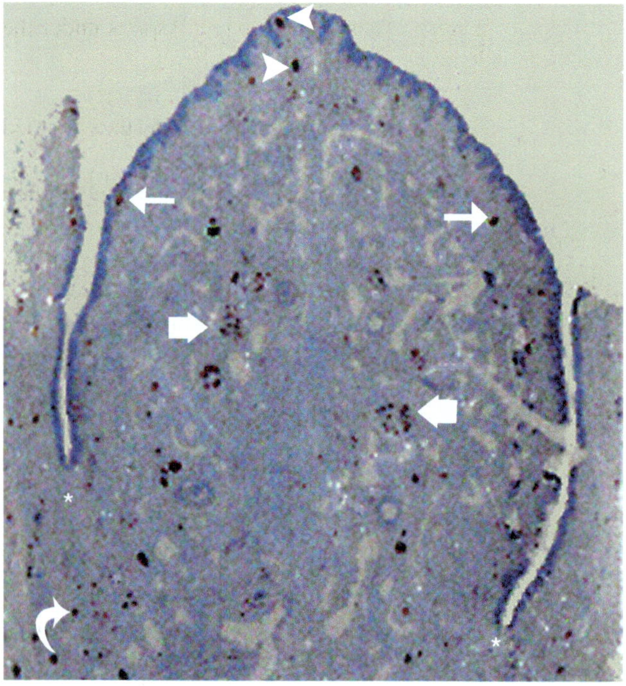

Fig. 9.11 Location of genital corpuscles (corpuscles of pleasure) in the glans clitoridis. Longitudinal section of the glans; PS100 immuno staining. *Arrowheads* distal superficial corpuscles ("corpuscles of the extremity"), *thin right arrows* lateral superficial corpuscles, *short large arrows* corpuscles of the middle part, *curved arrow* deep corpuscles. Note: The lack of tubercle on the middle part of the glans. The presence of corpuscles (*) at the neck of the glans

whether such systems are not "types of nerve impulse regulating organs, which either accumulate or absorb, such as relays inserted on certain nerve fibres".

6. Finally, the intra-corpuscular complication, the fibrillary tangle and the rounded shape of the corpuscles (allowing them to have equivalent receptor properties over their entire surface) demonstrate the perfect adaptation of the small clitoral surface to the reception of the slightest erogenous stimuli.
7. The presence of genital corpuscles on the hood, the clitoral prepuce, the labia minora, the vestibule and even, such as recalled by K.E. Krantz, in the labia majora and the mons pubis (mons veneris) shows that the clitoris is not the only excitable organ but that it remains the most excitable and the organ which is the most capable of generating erogenous excitation.[8]

[8] According to D. Perhaps Ohmori, "corpuscles organised in packs may be responsible for the sensations produced by friction or sliding, circumstances which occur with the greatest intensity in the external genitals". For our part, we believe that the possibility of spontaneous orgasms occurring during sport exercises, such as pedalling on a bicycle or climbing a rope, is due to the extreme abundance of genital corpuscles at the level of the clitoris associated with the corpuscles (scattered but present) of the external genitalia. An interesting study concerning "sexual pleasure induced during exercise (PSIE) and possible orgasms induced during exercise (OIE), in a group of women of all ages", was published recently and concluded that these situations were very frequent. The authors (D. Herbenick) have called this type of orgasm, induced by exercise, "coregasm".

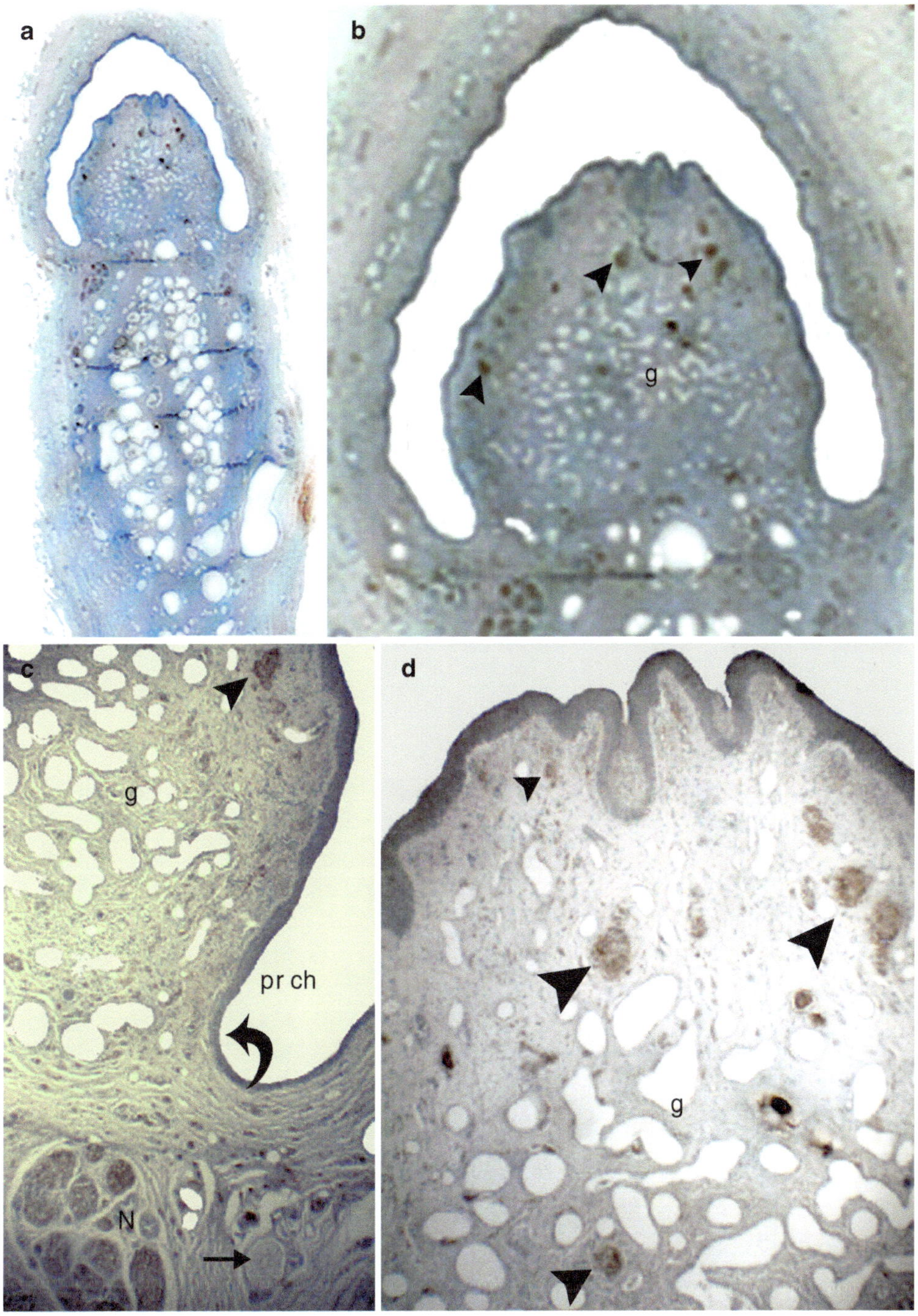

Fig. 9.12 The corpuscles of pleasure in the residual spongy part (lower half of the glans). (**a**–**d**) Transverse sections through the RSP, showing the corpuscles of pleasure in the glans (PS100 immuno staining). (**a**) photomicrograph (low power view), (**b**) photomicrograph (intermediate power view), (**c**, **d**) microscopic aspects (high power view). *g* glans, *N* terminal branch of the dorsal nerve of clitoris, *pr ch* preputial chamber, *curved black arrow* neck of the glans, *right black arrow* it shows a corpuscule of Pacini, b*lack arrowheads* corpuscules of pleasure (note on **d**, the different sizes of these corpuscles)

Overall, our entire study demonstrates, once again, the extraordinary complexity and abundance of sensory receptors in the clitoris, the exceptional richness of this organ in nerve terminations and the incredible density of genital corpuscles over a very small surface.

The clitoris is the most elaborate sensory organ of the female body. It is provided with "specific tactile sensitivity receptors", which are highly specialised, in connection with the sexual function. With its invaluable sensors, the clitoris has only one function: the reception of pleasure stimuli for a single purpose: to generate sexual excitation in women in all types of contexts (masturbation, foreplay, sexual intercourse).

10 The Vessels of the Bulbo-clitoral Organ

10.1 General

This vascularisation plays an extremely important double role. It ensures not only the normal nutritional physiological circulation of this organ but also the functional circulation of the erectile bodies, which allows the intumescence and clitoral erection phenomena during sexual arousal in women.

10.1.1 The Arteries

The arteries mostly originate from the branches of the internal pudendal artery (internal shameful artery according to previous authors), which is, itself, one of the branches from the division of the hypogastric artery. This pudendal artery follows the same pathway as the pudendal nerve. With the pudendal veins, it is routed with the nerve, on the lateral wall of the true pelvis, against the inferior insertions of the obturator internus muscle, inside a fibrous canal: the pudendal canal of Alcock. It is then divided into the perineal artery and the dorsal artery of the clitoris. The bulbar artery originates from the deep branch of the perineal artery, at the level of the posterior edge of the anterior perineum. The arteries of the bulbo-clitoral organ are thus the bulbar artery and the dorsal artery of the clitoris.

10.1.1.1 The Bulbar Artery

It detaches from the deep perineal artery (of which it often forms the terminal branch) at the level of the ischio-bulbar triangle (Figs. 12.1 and 13.2). It is routed obliquely towards the posterolateral part of the bulb and thus irrigates the greater vestibular gland. It is then divided into several rami, which approach the convex part of the bulb, via its lateral surface. The deep perineal artery also emits a bulbo-urethral collateral vessel, whose bulbar ramus also contributes to the vascularisation of the bulb.

10.1.1.2 The Dorsal Artery of the Clitoris

This artery is the terminal branch of the internal pudendal artery. It has the same ischio-pubic trajectory as the homologous nerve (see Chap. 8). Once it has left the canal of Alcock, it appears at the level of the lateral part of the infra-pubic canal, under the arched ligament of Lauth, wrapped in a sort of fibrous cone, under a similar cone, from which emerges the homologous nerve. These fibrous structures, whose protective role seems obvious, are formed by splitting of the transverse ligament of the pelvis (ligament of Krause), which itself is formed by coalescence, in front of the urogenital diaphragm, of the superior and inferior fascias covering this diaphragm. It is necessary to recall, such as we have already observed, that the arterial cone is always located under the nerve cone and that the 2 cones are adjacent and separated by a common fibrous lamina, a genuine fibrous "spacer", according to the expression employed by G. Paturet (Fig. 8.2). The artery then approaches the dorsal surface of the clitoris; extends along the homologous pillar, thus providing this pillar with the deep artery; then passes rapidly under the retrocrural fascia; and rises obliquely up to the angle of the clitoris. The artery then penetrates the base of the suspensory ligament (Fig. 12.3) and progressively passes on the dorsolateral part of the descending portion of the clitoral body. It is then routed towards the glans, inside the nerve, on either side of the superficial dorsal vein. Over this trajectory, it remains underneath the clitoral fascia, which separates it from the cell tissue of the prepuce. On the lateral walls of the descending portion, it provides branches perpendicular to its axis: circumflex branches. It then supplies the glans and its hood. Anastomoses exist between the branches of this artery and the distal rami of the superficial perineal artery (which vascularises the prepuce, the labia majora and labia minora). The vascularisation of the latter is particularly rich and thus explains the involvement of these formations in the turgescence phenomena and their extremely characteristic histological aspect.

V. Di Marino, H. Lepidi, *Anatomic Study of the Clitoris and the Bulbo-Clitoral Organ*,
DOI 10.1007/978-3-319-04894-9_10, © Springer International Publishing Switzerland 2014

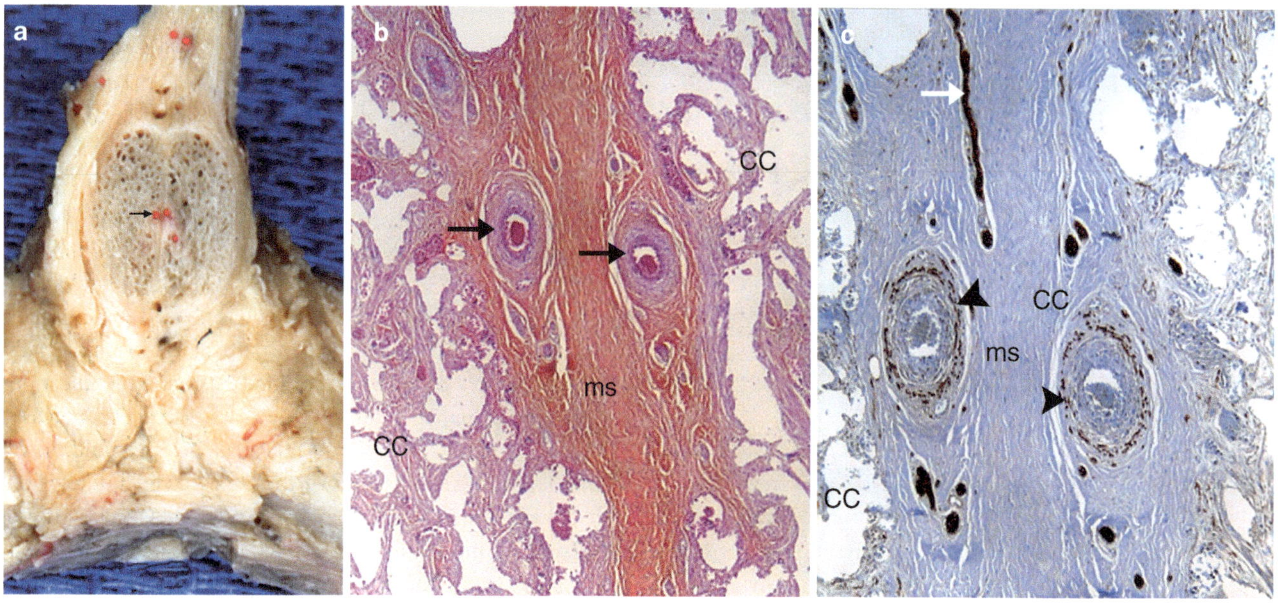

Fig. 10.1 The deep arteries of clitoris. (**a**) Situation of the deep arteries (*black arrow*) on a frontal section of the clitoral body, just after the joining of crura. (**b**) Microscopic aspect of the deep arteries, included in the fibrous tissue of the median septum (classical staining by HEAS). (**c**) Another aspect of the deep arteries: staining of the section by PS100. *CC* corpora cavernosa, *ms* median septum, *right black arrows* they show the deep arteries of clitoris, *black arrowheads* they show the vegetative nerve in the arteries' wall, *white right arrow* it shows a nervous ramus oriented in the axis of the median septum and innervating this structure. Note: on (**a**) the position latero-septal of the deep arteries, their conjunctive wrapping connected with the median septum; on (**b**) the intra-septal position of the arteries; on (**c**) the rich vegetative innervation of the arteries

10.1.1.3 The Cavernous Artery

English authors refer to this artery as the deep artery of the clitoris. It generally originates from the dorsal artery of the clitoris, shortly after the point where this vessel reaches the clitoral pillar. It can also originate from the pudendal artery itself, along the ischio-pubic trajectory of this vessel, which then seems to have 2 terminal branches, the dorsal artery of the clitoris and the cavernous artery. In the first case, it is very short. In the 2nd case, it is longer and can measure more than 2 cm. When it is a collateral vessel of the dorsal artery of the clitoris, it detaches from this vessel almost nearly at a right angle and penetrates the clitoral pillar via its lateral surface (Fig. 8.5). It rapidly divides inside the pillar into 2 rami, a proximal ramus, referred to as recurrent, which irrigates the pillar and the pre-angular ascending portion of the body, and a distal ramus, which follows the direction of the clitoral body, in the centre part of the corpus cavernosum. It can remain single or divide into 2 or 3 rami. In the corporeal corpora cavernosa, these cavernous arteries approach the median septum and remain in contact with it via a connection consisting of a fibrous crown, appendage of this septum.[1] The branches originating from the cavernous artery, and which have penetrated the right pillar, remain on the right of the septum. The branches originating from the artery, and which have penetrated the left pillar, remain on the left. In his dissections, Kobelt observed, with great frequency, a retro-cavernous anastomotic arch joining together the 2 right and left cavernous arteries. From this arch originate "2 small rami, which deeply penetrate the corresponding corpus cavernosum and which would be the true deep arteries of the clitoris". In any case, the cavernous arteries are of great importance for the operation of the corpora cavernosa, which is demonstrated not only by their size but also and especially by the very rich vegetative innervation of their wall (Fig. 10.1).

Each cavernous artery gives, in the ipsilateral corpus cavernosum, the helicine arteries. In turn these arteries branch into capillaries which vascularize the trabecular structures of the cavernous tissue. These helicine arteries are very specific: (a) they open directly in the sinusoidal spaces and (b) they have a structure adapted to cavernous physiology: Indeed, their wall has a particular longitudinal smooth muscle layer made of helical fibre bundles (hence their name), thickened intermittently and forming heightenings of the intima,[2] the "intimal cushions"[3] of von Ebner. These cushions modulate the arterial flow by serving as valves. Under

[1] They can also penetrate and be routed within the septum.

[2] See footnote 1.

[3] The helicine arteries are part of the "polsterarterien" of the German authors (G. Paturet).

the effect of parasympathetic nerves, these cushions relax, allowing the inflow of arterial blood and the rapid filling of the cavernous sinusoids prelude to the swelling and erection of clitoris.

10.1.1.4 The Arteries of the Hood and Prepuce

These are the arteries of the anterior part of the vulva, which therefore vascularize the anterior part of the labia majora and labia minora. These arteries are not only distal rami of the superficial perineal arteries but also rami from nearby arteries, which have crossed the genito-femoralis sulcus: external pudendal arteries (collateral branches of the femoral artery) and obturator artery (collateral branch of the internal iliac artery). Rich anastomoses exist between all these vessels and the small terminal rami of the dorsal artery of the clitoris, as well at the level of the hood of glans as of the clitoral frenula.

10.1.2 The Veins

They specifically receive venous blood from the draining of the corpora cavernosa and the sub-albugineal veins. Two venous circulations drain the clitoris: the deep venous circulation and the superficial venous circulation.

10.1.2.1 The Deep Vein of the Clitoris

It consists of the junction of the venules from the corpora cavernosa (Fig. 7.2), especially the venules which are routed along the corporeal septum and the venules which drain the pillars. This vein can be single or divide into 2 superimposed veins. It leaves the bulbo-clitoral organ at the level of the median part of the retro-crural region (Fig. 8.4). It has a significant size of around 2 mm in diameter. It is directly routed under the subpubic urethral plate or in a split, whose protective role seems obvious and through which it is possible to see it by transparency (Fig. 13.4). This vein perpendicularly receives other venous branches in a ladder-type configuration. It passes very rapidly under the ligament of Lauth, in the infra-pubic region, and reaches the infravesical venous plexus (plexus of Santorini),[4] which drains itself in the internal pudendal veins and vesical veins. Thus, the deep venous circulation of the clitoris ends at the pelvic venous networks. This is why a pelvic phlebography method by intra-clitoral injection of a contrast product has been described: **pelvivenography** (S. Petkovic), a method which can be used to obtain excellent pelvic phlebographies, in cases of thrombosis or compression by pelvic tumours.

[4]This infravesical venous plexus anastomoses with the urethral, vaginal and uterine venous plexuses.

10.1.2.2 The Superficial Dorsal Veins of the Clitoris

They are routed along the dorsal surface of the clitoral body, parallel to the homologous arteries, under the clitoral fascia. In most cases, they are multiple and arranged against the dorsal surface of the albuginea of the corpora cavernosa and, therefore, a little deeper than the above-mentioned arteries. They are generally positioned in the same horizontal plane but can also be placed according to an oblique plane (Fig. 12.3). They receive the veinlets, which drain the venous reticulum of the glans and the circumflex branches from the lateral surfaces of the body of the clitoris, branches whose role is to establish an anastomotic system between the corporeal veins and the veins of the intermediate network. The superficial dorsal veins cross through the base of the suspensory ligament and then drain themselves into the nearby venous networks, especially the obturator veins, such as demonstrated by Kobelt (Fig. 10.3). One of these veins, more dorsal than the previous veins and especially with the largest diameter (2–3 mm), is the median superficial dorsal vein. After having penetrated the suspensory ligament of the clitoris (Fig. 12.3), this vein anastomoses with the superficial venous network of the abdominal wall.

10.1.2.3 The Veins of the Intermediate Network

They ensure the communication, such as shown by Kobelt, between the veins, which detach at the level of the inferior surface of the corpus cavernosum of the clitoris and the veins arranged at the level of the superior surface of the spongy body. These parallel and vertical veins, organised as organ pipes, detach obliquely with respect to the corporeal axis. They have large diameters and are in contact with the thin arterial rami. They are often organised in 2 more or less symmetrical, right and left parallel rows (Fig. 5.18). These veins of the intermediate network connect the veins of the commissure of the bulbs and the veins of the infra-clitoral residual spongy part (ic rsp).

10.1.2.4 The Veins of the Spongy Bulbs and of the "ic rsp"

They are the draining pathways of the spongy body. For the bulbs, these veins are, in the front region, the veins of commissure of the bulbs and, in the back region, the "veins of the posterior bulge of the bulb" (Kobelt): The latter are drained, according to this author, by the venules of the major vestibular gland, in the lower rectal veins and therefore in the venous internal pudendal network. The veins of the ic rsp drain themselves not only via the communications with the intermediate venous network but also into the venules of the frenula and labia minora. All of the veins of the bulbo-clitoral organ have connections with:

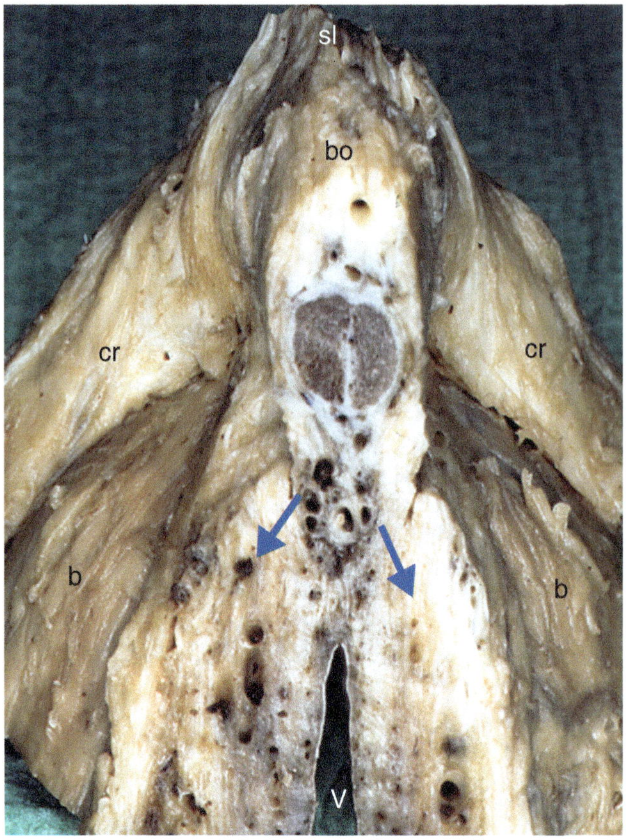

Fig. 10.2 Coronal section through the clitoral body and the vestibule showing the vascular connections (*blue arrows*) between the vessels of the RSP and those of the vestibular walls. *b* bulb, *bo* clitoral body, *cr* crus clitoridis, *sl* suspensory ligament, *V* vestibule

- On the one hand, the veins of the labia majora, the labia minora and the vestibule (all these veins end at the affluents of the curved end of the great saphenous vein, at the obturator veins and at the superficial perineal veins) (Fig. 10.2)
- On the other hand, the urethral veins and the veins of the peri-vaginal plexuses (Fig. 10.3)

10.1.2.5 The Veins of the Hood and Prepuce

They are drained essentially by the superficial dorsal veins of the clitoris. There is also an accessory drainage by the tributaries of the great saphenous vein.

10.1.3 The Lymphatic Vessels

The lymphatic vessels were not well known[5] or not known at all, but were then studied in a masterly way, at the beginning of last century, by 2 well-known anatomists: Cunéo and Marcille. In our era, there is renewed interest in these vessels due to their involvement in inflammatory and infectious phenomena (adenitis and lymphangitis), which can result from piercings (performed in dubious asepsis conditions) of the hood or the glans itself, and also in the frightening neoplastic pathologies affecting external female genitalia.[6]

The pathways draining the lymph of the glans clitoridis (lymphatic vessels of the glans) originate, as observed during our own injections, from a very thin and very tight plexiform network (Fig. 10.4). This network ends at 3 or 4 trunks, which pass under the clitoral fascia, parallel to the deep dorsal vein up to the suspensory ligament of the clitoris. These trunks anastomose in the base of the ligament, thus forming a true plexus from which are formed 2 groups of collectors: femoral collectors to the femoral ring (crural canal) and inguinal collectors to the inguinal ring (inguinal canal).

The **femoral collectors** pass under the femoral fascia, against the fascia of the pectinate muscle and end in the medial lacunar lymph node(s), located inside the common femoral vein, below or on the same level (lymph node of Cloquet) as the medial part of the vascular lacuna of the femoral ring. These collectors can cross the femoral ring and end at a lymph node located inside the external iliac vein, in the pelvis, just above the ring.

The **inguinal collectors** are very few (generally there is only a single collector). They penetrate the inguinal ring and pass under the round ligament. They drain into a lymph node located above the deep inguinal ring, pushed against the external iliac artery (external iliac lymph node).

As for the **pudendal lymphatic collectors**, it could be believed that they contribute to the drainage of the entire lymph of the bulbo-clitoral organ, when, in actual fact (according to research conducted by Cunéo and Marcille), they only perform the lymphatic drainage from the distal part of the urethra and spongy part (bulbs, in particular) of the organ to the hypogastric lymph nodes.

[5] We only knew that the lymphatic vessels of the hood and prepuce are dependent on the superficial inguinal lymph nodes, in particular the supero-medial lymph nodes (Rieffel).

[6] Given the structure of the bulbo-clitoral organ and tissues that compose it, three categories of primary malignant tumours can affect it: carcinomas, sarcomas and melanomas. We also find in the literature some cases of benign tumours, including haemangiomas, glomus tumours and neuromas.

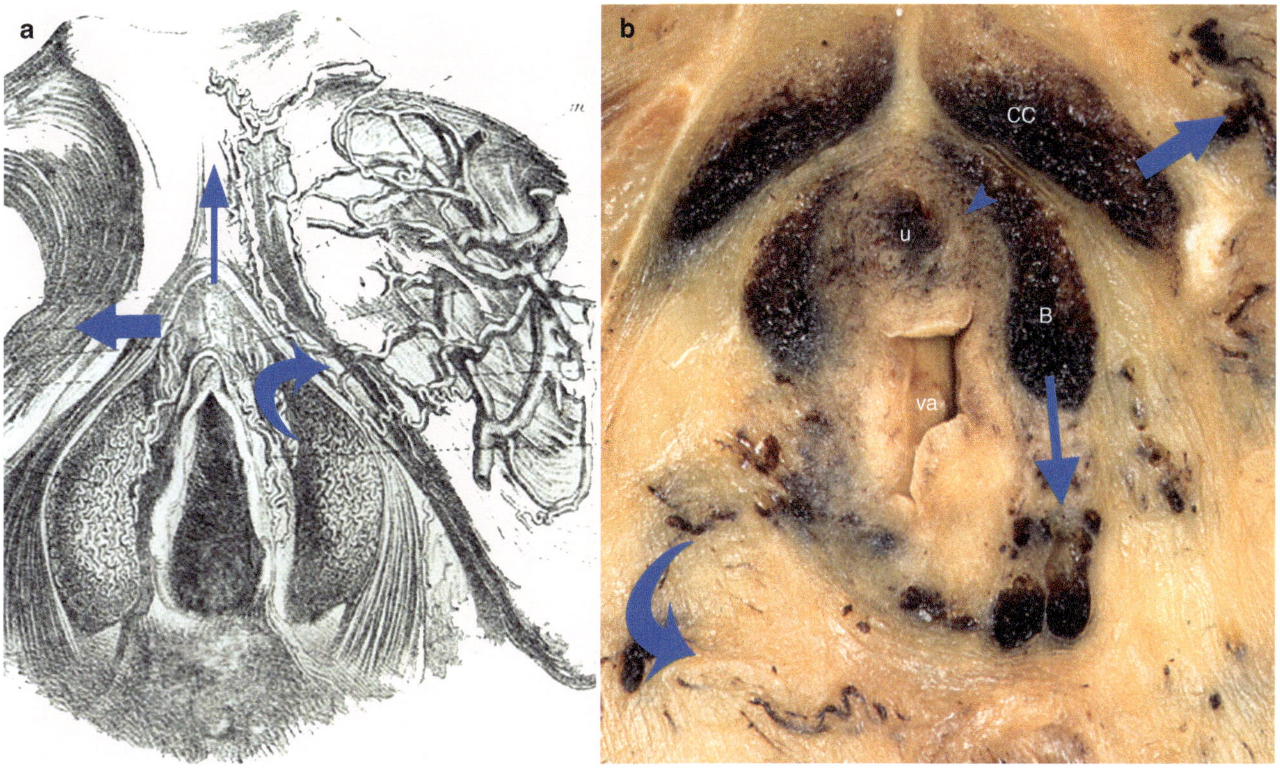

Fig. 10.3 Venous connections of female erectile bodies. (**a**) Drawing (from Kobelt) of a dissected vulva. *Large and short right arrow* connections with the tributaries of the great saphenous vein and femoral vein. *Thin right arrow* connections with the veins of the anterior abdominal wall. *Curved arrow* connections with the venous network of the obturator foramen. (**b**) Transverse section of the pelvis through female erectile bodies. *Large and short right arrow* connections with the tributaries of the internal iliac vein. *Thin right arrow* connections with the vaginal venous plexus. *Curved arrow* connections with the rectal venous plexus. *Arrowhead* connections with the urethral veins (draining to the vesical plexus and the internal pudendal vein). *B* bulb, *CC* corpus cavernosum, *u* urethra, *va* vagina

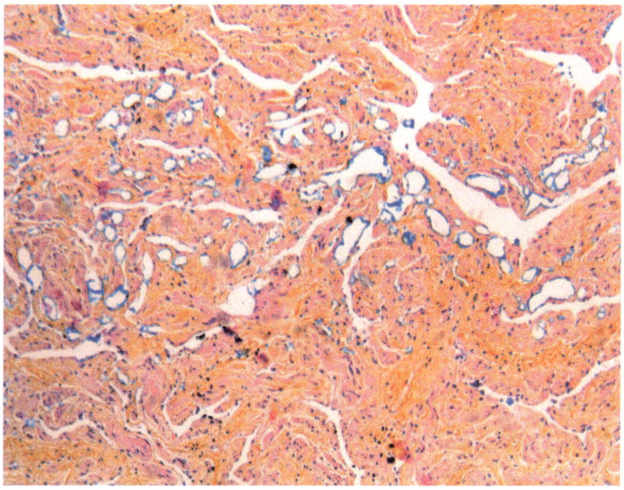

Fig. 10.4 Histologic aspect of the lymphatic vessels of the clitoris (coronal section of the glans clitoridis). Patent blue injection. Note the abundance of fine lymphatic vessels into contact with the vascular spongy network

11 Muscles of the Bulbo-clitoral Organ

These muscles (4 overall) are symmetrical, paired and striated muscles, which are divided into two groups: the group of the ischiocavernosus muscles (connected to the crura of the clitoral corpora cavernosa) and the group of the bulbospongiosus muscles (connected to the spongy bulbs). These muscles also exist at the level of man's perineum, in which case they are connected to the corpora cavernosa and to the spongy corpus of the penis.

These 4 muscles are located in the superficial region of the anterior perineum (urogenital perineum), between the superficial fascia of the perineum (fascia of Colles) and the inferior fascia of the urogenital diaphragm, i.e. the perineal membrane. They are innervated by branches from the perineal ramus of the pudendal nerve.

Their role is essential for the mobility of the clitoris and its erectile function.

11.1 Ischiocavernosus Muscles

Over time, they have been given very diverse names: tertius and quartus musculus (Colombo), clitoridis musculus (Fallope), clitoridis musculus tensioni dicatus (Laurent), superior rotundus (Riolan), ischio-clitoral (Dumas and then Testut), erector clitoridis (Cowper, Albinus, Soemmering), ischio-sub-clitoridis, ischio-urethralis (Chaussier) and collateralis sive penem erigens (Spigel). The current terminology is the terminology which had already been adopted by Winslow, Bichat, Portal, Boyden, etc., in their time.

According to traditional descriptions, the ischiocavernosus muscles resemble half-cones enveloping each of the 2 crura of the clitoris at the level of their inferior and medial surfaces. The insertion of each ischiocavernosus muscle is made on either side of the osseous or osteofibrous insertions of the clitoral crus, which it covers.

We have observed that reality appeared to be slightly different.

- Each crus clitoridis is completely surrounded by its ischiocavernosus muscle, which, furthermore, protrudes therefrom by 1 or 3 cm downwards. The clitoral ischiocavernosus muscle is not, such as often described, a small muscle of no importance, with much smaller dimensions than its penile counterpart. We even think that it is the opposite: if a ratio is established between the dimensions of this muscle in women and the dimensions of the thin and short crus, which it covers, it is observed that the clitoral ischiocavernosus muscle represents a powerful and bulky muscle, with major functional capacities. Such as already noted by L. Kobelt, "its length is of 8 cm, sometimes more still, so as to be compatible with the dimensions of the woman's pubic arcade".
- Another observation resulting from our experiment: The ischiocavernosus muscle has all of the characteristics of a penniform muscle, i.e. of a muscle "with less shortening possibilities but with the capacity to develop a greater force": tendon of origin, medial, very long and very solid, aponeurosis originating from this tendon and enclosing the crus, oblique muscular bundles, originating from the tendon and pushed against the inferior and medial surfaces of the aponeurosis, short terminal and tendinous fibres.
- The insertions of the ischiocavernosus muscle are particular (Fig. 11.1):

  It is firstly attached via its tendon of origin, onto the ischial tuberosity. It then takes, via its aponeurosis, osteofibrous attachments, which, on the one hand, are osseous over the ¾ of the ischio-pubic ramus, from the ischial tuberosity, and on the other hand fibrous on the adjacent part of the perineal membrane (thickened as an "attaching lamina of the corpora cavernosa").[1]
- From these origins is formed a not very thick, flattened muscle, rolled up like a cone (and not a half-cone, such as

[1]Due to the existence of 2 osseous insertions, Rouvière describes 2 bundles of origin for this ischiocavernosus muscle: an ischiatic bundle and an ischio-pubic bundle.

V. Di Marino, H. Lepidi, *Anatomic Study of the Clitoris and the Bulbo-Clitoral Organ*,
DOI 10.1007/978-3-319-04894-9_11, © Springer International Publishing Switzerland 2014

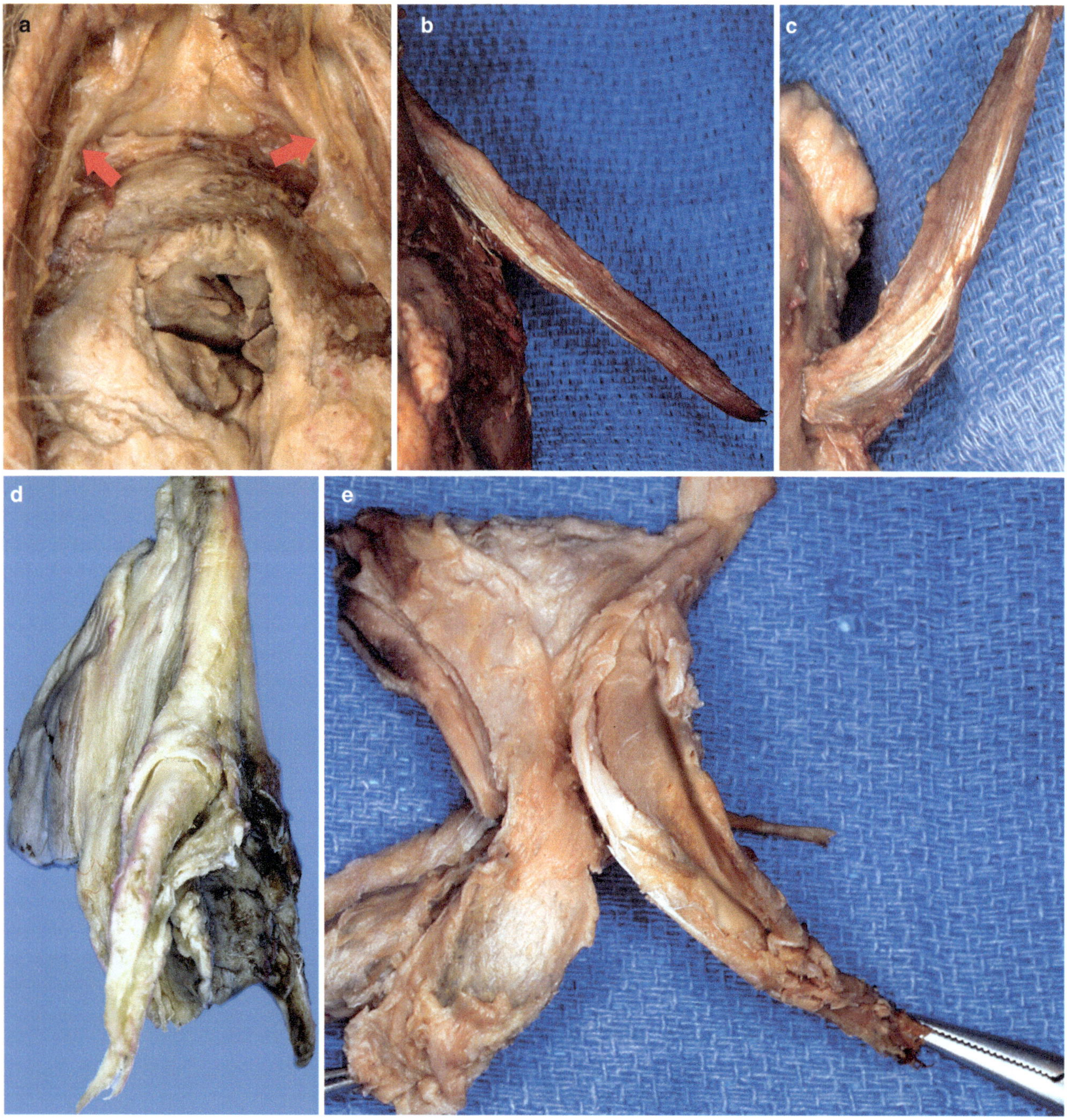

Fig. 11.1 Aspects of the ischiocavernosus muscles. (**a**) The two ischiocavernosus muscles on a perineal view of the anterior perineum (crura clitoridis dissected and pushed towards the pubis); the *red arrows* show the tendons of the two muscles. (**b**) The tendon of the left ischiocavernosus muscle on an anterior view of an anatomic specimen (notice the medial position of this tendon). (**c**) Inferior view of the same tendon (after raising of the left crus clitoridis). (**d**) Dorsolateral view of a bulbo-clitoral organ showing the left crus released after longitudinal opening of the ischiocavernosus fibro-muscular layer. (**e**) Left lateral view of a bulbo-clitoral organ, showing the left crus dissected and released from its fibro-muscular sheath (driven by a clamp)

described in too many cases) around the crus clitoridis. Thus, the body of the muscle truly encloses the erectile body so that the latter is not visible. In order to observe the crus, it is therefore necessary to longitudinally incise the lower surface of the musculo-aponeurotic envelope and release this crus from this pearly envelope (Fig. 11.1). This operation is delicate due to the fact that very tight adherences connect the albuginea of the crus and the internal surface of the aponeurosis of the ischiocavernosus muscle. When the crus is completely released, the hollow aponeurotic cylinder, inside which it was located, is clearly visible and appears shiny and pearly, completely intact and uninterrupted in its attachment region. There obviously are no muscle fibres on this internal surface. It is thus understood that insertions onto the perineal membrane are made via the external surface of the cone and not via the crus.

- Overall, the ischiocavernosus muscle is such as described by L Kobelt: "The ischiocavernosus muscle is not a lapped muscle, but a hollow muscle, shaped like a cone, which contains, in its cavity, the entire surface of the crus".
- The muscle fibres are arranged so as to form small superimposed, but irregular, oblique bundles (Fig. 11.2), and it is easy to define the capacity of these bundles to evacuate the blood contained in the sinuses of the crura.[2]
- The right and left muscles then converge towards the lateral surfaces of the clitoral body, and the muscular bundles give way to short terminal tendinous fibres. As for the tendon of origin, which is generally very long, it borders the muscle up to the point of convergence of the crura, where it is lost and sometimes exchanges tendinous fibres with the contro-lateral tendon.
- The termination of the ischiocavernosus muscles is also complex.

  The superficial bundles, joined by bundles from the homolateral bulbospongiosus muscle, will reach the lateral parts of the body of the clitoris in the angle region and end in the clitoral fascia. Some fibres become lost on the lateral surfaces of the suspensory ligament. Lastly, others are inserted on the superior surface of the clitoral body (a few fibre exchanges with the opposite side).

  The deep bundles are dorsal (they are only visible once the clitoris and its suspensory ligament have been unstuck from the pubic symphysis) and join the bundles on the opposite side, by passing as a bridge behind the crura to provide reinforcement in front of the symphysis, the retro-crural fascia and the subpubic urethral plate (Fig. 11.3).
- The role played by the ischiocavernosus muscles is determined from the anatomical concepts studied previously:

The ischiocavernosus muscles are lowering muscles for the clitoris. During intercourse, they lower the glans clitoridis against the dorsal surface of the penis (Testut).

They play an important part in the erection of the clitoris, as they make the blood from the crura flow back to the body of the clitoris. According to E. Gardner, they can even "contribute to maintaining the erection of the clitoris by compressing the crura and, as a result, delaying the blood returning from this organ".

11.2 Bulbospongiosus Muscles

Over time, they have also been given many names: primus musculus (Vesalus, Colombo), inferior sive urethram trahens (Spigel), musculus urethrae seu accelerator (Riolan), accelerator urinae (Cowper, Douglas), urethram dilatans (De Graaf), accelerator (Morgagni, Santorini, Albinus, Soemmering), bulbo-urethral (Chaussier) and bulbo-syndesmo-cavernous (Dumas).

- The insertions of the bulbospongiosus muscles are relatively large. They originate at the central tendon of the perineum (former perineal body of Savage or central fibrous core of the perineum according to French authors) and at the perineal membrane (inferior fascia of the urogenital diaphragm). There are also crossed fibres from the external anal sphincter, via the central tendon.
- The muscle fibres from these insertions, spread out from back to front and laterally.
- Large flattened muscular lamina on either side of the vaginal and urethral orifices. This muscular body has two parts (Fig. 11.2):
  - A peri-vaginal medial part, referred to as the constrictor muscle of the vagina[3] (Sabatier, Portal, Bichat), constrictor cunni (Albinus, Soemmering), clitoridis inferior latus *et* planus (Arantius), portio musculosa in externâ parte vaginae or vaginae musculi constrictorii (Riolan), sphincter vaginae (Verheyen), perineo-clitoral (Chaussier) and annulo-syndesmo-clitoridien (Dumas), while it only exerts a relatively modest constrictive action on the vaginal wall
  - A lateral part, which successively covers the inferior surface of the greater vestibular gland then that of the spongy bulb
- The termination of the bulbospongiosus muscle is even more complex than that of the ischiocavernosus muscle. Two bundles are to be considered (Fig. 13.2):
  - The fibres of the superior bundle (often consisting of medial infra-bulbar fibres) meet the fibres of the ischiocavernosus muscle (Figs. 11.2, 11.4 and 8.5) and end on the lateral edges of the body, in the clitoral fascia. Some fibres become lost on the lateral surface of the suspensory ligament. Other fibres of this

[2]Thanks to their orientation and their direction, the ischiocavernosus muscle fibres can really squeeze out the entire blood content of the clitoral crura as "a cloth which is being wrung out"!

[3]For a long time, it was considered as responsible for vaginismus (it was believed to be due to its spasmodic contraction). Such as we will see further on in this study, the introitus shrinking faculty belongs to the powerful pubo-rectal bundle of the levator ani muscle.

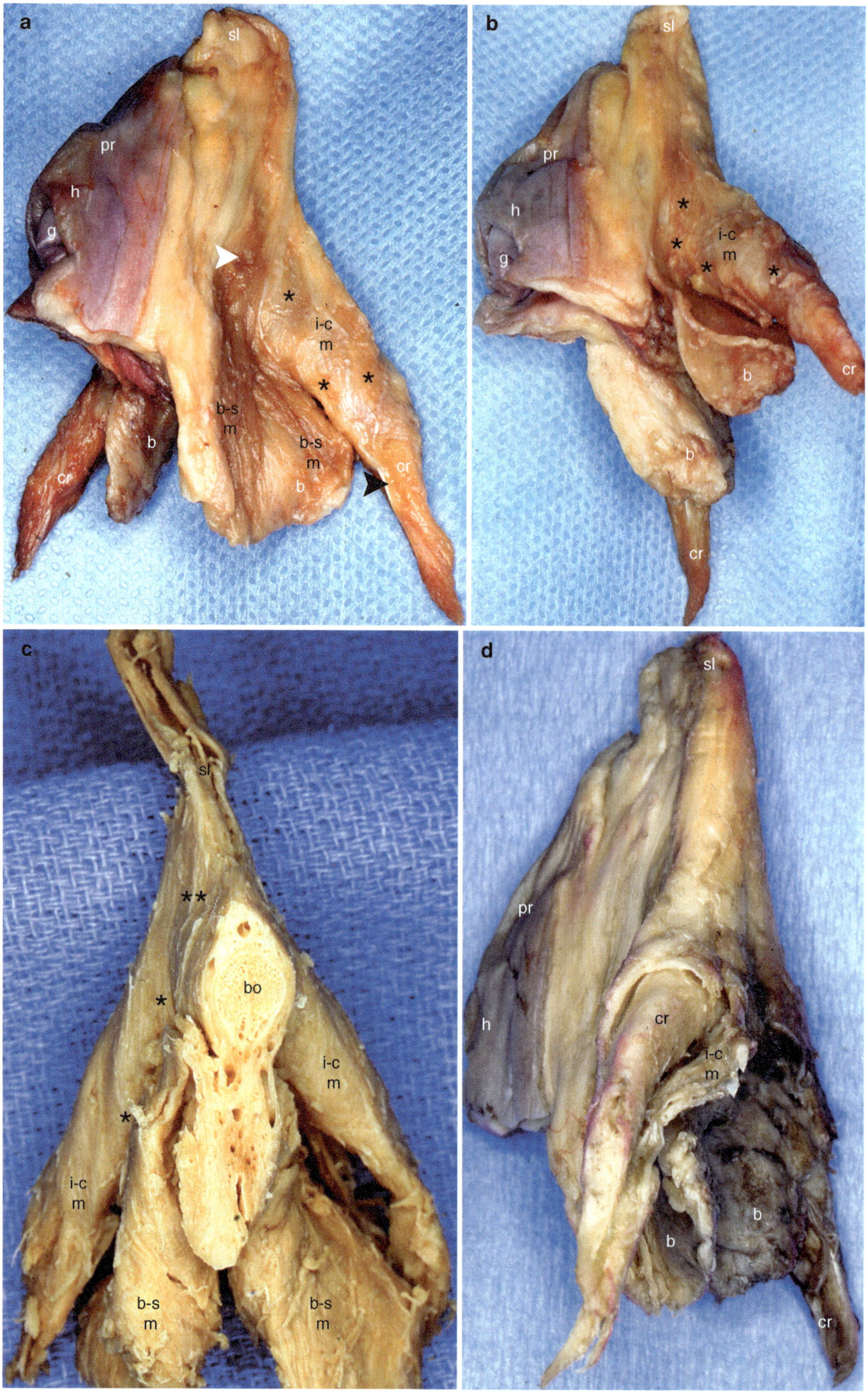

Fig. 11.2 The muscles of the bulbo-clitoral organ. (**a**) Lateral view; (**b**) inferior-lateral view; (**c**) ventral view with frontal section of the body; (**d**) dorsolateral view (i-c muscular layer opened longitudinally). *b* bulb, *bo* body, *b-s m* bulbospongiosus muscle, *cr* crus clitoridis, *g* glans, *h* hood, *i-c m* ischiocavernosus muscle, *pr* prepuce, *sl* suspensory ligament, *black arrowhead* it shows the tendon of the i-c muscle, *white arrowhead* it shows a bundle of the bulbospongiosus muscle reinforcing the i-c muscular layer, *oblique bundles of the i-c muscle, **muscle bundles towards the sl

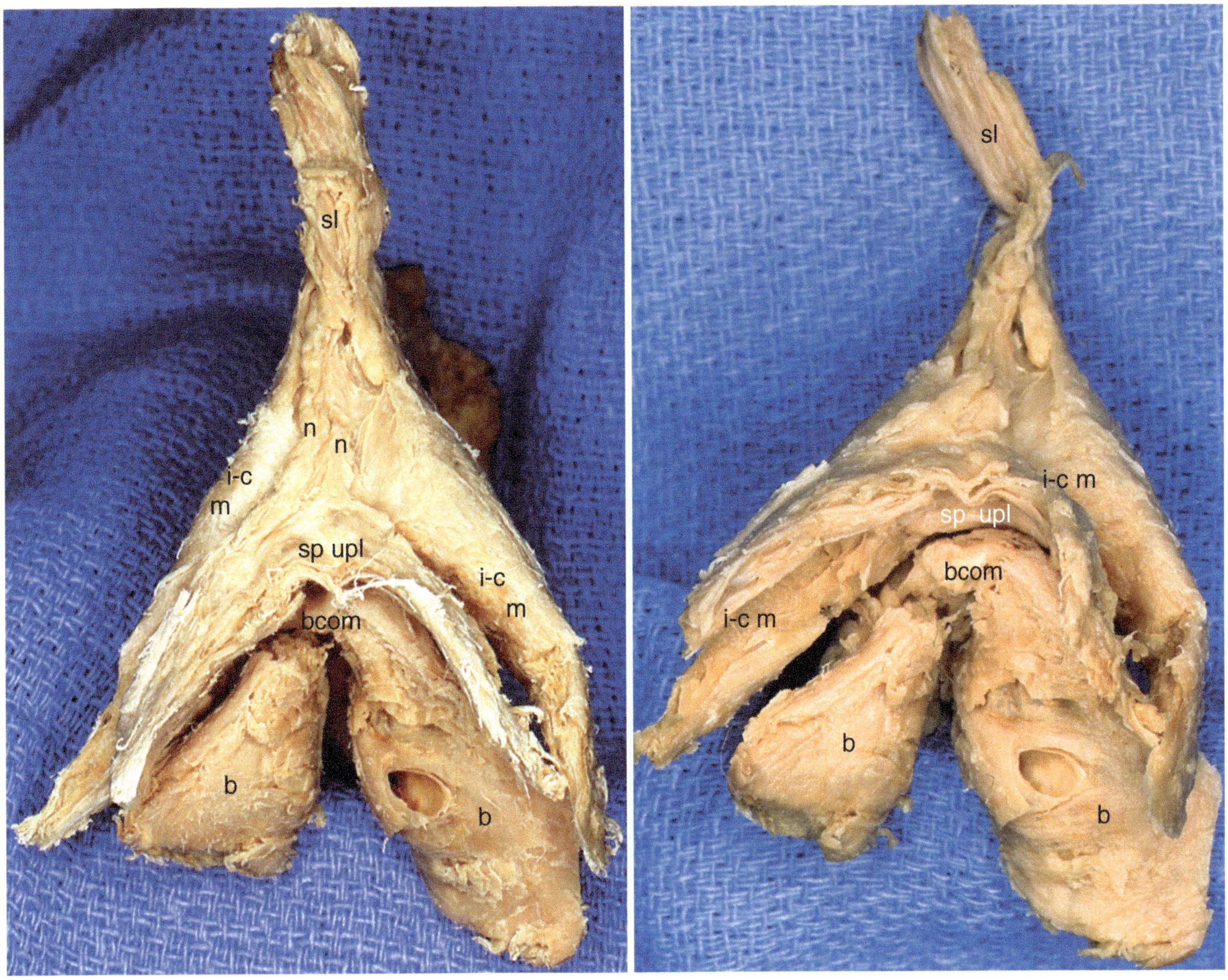

Fig. 11.3 Dorsal views of the bulbo-clitoral organ showing the subpubic urethral plate. *b* bulb, *bcom* bulbar commissure, *i-c m* ischiocavernosus muscle, *n* dorsal nerve of clitoris, *sl* suspensory ligament, *sp upl* subpubic urethral plate (*black legend* superior surface of the plate, *white legend* urethral surface)

superior bundle reach the back of the clitoris to merge with the fibres of the counter-lateral homonymous muscle and form a supra-venous strap, capable of compressing the large superficial dorsal vein of the clitoris and its satellite veins. It is this strap which is referred to as the **compressing muscle of the dorsal vein of the clitoris** (muscle of Houston according to previous authors).

– The fibres of the inferior bundle (often consisting of lateral infra-bulbar fibres) end under the commissure of the bulbs, in the region which separates it from the urethral duct. They have not been studied in depth by the various authors who have examined these muscles. In order to clarify their trajectory, it is necessary to perform a microscopic observation of anatomical sections as we have done in the past. It is then noted that this bundle detaches from the peri-bulbar muscular body and extends medially above the commissure of the bulbs (Fig. 11.4). Then the striated muscle fibres penetrate between the spongy caves to finally be positioned above the urethra, merge with the muscle fibres of the opposite side (Fig. 13.6) and, in fine, will be arranged as a **distal sphincter of the female urethra** (compressor urethrae muscle of ancient authors).

- The action of the bulbospongiosus muscles is also explained by their anatomy:

1. They play an active role in the erection of the clitoris:

   By making the blood from the posterior part of the bulbs to flow back to the pars intermedia (intermediate network) and the body of the clitoris

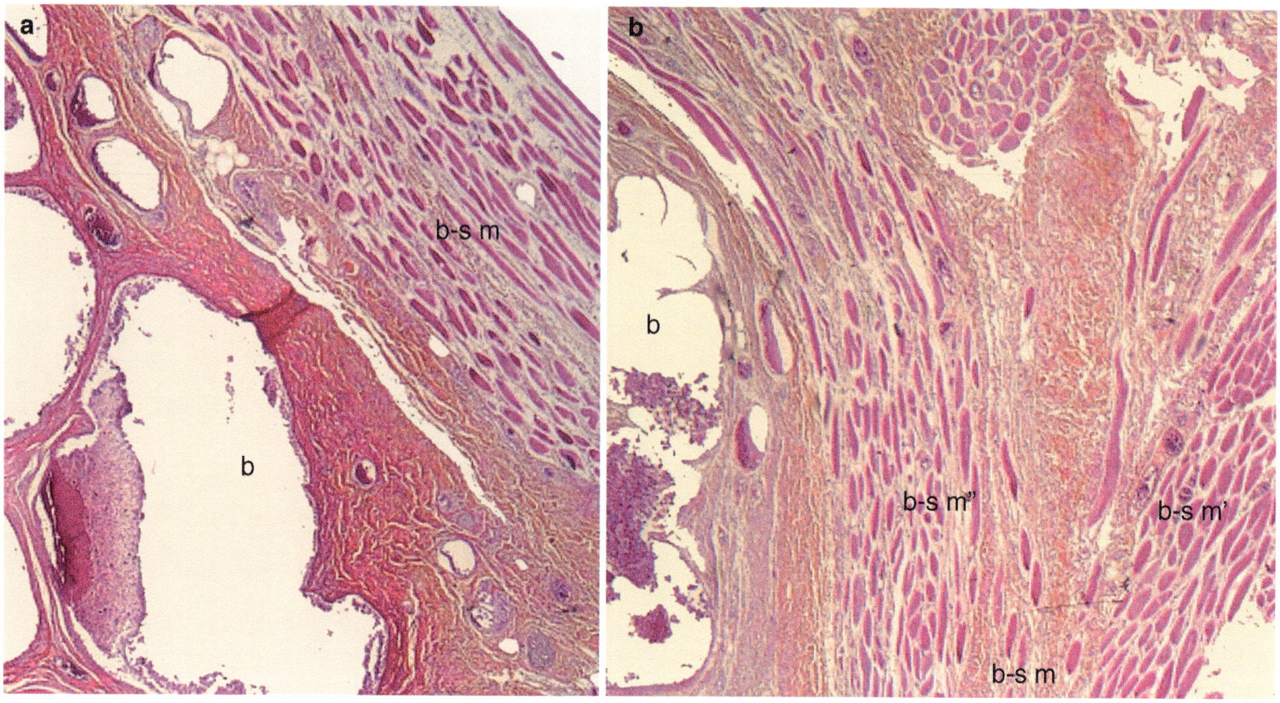

Fig. 11.4 Microscopic aspects of the bulbospongiosus muscle's bundle. (**a**) The bulbospongiosus muscle (*b-s m*) covering the spongy bulb (*b*). (**b**) The termination of the bulbospongiosus muscle in two bundles: an inferior bundle *b-s m''*, a superior bundle *b-s m'*

By, at the same time, blocking the venous leakage by compressing the dorsal vein (action of the muscle of Houston)

By contributing to the lowering of the clitoris

2. They also play a part in the contraction of the vaginal introitus, but this contribution is modest and they do not deserve to be called "sphincter of the vagina", such as some authors wanted to call them.
3. Lastly, they exert an action on vaginal humidification by compressing the major vestibular gland and causing its secretion.

12 The Suspensory Ligament of the Clitoris

This ligament is a fibro-elastic and fatty formation, which connects the bulbo-clitoral organ to the pubic symphysis and to the infra-umbilical linea alba. In spite of its volume and size, few authors have actually studied this formation and most of the authors, who have written about the clitoris, have simply pointed out the existence of this formation or provided erroneous representations, such as a minuscule symphysis fibrous attachment. It is Poirier and Charpy, in 1901, who provided the first anatomical description of this ligament.

At first sight, the shape of the ligament is that of a quadrangular pyramid, with a clitoral base and a prepubic and sagittally flattened top. The ligament insertions at the level of the clitoris protrude ahead of the elbow and overflow onto the descending part of the body and may even reach the glans. These insertions are present on the corporeal albuginea and spread to overflow over the lateral sides of the latter. Certain fibres of the ligament meet on the inferior surface of the clitoral body to form a sort of hammock. Other fibres extend laterally on either side of the clitoral body and are lost in the derma of the labia majora (Fig. 12.1). Some fibres also reach the medial part of the bulbospongiosus muscle. The ligament extends with the fibrous tissue of the retro-crural fascia and is connected, with the latter, to the pre-symphysis fibrous coating, at the level of the crura and of the ascending portion of clitoral body. From this clitoral base, the ligament pyramid can be described as follows (Fig. 12.1):

- Two large triangular lateral surfaces, whose base measures on average 2.3 cm and extends from the retro-crural fascia to the junction between the body and the glans clitoridis.
- A ventral surface corresponding to the thickness of the ligament (3–4 mm), which becomes wider at the level of its clitoral attachment (around 1 cm large).
- A dorsal surface extending without any boundary line with the retro-crural fascia. When this fascia is absent or not very developed, its base measures around 8 mm but when in contact with the pubic symphysis, the ligament insertion spreads out as a triangular surface.
- A superior end, which gradually becomes tapered and loses itself on the linea alba, above the pubic insertions of the latter.

The ligament has an overall length of 8–9 cm.

Such as already described by Poirier and Charpy, it is possible to distinguish two parts of the suspensory ligament of the clitoris:

- **A deep part**, attached to the angle of the clitoris and the back of the first centimetres of the body. This part is attached to the retro-crural fascia, on the caudal half of the anterior surface of the pubic symphysis, and its fibres mix with those of the pre-symphysis fibrous coating (Fig. 5.3).[1]
- **A superficial part**,[2] covering the preceding part, attached to the descending portion of the body of the clitoris, passing in front of the symphysis, then spreading out and ending laterally at the level of mons pubis (mons veneris) in the fibrous parts of the subcutaneous fascia of Camper and, for its median part, over the last caudal centimetres of the infra-umbilical linea alba (Fig. 12.1).

  It is interesting to observe that this superficial part is significantly long and especially that it has, such as already observed by M.A. Rees et al., a multiplanar configuration (such as the entire bulbo-clitoral organ). The attachments to the mons pubis are made according to a frontal plane (that of the abdominal wall) but then, imperceptibly, the fibres of this superficial part are arranged on a sagittal plane to join the clitoris (Figs. 12.1 and 12.2). It should also be noted that the attachments to the mons pubis are very spread out (8 cm on average). As regards the length of the superficial part, from the mons pubis up to the clitoris, it is difficult to give a figure because, in more than half of

[1] From a surgical viewpoint, the deep part of the suspensory ligament and the retro-crural fascia detach relatively easily from the pre-symphysis fibrous coating, thus providing numerous accesses (urethra, bone pubis, etc.) for surgical purposes.

[2] This superficial part of the ligament was unknown for a long time. Its description by Charpy, in 1901, was largely ignored by the authors of anatomy books.

V. Di Marino, H. Lepidi, *Anatomic Study of the Clitoris and the Bulbo-Clitoral Organ*,
DOI 10.1007/978-3-319-04894-9_12, © Springer International Publishing Switzerland 2014

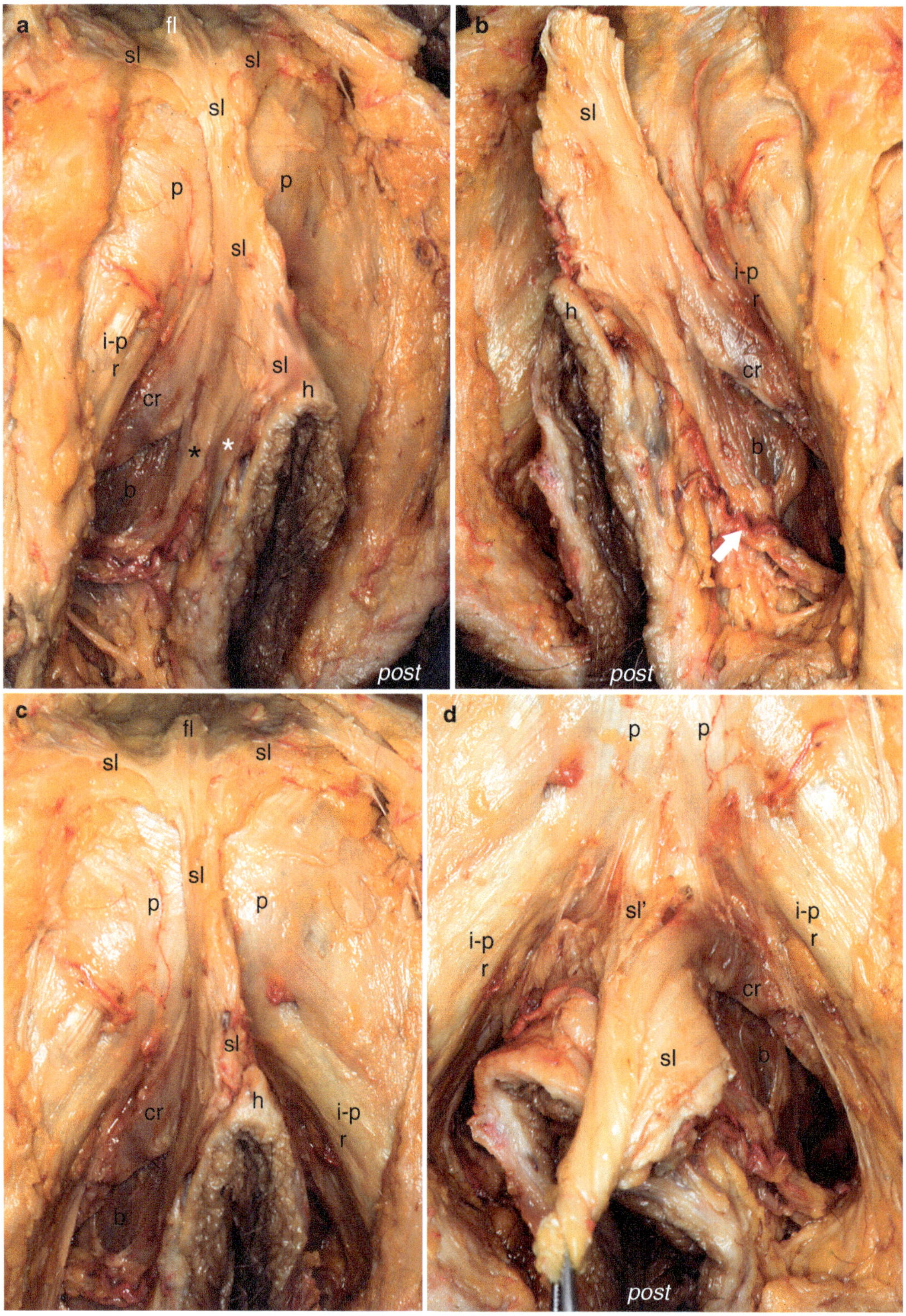

Fig. 12.1 Four pictures of a prepubic dissection to demonstrate the position of suspensory and fundiform ligaments. (**a**, **c**) Right lateral view and anterior view showing the suspensory ligament in place. (**b**, **d**) Left lateral views showing the suspensory ligament partially detached. *b* bulb, *cr* crus, *fl* fundiform ligament, *h* hood, *i-p r* ischio-pubic ramus, *p* pubis, *sl* suspensory ligament, superficial portion, *sl′* suspensory ligament, deep portion. The *White asterisk* shows ligamentary fibres extending in the derma of labia majora. The *Black asterisk* shows ligamentary fibres reaching the medial part of the bulbo-spongiosus muscle. The *white arrow* shows the neuro-vascular bundle (bulbar vessels and bulbar nerve).

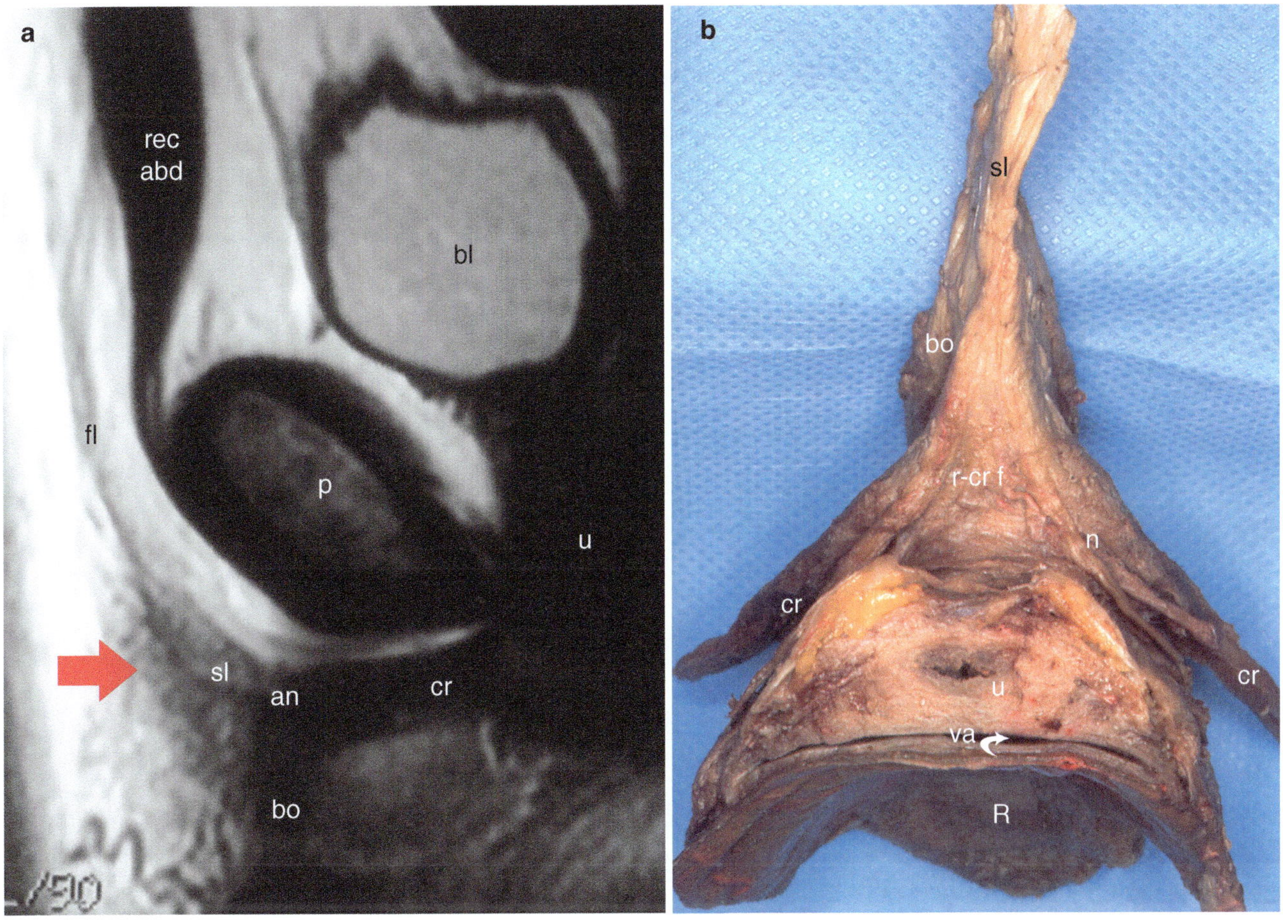

Fig. 12.2 Aspects of the suspensory ligament of clitoris. (**a**) NMR median sagittal section of the pelvis showing the clitoris and his suspensory ligament (*red arrow*). (**b**) Dorsal view of the suspensory ligament on a pelvic dissection. *an* angle of clitoral body, *bl* bladder, *bo* clitoral body, *cr* crus clitoridis, *fl* fundiform ligament, *n* dorsal nerve of clitoris, *p* pubic symphysis, *R* rectum, *r-cr f* retro-crural fascia, *rec abd* rectus abdominis muscle, *sl* suspensory ligament, *u* urethra, *va* vagina (white curved arrow showing vaginal cavity)

the cases, we could observe a median cranial extension of this superficial part in the form of a thin 1-cm wide fibrous cord, of variable length, in front of the infra-umbilical linea alba, in the direction of the umbilicus, which it sometimes reaches. However, in most cases, this fibrous condensation becomes thinner and then merges with the white line. It is this formation, which has previously been described by authors as the **fundiform ligament of the clitoris** (Fig. 12.1). Lastly, it should be noted that the superficial and deep parts of the division are quite often visible on the median sagittal MRI sections of the abdominal wall (Fig. 12.2).

The structure of the ligament explains its solidity: It consists of resistant collagen packs, with which are mixed numerous elastic fibres and fatty tissue (Fig. 12.3). The microscopic examination is particularly interesting because it shows an accordion-shaped fold of the collagen bundles. This configuration obviously forms a lengthening reserve for a ligament, which must allow clitoral mobility, even of small amplitude. At its base, the collagen bundles extend with the corporeal albuginea bundles and the cavernous septum bundles. Numerous neurovascular components, following the dorsolateral parts of the clitoral body, cross through this base (Fig. 12.3): the 2 arteries and the 2 dorsal nerves of the clitoris, or the median dorsal part, such as the main large dorsal vein of the clitoris, always with 2 or 3 satellite veins.[3] However, while the veins cross directly through the fibrous framework, which maintains them open, the nerves pass through small islands of fatty tissue, thus providing them with a certain amount of freedom. The arteries, as for them, remain surrounded by a protective fibrous sheath. The collagen packs become denser and exchange their fibres with

[3] All the veins of the suspensory ligament communicate with the sinuses of the cavernous bodies, especially via emissary veins crossing through the peri-cavernous albuginea (Fig. 7.1).

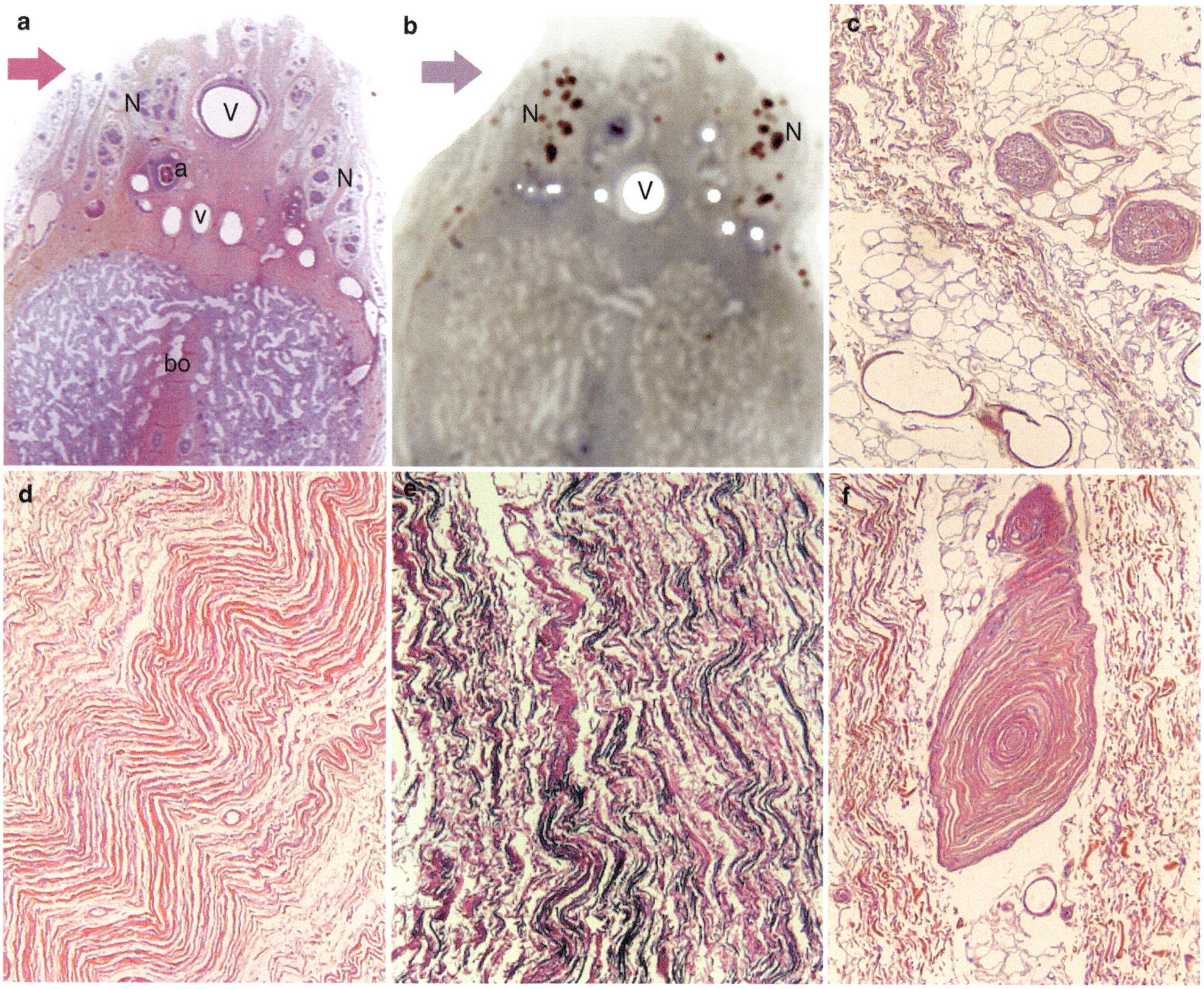

Fig. 12.3 Histological aspects of the suspensory ligament of clitoris. (**a**) Photomicrograph of a frontal section of the base of suspensory ligament (*sl*) (*magenta arrow*): standard staining by HEAS. *a* artery, *N* nerve, *V* principal dorsal vein, *v* accessory dorsal veins, *bo* clitoral body. (**b**) Photomicrograph of a frontal section of the base of suspensory ligament (*mauve arrow*): PS100 staining. (**c**) Nerves through the suspensory ligament within islands of adipose tissue. (**d**) Curious kinks in collagen fibres of ligament. (**e**) Staining by orcein, highlightening the elastic fibres (*black coloured*). (**f**) A great Pacinian corpuscle in the sl, among the collagen fibres

those of the prepubic fibrous coating at the top of the ligament, i.e. in the tapered part. It should also be noted that extremely developed venous networks cross through the ligament longitudinally. These networks establish anastomoses between the clitoral veins and the median veins of the anterior abdominal wall (which form an additional drainage pathway for clitoral venous blood).

The question of the role played by the suspensory ligament is of primary importance: Why such a large ligament for such a small organ? The answers are determined from anatomical observation:

- The first role is a support role for the clitoris which is obvious when observing the "ligament hammock" arranged not only under the elbow (or angle) but also under the entire descending portion of the clitoris. This support is static during the quiescent phase. It becomes active during intercourse and "contributes to the stability of the clitoris" (M.A. Rees et al.), thus preventing the clitoris from hindering the penis' back-and-forth movements. Abdominal musculatures are involved in this active role (via the insertion of the superficial bundle of the ligament on the linea alba). Musculature of erectile bodies, via fibres of the ischiocavernosus and bulbospongiosus muscles, which mix with the collagen bundles of the suspensory ligament, is also involved.

- The second role is a protective role for the components which come from or go towards the clitoris, by crossing it. This role especially involves providing an additional protection for the dorsal nerve of the clitoris, whose trajectory is particularly complex at this level, with the rise of the angle of the clitoral body and whose contribution of the vegetative contingent, via the arrival of the thin and fragile cavernous nerve, must be carefully protected. In this protection, it is necessary to recall the very important role played by the fatty tissue, whose small islands within the collagen tissue are preferential areas of transit for the nerves.
- Lastly, it should be recalled that the ligament (and especially its base) has a particularly high content in Pacini's corpuscles. This is most certainly due to the proximity of the dorsal nerves of the clitoris, but also to a potential role played by the ligament as a specialised receptor, a starting point for proprioceptive impulses. This suspensory ligament may therefore be more than just a simple support structure. It may even contain specific sensors (neurotendinous spindles).

It will therefore be necessary to conduct more research work in this field in order to improve our understanding of this ligament, which has always intrigued researchers.

13 Relationships of the Bulbo-clitoral Organ

13.1 General

The topographical anatomy of the bulbo-clitoral organ was already partially covered in the first chapters, with its morphology and structure having been observed in situ, i.e. within the external feminine genitalia. However, several points remain to be clarified and very specific connections have yet to be studied. We will discuss the connections with the vestibule, labia and greater vestibular glands; the connections with the urethra, the vagina and the perineal membrane; and finally, the relations with the pelvic diaphragm (levator ani muscles).

13.1.1 Connections with the Vestibule

13.1.1.1 Position of the Bulbs

These connections mainly involve the bulbs for which we have already examined the naming difficulties. Are the spongy bulbs pressed against the walls of the vestibule or do they go deeper? Against the terminal urethra? Against the end of the vaginal duct?

L. Kobelt, who first gave an exact and complete description of the bulbo-clitoral organ, was also the first to replace the older terminology (plexus retiform, plexus reticular) with the name "vestibular bulbs". But a semantic ambiguity should immediately be noted as this author explains why he calls them as such since the formations "are not located around, but at the entry to the vagina". Is this enough to abandon the terminology "vaginal bulbs"? The reality is in fact more complex and only frontal sectioning of the external genitalia makes it possible to reveal the details of bulb topography: The result of these sections is incontrovertible. The bulbs are directly connected not only to the urethral and vaginal orifices but also to the parts preceding it, i.e. the vestibule and the base of the labia minora and what follows the entire distal part of the vagina (Fig. 13.1) and the end of urethra. In other words, the terms "vestibular bulbs", "vaginal bulbs" and "urethral bulbs" all have an element of truth and remain perfectly valid. However, it is the term "vestibular bulbs" that was retained by the Nomina Anatomica. We would have preferred "spongy bulbs in women" (or female spongy bulbs) which is more general and less subject to discussion. Please note, however, that the terminology now present in the nomenclature is in accord with Kobelt, thus paying homage to this author. As for choosing which of the adjectives "urethral" or "vaginal" is best suited for describing the bulbar topography, let us agree that there is no cause for debate. The sections clearly show that the two adjectives are true since the bulbs border the external orifices of both ducts. However, for the urethra, 3/4 of the circumference of its external orifice is lined by spongy bulbar tissue, whereas for the vaginal orifice, only the lateral parts are lined by this tissue. It should also be noted that the bulbs are of varying length and often asymmetrical, such that the lateral walls of the vagina may not be entirely lined by the bulbs or one of the walls may be lined to a greater extent than the other! Finally, let us recall the terminology proposed by H.E. O'Connell et al.: "clitoral bulbs", which cannot be supported from a semantic point of view given that the clitoris and bulbs are two different parts of a whole. In the interest of accuracy, one should therefore refer to "bulbs of the bulbo-clitoral organ" or "genital bulbs" which are easier to express.

13.1.1.2 The Greater Vestibular Glands

On either side of the vestibule, the spongy bulbs connect with the two greater vestibular glands or Bartholin's glands.[1] These glands are well known to gynaecologists and their patients due to their frequent infection (bartholinitis may be the manifestation of a sexually transmitted infection, notably the infection caused by gonococcus: gonorrhoea). These muciparous glands are topographically in contact with the posterior extremity of the bulbs. The glands have an average diameter of 15 mm, a more or less spherical, slightly

[1] The two Bartholin's glands are equivalent to the bulbo-urethral glands (Cowper's glands) in men.

V. Di Marino, H. Lepidi, *Anatomic Study of the Clitoris and the Bulbo-Clitoral Organ*,
DOI 10.1007/978-3-319-04894-9_13, © Springer International Publishing Switzerland 2014

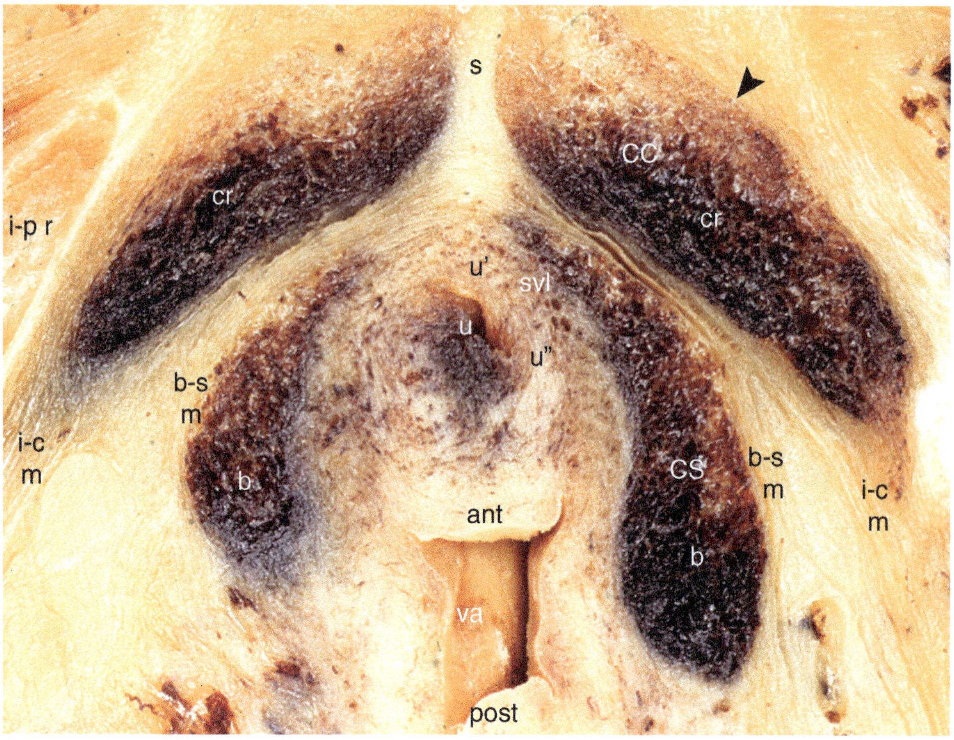

Fig. 13.1 Relationships between both the urethra and vagina with female erectile bodies. Tranverse section of female pelvis through the anterior perineum. *ant* anterior vaginal wall, *b* bulb, *b-s m* bulbo-spongiosus muscle, *CC* corpus cavernosum, *cr* crus clitoridis, *CS* corpus spongiosum, *i-c m* ischio-cavernosus muscle, *i-p r* ischio-pubic ramus, *post* posterior vaginal wall, *s* septum of the crura, *svl* superficial vascular layer (of urethra), *u* urethra, *u'* anterior wall (of urethra), *u''* lateral wall (of urethra), *va* vagina, *Black arrowhead* it points the clitoral albuginea. Note the close relationship between the bulbs and the urethro-vaginal pyramid

flattened shape, sometimes ovoid, with an irregularly mammillated surface. Their firm consistency allows them to be easily identified upon palpation during dissection. However, strong adherence of the gland to the adjacent cell tissue renders its release difficult. This notion is also demonstrated during surgical release in view of exeresis of this gland, which is made all the more difficult due to the significant inflammatory adherences. The gland has very tight connections with the swollen posterior part of the bulb (Fig. 13.2), connections that the surgeon must be very aware of so as to avoid often serious haemorrhaging due to injury to the homolateral spongy bulb. The gland may be located to the rear and simply tangent to the convexity of the bulb's swollen extremity. More often, it is inserted into this extremity, carving out a sort of bulbar cavity in which it is housed. The vestibular gland may therefore be described as follows: an anterior extremity directed towards the external orifice of the vagina; a posterior extremity connected to the superficial transverse muscle; one medial vulvar surface that is more or less concave; a convex lateral surface facing, but at a distance from, the initial tapered pillar of the clitoris; a superior edge pressed against the inferior fascia of the urogenital diaphragm; and an inferior edge in contact with the superficial transverse muscle. The bulbospongiosus muscle is applied to the inferior edge and then to the inferior surface of the bulb, as seen previously. The excretory duct[2] measures 1–2 cm. Its diameter can reach 2 mm. Slanted forward and inward, it opens into the vestibular sulcus[3] along the insertion of the hymen or its carunculate remnants (carunculae hymenales), at the junction of the anterior 2/3 and posterior 1/3 of the vestibule. The secretion produced by this gland (tubulo-alveolar gland from a histological perspective) is a transparent or opalescent, runny mucus. It is secreted during coitus and contributes to lubrication of the walls of the external genitalia.[4] The neurovascular connections of the greater vestibular gland are lateral. The neurovascular bundle (Fig. 13.2) engaged above the lateral surface of the gland includes rami of the deep perineal artery (bulbar artery), voluminous veins and the deep ramus of the pudendal nerve perineal branch.

13.1.2 Connections with the Urethra

These connections are exceptionally important and had surprising consequences from an anecdotal point of view (see Chap. 1). Despite the shortness of this duct, we now know the various parts of the feminine urethra well (Fig. 13.3) and notably the part that is connected to the bulbo-clitoral organ: It is the distal part of this duct, immediately beyond the posteroin-

[2]This excretory duct may be double.

[3]The vestibular (or labio-hymeneal) sulcus separates the hymen (or its remnants, carunculae hymenales), from the labium minus. The limit between the vestibular epithelium and the epithelium of the medial surface of the labia minora is called "Hart's line" (Fig. 4.2).

[4]The secretion is also produced by many very small "lesser vestibular glands" situated between the urethral and vaginal orifices and opening into the vestibule.

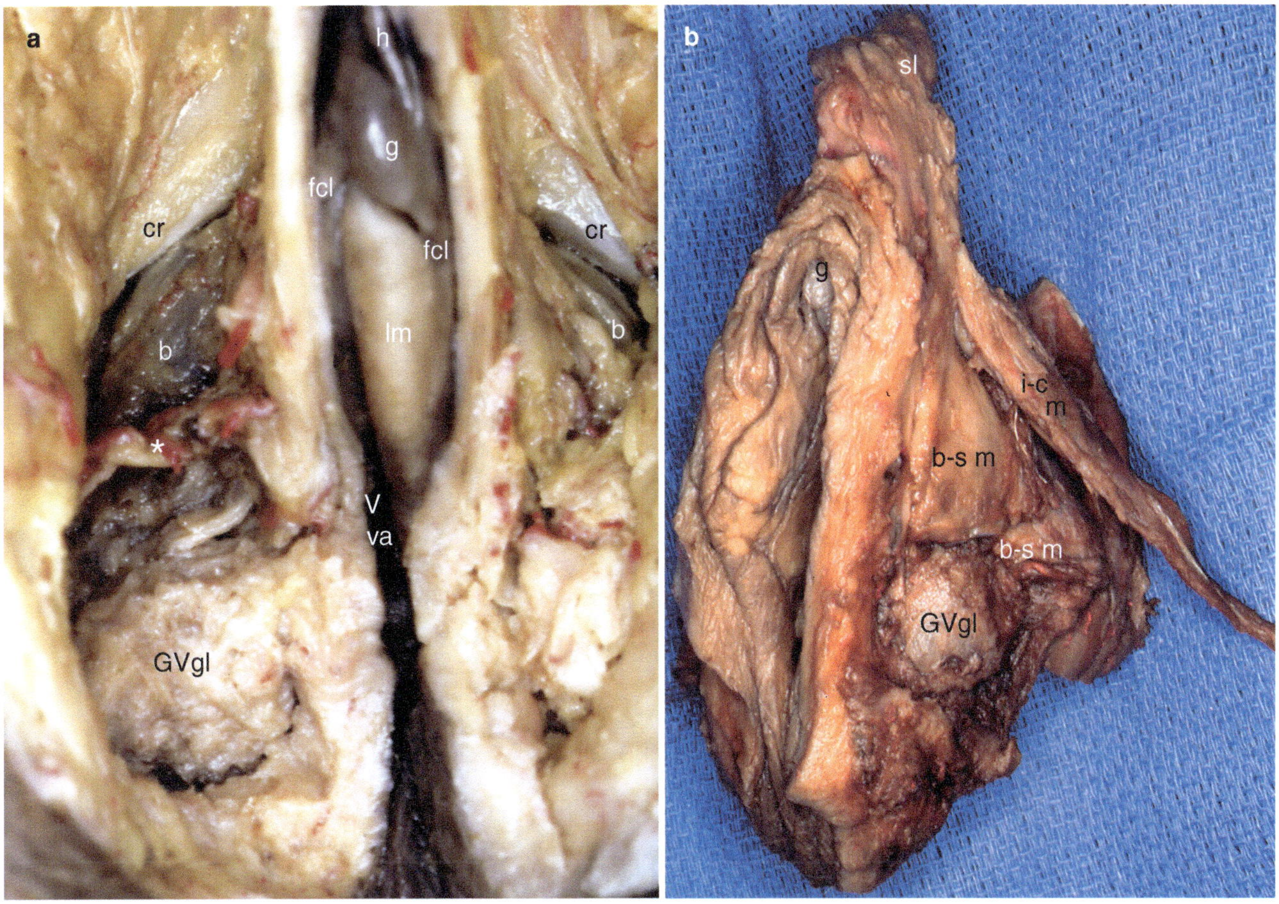

Fig. 13.2 Two dissections of the anterior perineum to show the close relationship between the bulbs and the greater vestibular glands (Bartholin's glands). (**a**) Anterior view showing the right great vestibular gland behind the right bulb; notice the neurovascular bundle* (bulbar artery and rami from the pudendal nerve's perineal branch) engaging obliquely in front of the gland. (**b**) Lateral left view showing a left great vestibular gland, behind the left bulb covered by the bulbospongiosus muscle. *b* bulb (with bulbospongiosus muscle), *b-s m* bulbospongiosus muscle (*black legend* medial part of the muscle, *white legend* lateral part), *cr* crus of clitoris, *fcl* frenum (frenulum clitoridis), *g* glans clitoridis, *h* hood, *lm* labium minus (lesser lip), *GVgl* greater (or major) vestibular gland (Bartholin's gland), *Vva* vestibule of vagina

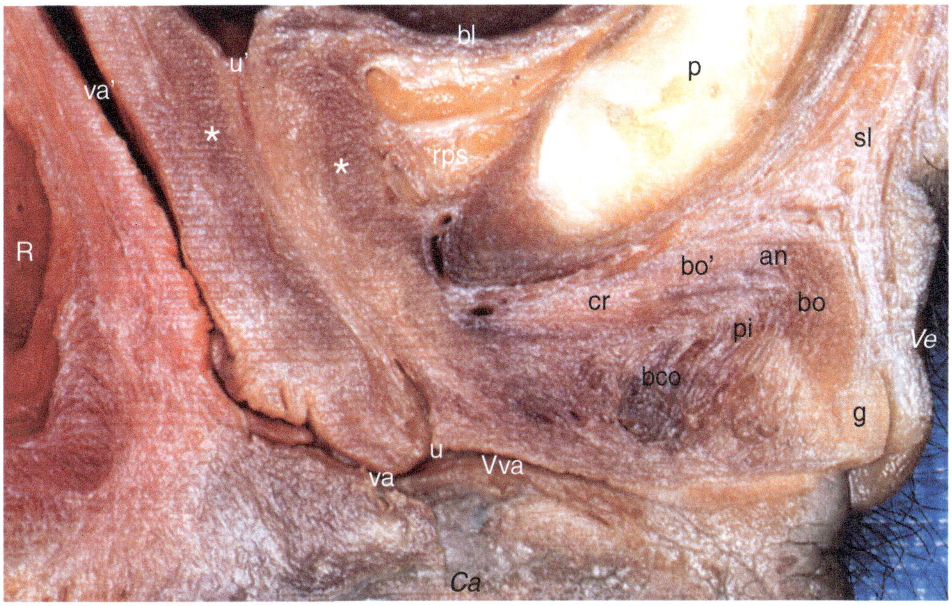

Fig. 13.3 Part of a median sagittal section through the female pelvis showing the location of the bulbo-clitoral organ and its immediate surroundings. *Ve* ventral, *Ca* caudal, *an* angle (elbow), *bco* bulbar commissure, *bl* bladder, *bo* clitoral body (descending part), *bo′* clitoral body (ascending part), *cr* crus clitoridis, *g* glans, *pi* pars intermedia of Kobelt, *p* pubic symphysis, *R* rectum, *rps* retro-pubic space, *sl* suspensory ligament, *u* external urethral orifice, *u′* internal urethral orifice, *va* vaginal orifice (introïtus), *va′* vagina, *Vva* vestibule of vagina, *white asterisks* external sphincter urethrae

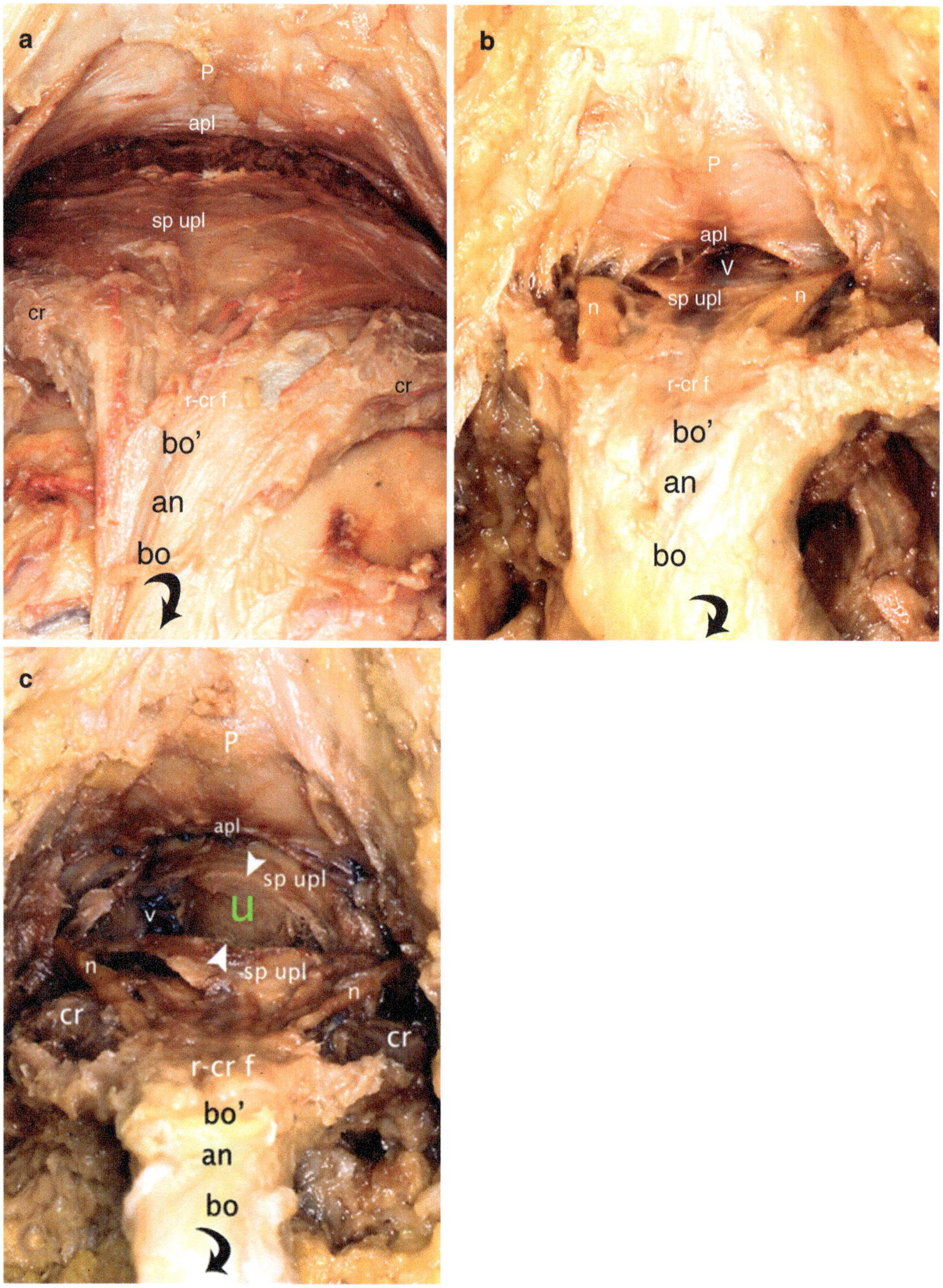

Fig. 13.4 Dissections of the pre- and infra-pubic areas to show the relationship between the dorsal parts of the bulbo-clitoral organ and the urethra (bird's eye views). The clitoral body has been retracted back and down (*black curved arrow*). (**a**, **b**) Gradual release of the subpubic urethral plate. (**c**) Transversal incision of the subpubic urethral plate highlighting the urethra. *an* angle, *apl* arcuate pubic ligament, *bo'* ascending part of the clitoral body, *bo* descending part of the clitoral body, *cr* crus clitoridis, *n* dorsal nerve of clitoris, *P* pubic symphysis, *r-cr f* retro-crural fascia, *sp upl* subpubic urethral plate, *u* (*green*) urethra, *V* deep vein of the clitoris (seen by transparency through the sp upl), *v* right latero-urethral venous plexus, *white arrowheads* they show the two banks of the subpubic urethral plate cut transversally

ferior edge of the pubic symphysis and more accurately immediately beyond the passage through the urogenital diaphragm. This is therefore the extra-pelvic portion of the urethra. This part, called the "distal urethra", very thoroughly evaluated by J.O.L. Delancey, corresponds to the terminal segment extending from 79 % of the total length of the duct to the external orifice (situated at "100 % of the length"). If we take the average length of the female urethra to be 3 cm (extremes: 25–44 mm), the distal urethra only represents 0.8 cm! Serial sectioning of this short segment, perpendicular to the urethral axis, allowed us to gain very precise information to complete the general notions that appeared during dissection.

We were able to observe deep and superficial connections during dissection.

Observation of deep connections first requires dissection of the bulbo-clitoral organ. The organ has to be pushed back and down with a fair amount of force (thus sacrificing the two thin cavernous nerves) in order to access the thin subpubic urethral plate (or supra-urethral lamina). This lamina continues from the retro-crural fascia (see Chap. 5). The underlying urethral duct can be felt through the lamina with a fingertip. As soon as the transversal incision is made in the lamina, the urethra appears (Fig. 13.4). The duct will then pass through the urogenital diaphragm behind the transverse

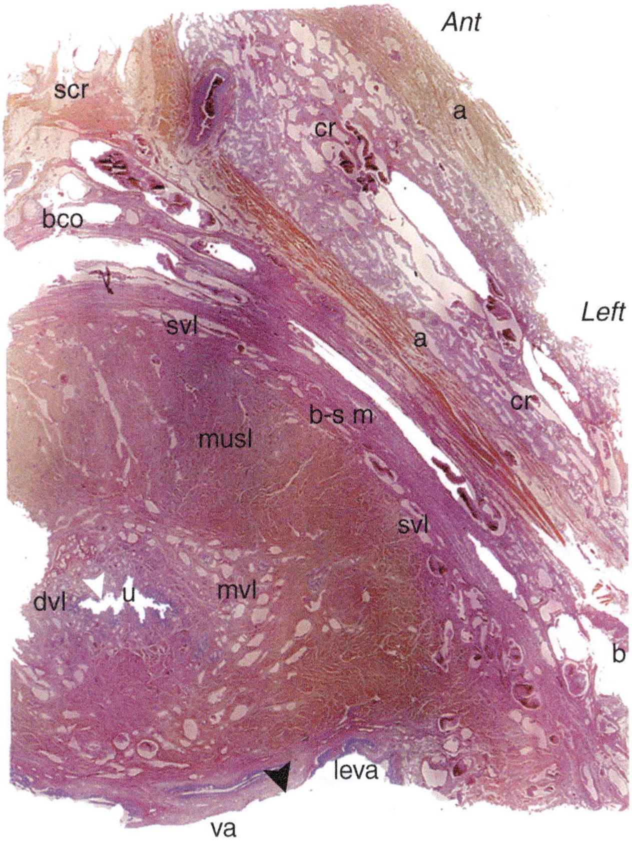

Fig. 13.5 Relationship between the urethra and female erectile bodies. Transverse section (Section 3) (perpendicular to the urethral axis) through the crura clitoridis. *a* albuginea, *b* bulb, *bco* bulbar commissure, *bs m* bulbospongiosus muscle, *cr* crus clitoridis, *dvl* deep vascular layer (layer of small submucosal vessels), *leva* lateral edge of vagina, *musl* muscular layer (thick layer of smooth muscle fibres), *mvl* middle vascular layer (bigger vessels than those of dvl), *scr* septum of crura, *svl* superficial vascular layer (large vessels communicating with the spongy lacunar cavities), *u* lumen of the urethra, *va* vagina, *black arrowhead* it shows the location of the vaginal mucosa on the anterior wall of vagina (this mucosa can be seen just above the left lateral edge of vagina), *white arrowhead* mucous membrane of the urethra

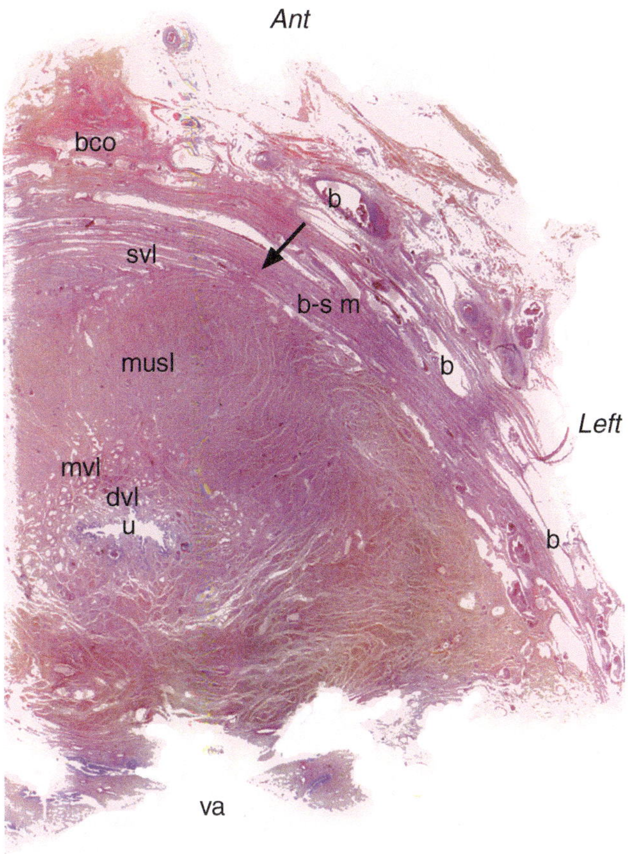

Fig. 13.6 Relationship between the urethra and female erectile bodies. Transverse section (Section 2) (perpendicular to the urethral axis) through the bulbar commissure. *b* bulb, *bco* bulbar commissure, *bs m* bulbospongiosus muscle, *dvl* deep vascular layer (layer of small submucosal vessels), *mvl* middle vascular layer, *musl* muscular layer (thick layer of smooth muscle fibres), *svl* superficial vascular layer (large vessels communicating with the spongy lacunar cavities), *u* urethral lumen, *va* vagina, *black arrow* it shows the striated muscle fibres of the bulbospongiosus muscle, getting between the spongy bulb's vascular cavities and merging with the muscle fibres from the opposite side, so as to form a distal sphincter : the "compressor urethrae muscle" of ancient authors

perineal ligament. It is now becoming perineal. It appears as a cylindrical tube, slightly widened, with a thick wall.

Observation of superficial connections is simpler. The previously dissected bulbo-clitoral organ is left in place. We can see the final mm of the distal portion of the urethra and the external urethral orifice above the external vaginal orifice. The connections are also easy to identify: The urethra in this terminal segment is surrounded by the spongy tissue on 3/4 of its circumference (lateral surfaces bordered by medial surfaces of the bulbs and superior edge, covered by the commissure of the bulbs).

If we lift the entire bulbo-clitoral organ, we are just beyond the limit between the pelvic urethra and the perineal urethra, as the urethra has just started to cross the urogenital diaphragm (Fig. 13.10).

Three levels of histological sections perpendicular to the axis of the perineal urethral duct seem worth considering (Figs. 13.5, 13.6 and 13.7): Section 3 (the deeper) goes through the crura clitoridis; the middle, Section 2, only goes through the spongy structures; and the most superficial, Section 1, is tangent to the external urethral orifice.

- On Section 3, we can see a crus clitoridis from front to back (cavernous tissue surrounded by its albuginea) and the commissure of the bulbs (commissura bulborum) lined in front and back by elements of the bulbospongiosus muscle with striated muscle fibres slipping in between the vascular crevices of the spongy tissue. This spongy tissue covers the anterior surface of the urethra.

  From the centre to its periphery, the urethral Section 3 shows a urethral lumen bordered by its epithelium

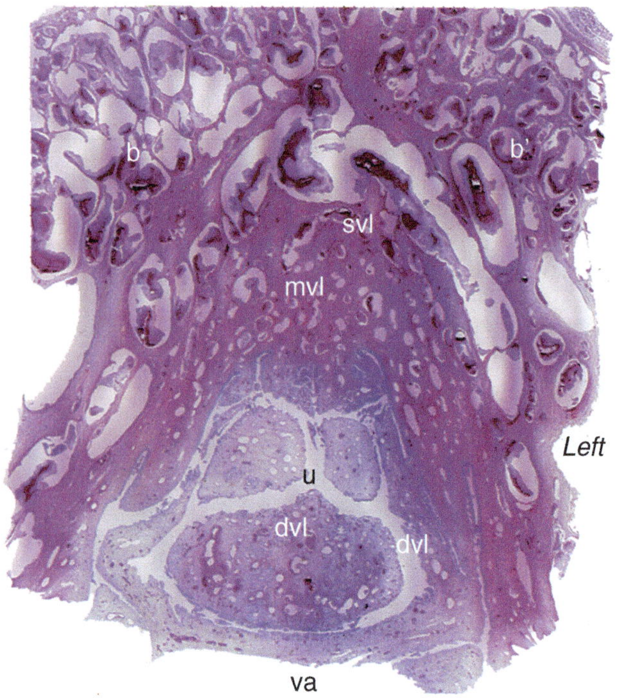

Fig. 13.7 Relationship between the urethra and female erectile bodies. Transverse section (Section 1) (perpendicular to the urethral axis) through the end of the urethra, near its external orifice. *b* bulb (right), *b′* bulb (left), *dvl* deep vascular layer (layer of small submucosal vessels), *mvl* middle vascular layer, *svl* superficial vascular layer (large vessels communicating with the spongy lacunar cavities), *u* urethral lumen (tangential to the external urethral orifice), *va* vagina (tangential to the vaginal orifice)

(squamous, multistriated, non-keratinised epithelium), a peri-epithelial layer rich in small vascular lumina (evoking a sort of erectile tissue), then a new vascular layer with vessels of diameters greater than those of the previous layer, a very thick muscularis (with an internal layer of smooth muscle fibres arranged longitudinally and an external layer arranged in a circular pattern) and a last, thinner vascular layer with elongated vessels of greater dimensions than those of the deep vascular layers, but smaller than those of the neighbouring spongy tissue vessels. Behind the urethra, we can finally observe until its lateral edge the thin anterior vaginal wall with its mucous membrane.

- The urethral Section 2 no longer displays cavernous tissue. It only goes through the spongy tissue (bulb and commissure of the bulbs). The striated muscle fibres from the bulbospongiosus muscle are much more abundant and, behind the commissura bulborum, they make up a sort of sphincter that compresses the anterior median part of the urethra. The urethra has a similar appearance to that observed on Section 3. However, the vessels of the two deep vascular layers are less abundant and smaller. On the contrary, the vessels of the superficial layer are larger and already communicate with the spongy vascular lacunae.
- The very superficial urethral Section 1 (at the external orifice), like the previous section, only includes spongy tissue which is particularly abundant (the section passes through meanders of the lumina, bordered in places by the remnants of the epithelium). The thick urethral muscularis cannot be seen on this section. However, the layer of medium-sized vessels is particularly developed and occupies the anterior sinus formed by the junction area of the two bulbs. In the most anterior part of this vascular layer, larger vessels communicate extensively with the spongy vascular lacunae. The role of this well-developed vascular cushion, capable of storing a large quantity of blood, if needed, has yet to be understood.

Another important connection is the **distance separating the summit of the glans and the external urethral orifice**. This distance, which is variable (Fig. 13.8), is the median of the urethral vestibular triangle (of which the summit is the clitoral glans, the two sides are the labia minora and the base is the horizontal line going through the external urethral orifice). Most authors evaluate this distance at an average of 2.1 cm (the extremes found during our study were 1.8 and 3.6 cm). If we believe the publication of a certain A. Narjani in 1924 (a publication that did not include any real statistical study), this distance is predictive of a woman's aptitude to experience orgasm. In fact, Narjani was the pseudonym used by the princess Marie Bonaparte,[5] very concerned with her own frigidity. To publish the measurements, she took on 200 women to prove her reasoning, providing figures on the CUMD distance (distance between the clitoral glans and the centre of the urinary meatus[6] or "clitoris-urinary meatus diameter"). The author classified women into three categories depending on this distance: If the distance is less than 2.25 cm (**paraclitoral** women; 70 % of cases), it is very probable that the woman will easily orgasm during sexual intercourse. If the distance is greater than or equal to 2.5 cm (**teleclitoral** women; 20 % of cases), orgasm during sexual intercourse will be exceptional or even impossible. The remaining 10 % (**mesoclitoral** women) have varying chances of orgasm. Marie Bonaparte explained that in fact, it is the distance between the clitoral glans and the external vaginal orifice (immediately behind the external urethral orifice) that plays a major role. According to her, if this distance is too great, the clitoris cannot be effectively stimulated by the male penis

[5]Granddaughter of one of Napoleon I's nephews, prince Pierre-Napoleon Bonaparte. She married prince Georges of Greece in 1907. She met Freud in 1925, underwent psychoanalysis with him and would become his student, friend and later his representative in Paris. She was grateful to him and enabled Freud and his family to leave Austria, under Nazi rule at the time. She also saved many Jewish intellectuals.

[6]The term "urinary meatus" used by Marie Bonaparte is not suitable. The author is actually referring to the "external urethral orifice" (term from the anatomical nomenclature!).

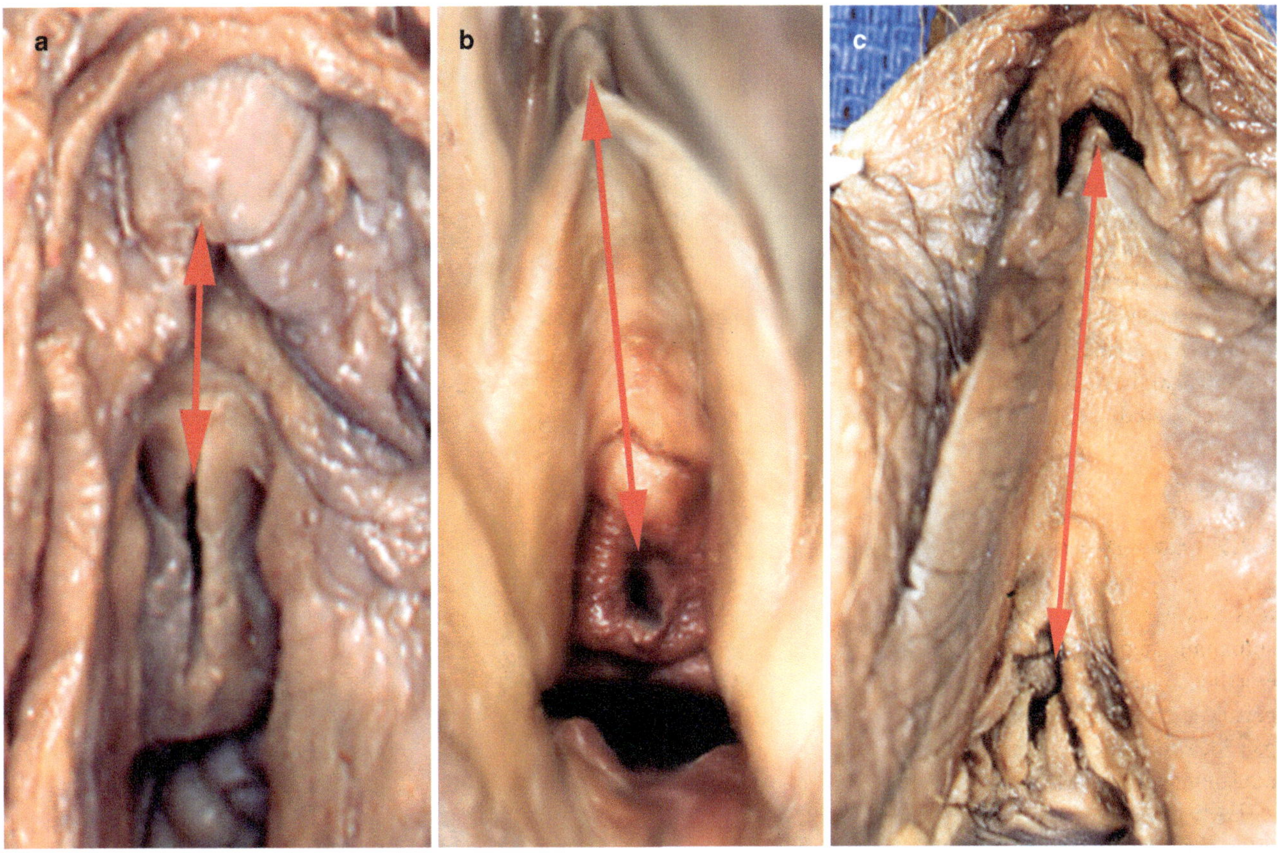

Fig. 13.8 Three examples of the extreme variability of distance between the glans clitoridis and the external urethral orifice (distance CUMD *clitoris-urinary meatus* according to Narjani). (**a**) Very short distance; (**b**) great distance; (**c**) very important distance. Notice on case **c**, the very particular and abnormal external urethral orifice with a hair hood

during coitus. Marie Bonaparte then went on to develop the Halban-Narjani procedure with the famous Austrian surgeon Joseph Halban, which consists in transecting the suspensory ligament of the clitoris to reposition the clitoris closer to the external vaginal orifice. Very convinced and courageous, Marie Bonaparte underwent the surgery herself. Not having received a successful result, yet still just as convinced, she went on to have the operation two more times… again without success (and surely with permanent damage resulting from the sectioning of nerves)! The results were the same for the four other patients who underwent the surgery. A few years later, Marie Bonaparte, convinced by the teachings of Freud, who was now her mentor, would radically change her viewpoint, affirming in 1933 that the CUMD had nothing to do with aptitude to orgasm and that frigidity only resulted from a psychoanalytic process. Despite this change in opinion, the initial theory of Marie Bonaparte remains current and has been adapted and studied by many authors since. An exhaustive, remarkable study co-signed K. Wallen and E.A. Lloyd addresses the question. The two authors conclude that statistical studies with very precise genital measurements could provide interesting conclusions. These authors note that undeniably, there are differences in individuals' aptitude to orgasm and that there are also major variations in the arrangement of external female genital organs (Fig. 13.8). There are surely correlations between these physiological and anatomical modalities. The two authors also ask what causes differences in positioning of the clitoral glans. Adapting the argument already offered by Marie Bonaparte, they mention the role of hormonal impregnation of the female foetus, itself a reflection of the hormonal impregnation of the pregnant woman carrying it. If prenatal exposure to androgens is low, the rostral migration of the genital tubercle does not take place and the clitoris remains near to the external orifice of the vagina. Inversely, in the case of prenatal exposure to androgens, the genital tubercle initiates migration, the resulting clitoris is slightly more developed and the CUMD distance is greater!

The embryological approach is also used to study what Pozzi, a famous French surgeon, called "**la bride masculine**" ("**the male bridle**")[7] and what Félix Léon Jayle would simply refer to as the "bandelette urétro-clitoridienne" (urethro-clitoral

[7] In fact, Neubauer had already included the so-called male bridle in his anatomical drawings as early as 1784 but had not made any comments on his observation of it.

band) in his work "l'Anatomie Morphologique de la Femme" (Morphological Anatomy of the Woman) since it linked the glans clitoridis and the external urethral orifice. In 1884, Pozzi described it as a triangular mucous membrane thickening with an anterior summit at the apex of the glans and a base that forks at the external urethral orifice into two rami that follow the lateral walls of the orifice, then meet under this orifice and finally diverge to "apparently continue with the hymen membrane". This bridle was, according to the author, clearly visible in the female foetus or young girl. He explained that "in the adult, virgin woman, this arrangement can be seen without difficulty, though it is less visible than in young girls. In deflorated women, and especially in multipara, the bridle sometimes becomes virtually invisible". Pozzi thought that this bridle was a vestige of the penile portion of the urethra in men (hence the name "**male bridle**"), but this was not quite accurate. If this thickening is indeed an embryological remnant, it corresponds not to the penile urethra, but to the spongy tissue that would have surrounded this penile portion of the urethra if the foetus had evolved to be male. In other words, this bridle corresponds to a median relief of what we designated as the "IC RSP". The conclusion from A.W.M van Turnhout et al.'s remarkable work can be cited as proof: these authors histologically studied the parts of the male bridles resected during metoidioplasties[8] performed in their hospital, and they found typical spongy tissue every time! In our study, we have highlighted this male bridle several times, once with formation of the small median groove, as observed by F.L. Jayle (Fig. 13.9).

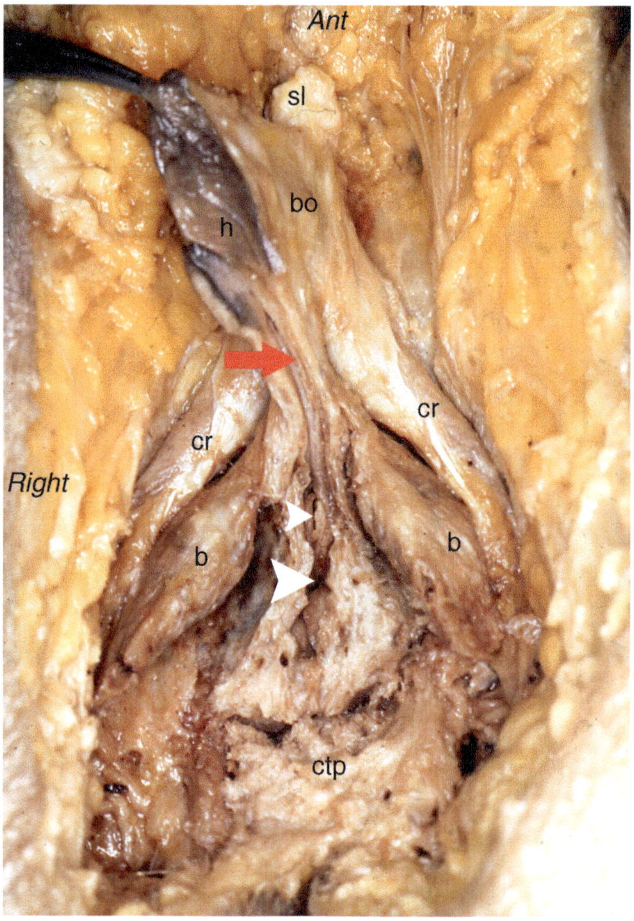

Fig. 13.9 The male bridle ("bride masculine" of French authors) on a perineal view of the female external genital apparatus, after dissection of the erectile bodies (the forceps tracts the prepuce back and right). *b* bulb, *bo* clitoral body, *cr* crus clitoridis, *ctp* central tendon of perineum, *h* hood, *sl* suspensory ligament, *great white arrowhead* vaginal orifice (introïtus), *little white arrowhead* external urethral orifice, *red arrow* male bridle (urethro-clitoral tractus)

13.1.3 Connections with the Perineal Membrane

The pillars (crura) of the bulbo-clitoral organ are applied against the inferior surface of the "**perineal membrane**", i.e. the fascia that lines the inferior surface of the deep transverse muscles of the perineum (former inferior fascia of the urogenital diaphragm). As we saw during the description of this organ, the erectile bodies are attached to this structure, which represents the floor of the deep space of the perineum (and the roof of the superficial space!).

The crura clitoridis attach, by way of their musculofascial envelope, to the lateral edges of the perineal membrane caudal surface, along the ischio-pubic rami (into which they are inserted as well). The perineal membrane is thickened here and the fibrous attachments are very solid. They are formed between the fibrous envelope of the cavernous pillars and the thickened membrane, which has become the "attaching lamina of the corpora cavernosa".

When we release the pillar, we can observe that this manoeuvre is difficult and that the bistoury "squeaks" upon contact with a dense fibrosis! The two spongy bulbs are also attached to the caudal surface of this perineal membrane. Their insertions are near to the most distal part of the urethra and the vagina, on the medial parts of this membrane, along the lateral edges of the urethral and vaginal orifices. Here again, the adherences are solid, even if less so than for crura clitoridis. Let us repeat finally that the perineal membrane is nothing more than the inferior fascia of what was called the "urogenital diaphragm" and a certain trend aims to erase it from anatomical texts. Traditionally, this diaphragm is made up of the deep transverse muscle and two fascias: one inferior, the inferior fascia of the urogenital diaphragm (perineal membrane), and the other superior, superior fascia of the urogenital diaphragm. Currently it is common for authors to contest that the existence of the superior fascia, which granted, is less clearly defined, and the same is true for the deep transverse muscle (T.A. Stein et al., P. Mirilas et al., W. Dorschner et al.). We affirm, after anatomical-histological

[8] Metoidioplasty or metaoidioplasty is a phalloplasty (plastic surgery aiming to construct a phallus in view of sex reassignment).

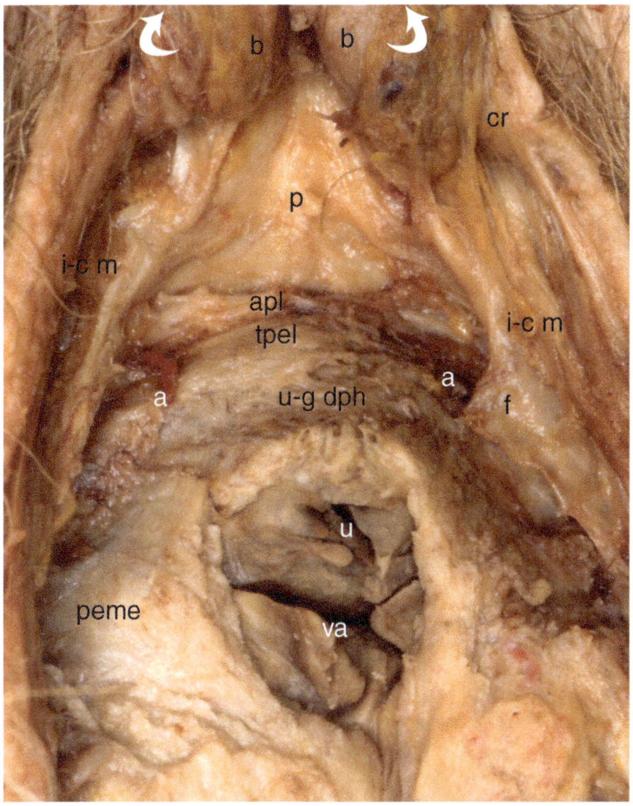

Fig. 13.10 Relationship between the ischiocavernosus muscles and the urogenital diaphragm (dissection of female anterior perineum). The bulbs and the crura of clitoris have been released from their attachments and pulled against the pubis (*white curved arrows*). *a* dorsal artery of clitoris, *apl* arcuate pubic ligament, *b* bulb, *cr* crus clitoridis, *f* fascia encasing ischiocavernosus muscle, *i-c m* ischiocavernosus muscle, *peme* perineal membrane, *tpel* transverse perineal ligament, *u* external urethral orifice, *u-g dph* urogenital diaphragm, *va* vaginal orifice

verifications, that an unchanging musculo-aponeurotic structure exists. The urethra and the vagina pass through it (Fig. 13.10). It is perfectly dissectible, and it plays an undeniable role in urinary continence. This structure, 2–3 mm thick, is lateral to the aforementioned ducts and is formed by two thick fascias (under microscope: thick bundles of collagenous fibres) between which we can see a muscular structure, made up of an abundance of striated muscle fibres in a transverse orientation (Fig. 13.10). These striated muscle fibres must not be confused with the urethro-vaginal sphincter, which belongs to the striated sphincter of the urethra![9] Serial sections of the urogenital diaphragm show that in front, the striated muscle fibres become rare and then disappear such that the two fascias join together as one thick ligament: the transverse perineal ligament (Fig. 13.10). This is the ligament that passes as a bridge to the front of the distal urethra. This is also the ligament that, along with the arcuate pubic ligament, delimits the infra-pubic space where the dorsal and deep veins of the clitoris and the cavernous nerves pass through. In its entirety, the musculo-aponeurotic structure observed consistently at the anterior perineum, if dissected minutely and attentively, presents the shape of a horseshoe, tying the urethro-vaginal assembly "from the front". The crura clitoridis and the spongy bulbs are attached to the caudal surface of this musculo-aponeurotic structure.

[9] It is the caudal part of the striated sphincter of the urethra, located above, pressed in a horseshoe shape against the ventral surface and the lateral surfaces of the urethra, which should be called the "urethro-vaginal sphincter".

13.1.4 Relations with the Pelvic Diaphragm

Relations with the **pelvic diaphragm** are worth mentioning, even if this fibro-muscular layer is deep and more cranial than the perineal membrane to which the erectile bodies are attached, as we have just seen. The **puborectal bundle of the levator ani muscle** (main, very powerful and very active muscular component of the pelvic diaphragm) is tangent to the lateral walls of the urethral and vaginal distal extremities.

Let us remember that the pelvic diaphragm is a musculo-aponeurotic assembly that closes at the bottom of the pelvic cavity and on which the pelvic viscera rest, as on a hammock. It comprises four muscles: the two levator ani muscles, which make up the majority of the muscular plane, and the two coccygeus muscles located behind the levator ani. All of these muscles are covered with fascias, the **superior fascia** and the **inferior fascia of the pelvic diaphragm**. Two large orifices, the urogenital hiatus (formerly the urogenital opening) and the anorectal hiatus, enable the pelvic viscera to cross this diaphragm and pass from the pelvic cavity to the perineum. For both sexes, the anorectal hiatus allows passage of the terminal part of the intestine or rectum.[10] In women, the urogenital hiatus allows passage of the urethra and the vagina. The two levator ani muscles are even and symmetrical and comprise two muscular layers: one cranial layer formed by three bundles—pubococcygeal, iliococcygeal and ischiococcygeal—and one caudal layer formed mainly by a so-called puborectal bundle which is of major importance in pelvic biomechanics as it ties the rectum from behind. This puborectal bundle (Figs. 13.11 and 13.12) follows the lateral wall of the vagina and has very tight connections with the vagina.[11] It is presented as a muscular, ribboned bundle, slightly twisted and very solid, that links the **central tendon of the perineum** to the angular surface of the pubis to which it is fastened. Retrovaginal fibres cross with those on the opposite side to form a true hammock that supports the vagina. The other retrorectal fibres determine the **anorectal angle**, of major importance in terms of function,

[10] The levator ani muscle marks the level where the pelvic rectum continues as the anal canal (former perineal rectum).

[11] The puborectal bundle can be easily perceived on the lateral walls of the vagina during vaginal examination. During examination with the aim of correcting a prolapse, the gynaecologist may evaluate the quality and tonus of the muscle by asking the examined patient to "squeeze" the examiner's fingers.

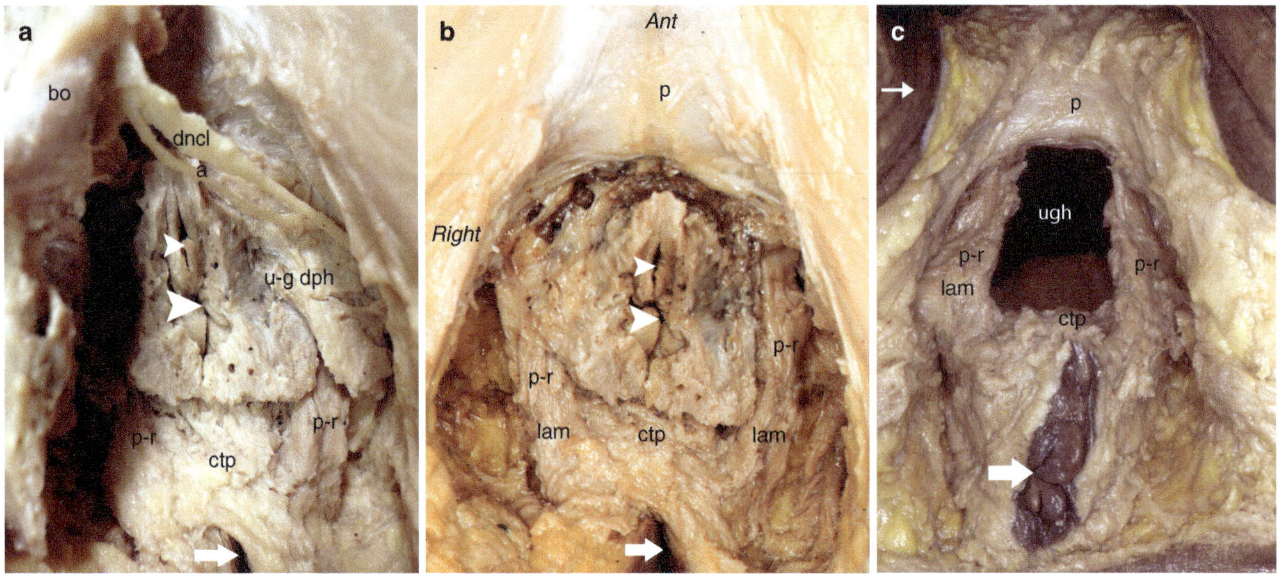

Fig. 13.11 Relationship between both the bulbo-clitoral organ and the urethro-vaginal pyramid with the levator ani muscles (pelvic diaphragm). (**a**) Perineal view of female pelvic floor. The bulbo-clitoral organ, released from its attachments, has been moved backwards and on the right, drawing the left dorsal artery and the left dorsal nerve of clitoris. The urogenital diaphragm (perineal membrane) is always present below the pubo-rectalis (bundle of the levator ani muscle). (**b**) Same perineal view of female pelvic floor after the removal of both the bulbo-clitoral organ and urogenital diaphragm. (**c**) Perineal view of female pelvic floor highlighting the urogenital hiatus (hiatus of levator ani muscles), after removal of bulbo-clitoral organ, urogenital diaphragm and "urethro-vaginal pyramid". *a* dorsal artery of clitoris, *bo* body of clitoris, *ctp* central tendon of perineum, *dncl* dorsal nerve of clitoris, *lam* levator ani muscle, *p* pubic symphysis, *pr* pubo-rectalis (bundle of levator ani), *u-g dph* urogenital diaphragm, *ugh* urogenital hiatus (hiatus of levator ani muscles), *great white arrow* anus, *small white arrow* genitofemoralis sulcus, *great white arrowhead* vaginal orifice, *small white arrowhead* external urethral orifice

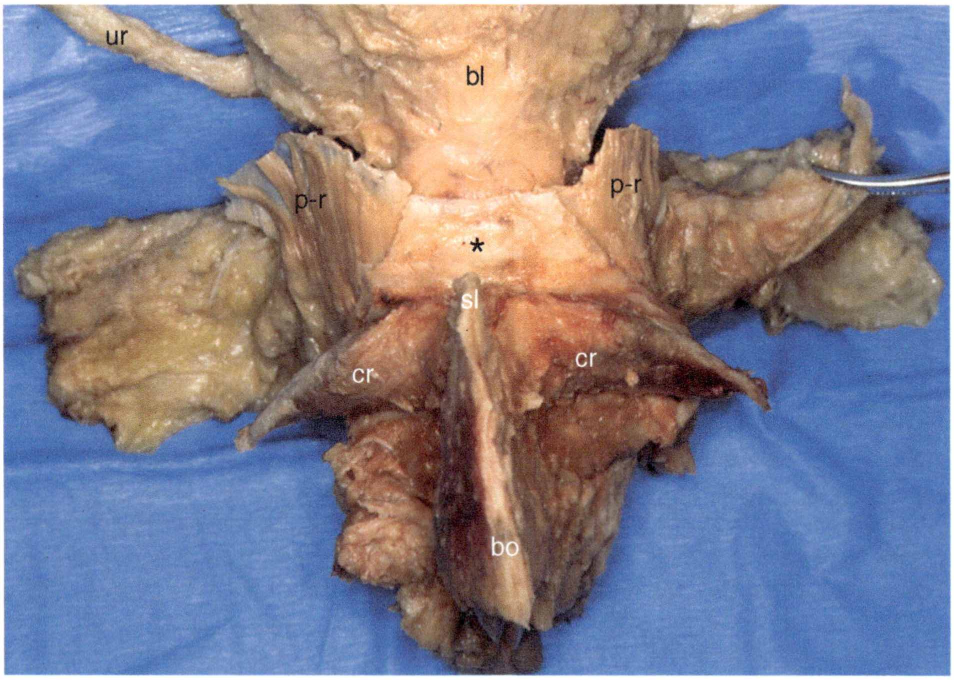

Fig. 13.12 Relations between the crura clitoridis and the pubo-rectalis muscles exposed on a "pelvi-perineal" dissection (*specimen removed from the pubic symphysis*). *bl* bladder, *bo* clitoral body (descending part), *cr* crus clitoridis, *pr* pubo-rectalis muscle (bundle of the levator ani muscle), *sl* suspensory ligament, *ur* right ureter, *black asterisk* fibrous coating of the pubic symphysis's posterior surface; the clamp lifts the left paracervix (Mackenrodt's ligament)

and they participate in anorectal continence and defecation along with the external anal sphincter (the deep part of which the fibres are joined to). Topographically, it is the perineal membrane and the deep transverse muscle that separate the female erectile bodies from the ischioanal fossa (spaces located under the levator ani muscles) and from their ventral extensions. The pelvic-subperitoneal space above the pelvic diaphragm represents a more distant connection.

14 Imagery of the Bulbo-clitoral Organ

14.1 General

Significant knowledge of the bulbo-clitoral organ has been acquired thanks to the enormous progress made in relation to new medical imagery techniques, which have been used to explore the pelvis and especially the female pelvis. X-Ray computed tomography and then NMR (nuclear magnetic resonance) are at the forefront of these new exploration techniques and are mutually complementary. At the same time, great progress has been achieved in medical echography, which has not only become a routine and essential process for pregnancy monitoring but also a sophisticated method for studying pelvic organs and more particularly female genitalia.

14.1.1 Computed Tomography (CT Scan Imagery)

The images obtained by X-Ray CT appeared, as soon as this process was developed, to be extremely interesting as they could be compared to (and almost superimposed on) the cadaveric sections prepared by anatomists. The progress made since then, in only 20 years, is considerable and at present the 64-slice scanners, due to their speed of acquisition and to the multiplicity of the cross-sections, can be used to reconstruct (CT image reconstruction) the organs under study in 3D and to generate three-dimensional images.

The bulbo-clitoral organ appears to be extremely complex and complete on the most superficial sections of the pelvis, especially on the sections made in the planes of the anterior perineum. On such sections, which are strictly comparable to anatomical sections (compare Figs. 14.1b and 13.1), all the parts forming the bulbo-clitoral organ are clearly observed. However, this observation varies according to the levels of the sections:

- On sections tangent to the pelvic outlet but not concerning this bony part (Fig. 14.1a), it is especially possible to observe 2 spongy bulbs bordering the urethral and vaginal ducts and then further ahead, in the extension of the commissure of the bulbs, the entire IC RSP (infra-clitoral residual spongy part) up to the glans.
- On sections concerning the ischio-pubic rami (Fig. 14.1b), the bulbs are almost no longer visible. It is possible to assume the existence of the commissura bulborum overhanging the

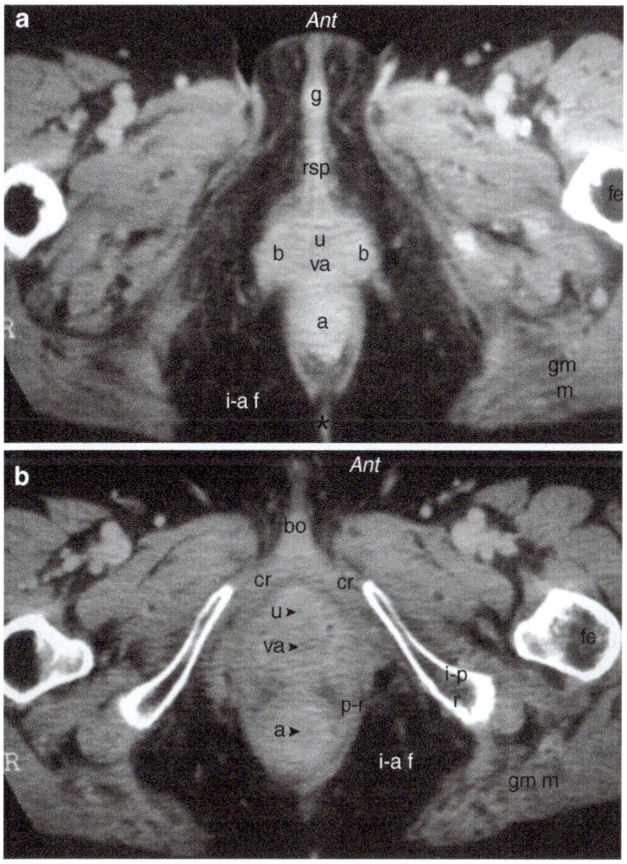

Fig. 14.1 CT scan sections of the bulbo-clitoral organ (scanner of female perineum). (**a**) Section concerning the spongy parts. (**b**) Section concerning the cavernous parts. *a* anus, *a* (*with arrowhead*) anal canal, *b* bulb, *bo* clitoral body, *cr* crus clitoridis, *fe* femur, *g* glans clitoridis, *gm m* gluteus maximus muscle, *i-a f* ischio-anal fossa, *i-p r* ischio-pubic ramus, *p-r* pubo-rectalis muscle (levator ani muscle), *rsp* residual spongy part, *u* external urethral orifice, *u* (*with arrowhead*) urethra, *va* vaginal orifice, *va* (*with arrowhead*) vagina, *ano-coccygeal ligament

V. Di Marino, H. Lepidi, *Anatomic Study of the Clitoris and the Bulbo-Clitoral Organ*,
DOI 10.1007/978-3-319-04894-9_14, © Springer International Publishing Switzerland 2014

urethra, but the IC RSP (infra clitoral residual spongy part) is no longer visible. On the other hand, the crura of the clitoris, applied to the lower rami of the pubis and then backed against the front of the pubic symphysis, are quite visible. On these same sections, the descending portion of the clitoris and the glans are not very visible, which is due to the elbow of the clitoral body and the triplanar architecture of the bulbo-clitoral organ.

- It should finally be recalled that the overlying sections provide a good overview of the pubo-rectal bundles of the levator ani muscles, thus limiting the hiatus of the levators ("urogenital hiatus").

14.1.2 Nuclear Magnetic Resonance (NMR Imagery)

The images of the female pelvis obtained by means of the NMR technique are extremely interesting: They are generally used to identify more anatomical details than those observed on CT images and to highlight them subject to using sequences, which are the most compatible with the target observation. For the bulbo-clitoral organ, it is incontestably the weighted T2 sequences, which are the most interesting, because they provide very contrasted images, which can be modulated by means of fat saturation techniques (removal of the fat signal). Furthermore, the possibility of obtaining sections of the pelvis in the 3 dimensions of space is particularly useful for studying the bulbo-clitoral organ, whose triplanar architecture is already known.

Anatomical structures of the external genitalia can be highlighted by performing an initial intravenous injection of gadolinium (this process can be combined with the fat saturation technique to obtain striking contrasts) for T1 sequences as well as for T2 sequences.

- The median or paramedian sagittal sections, which are used for exploring the internal genitalia and all of the pelvic organs, provide a good overview of the clitoris (angle, ascending and descending portions of the body) and the glans (Fig. 12.2a). On these median section drawings, the crura of the clitoris and the bulbs, in a more lateral position, are not (or only slightly) visible.
- The coronal sections are used for observing the glans, the vestibule, the labia minora and labia majora.
- The axial transverse-oblique sections are the most interesting because they provide images, which are perfectly similar to the sections obtained in the anatomy laboratory (Fig. 14.2): They show the erectile structures overhanging, in a horseshoe configuration with a posterior concavity, the external superimposed orifices of the urethra and vagina. The 2 crura connected to the ischio-pubic rami are perfectly visible, with their junction area and then their apposition to form the body of the clitoris. Furthermore, NMR

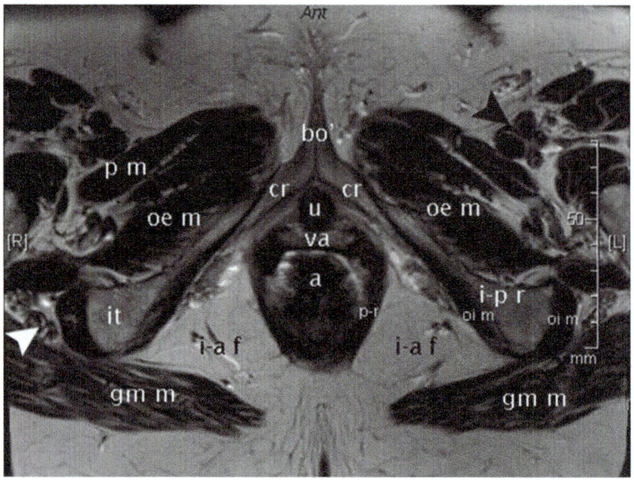

Fig. 14.2 NMR section of female pelvis through the lower third of the ischio-pubic rami (weighted T2 sequence. Very thin section). It shows the two crura clitoridis joining to form the clitoral body. *a* anal canal, *bo′* pre-angular part of the clitoral body (ascending part), *cr* crus clitoridis, *gm m* gluteus maximus muscle, *i-a f* ischio-anal fossa, *i-p r* ischio-pubic ramus, *it* ischial tuberosity, *oe m* obturator externus muscle, *oi m* obturator internus muscle, *pm* pectineus muscle, *p-r* pubo-rectalis muscle (part of the levator ani muscle), *u* urethra, *va* vagina, *white arrowhead* right ischiatic (sciatic) nerve, *black arrowhead* left femoral artery, left femoral veins and left femoral nerve

can be used to observe the median septum separating the 2 corpora cavernosa merged together as a "clitoral corporeal cylinder". The bulbs are also very visible on either side of the urethro-vaginal "block". All these erectile structures are clear, white and luminous, whereas the other perineal structures seem to be grey or black with a low intensity (this is the case of the urethra or the vagina[1] or of the greater vestibular glands, observable at the rear of the bulbs). It should be noted that the external urethral orifice and the vaginal orifice are hardly visible. As for ischiocavernosus, bulbo-spongiosus and superficial transverse muscles, they contrast perfectly with the erectile bodies due to their black aspect corresponding to a hyposignal. Lastly, it is possible to observe the pudendal vessels and their division into a perineal branch and dorsal branch of the clitoris, by means of the fat saturation technique (H.E. O'Connel and J.O.L. Delancey). Comparative studies carried out on non-menopaused women and menopaused women (D.D. Suh et al.) did not show significant differences between the 2 groups as regards the bulbo-clitoral organ (a slight reduction in size of the bulbs in menopaused women at the most). The differences, which are all related to the reduction in oestrogen secretions, are observed "at the labia (reduction in

[1] These ducts are highlighted if the fat saturation technique is also used. However, this highlighting remains less significant than that of the components of the bulbo-clitoral organ.

size), the vaginal epithelium (thinning), the vaginal canal (narrowing) and the cervix (atrophy)".

Overall, pelvic NMR imagery is used not only to perform a specific study of the bulbo-clitoral organ but also to provide an excellent vision of the entire female genitalia and to perform a complete study of its relationships (sections in the 3 dimensions of space) and a structural study of the various pelvic organs and of the anatomical sites inside which these organs are located. It should be noted that although NMR remains ideal for a topographic anatomical study of the bulbo-clitoral organ, CT provides a better understanding of the descriptive anatomy of this organ, such as can be observed on scanner images (Fig. 14.1).

NMR preceded by the intravenous administration of gadolinium is, on the other hand, an ideal method for studying the bulbo-clitoral organ during a sexual arousal phase. The highlighting achieved at the level of all the erectile bodies makes it possible to explore the modifications occurring in this context: increase in dimensions and, especially, in the diameters of the erectile bodies and intense blood flow congestion with filling of the cavernous and spongy lacuna (K.R. Maravilla et al.).

14.1.3 Ultrasonic Echography (Imagery by Ultrasound)

It is the merit of French echographists (O. Buisson and P. Foldes) to have used ultrasound to explore the bulbo-clitoral organ. They therefore obtained, with this easy-to-use and low-cost technique, very good quality images, which are, especially using the technique of ultrasonic CT, also comparable to anatomy data. They could also easily measure the diameters and lengths of the constituent parts, in particular the dimensions of the clitoris. However, the bulbs remain difficult to observe if there is no stimulation. From a topographic viewpoint, it is the coronal sections, which provide the best anatomical approach: they are especially used to view the commissure of the bulbs, overhanging the urethra (which, itself, is visible), and the anterior wall of the vagina (hypoechoic according to the authors' experience).

Thereafter, the authors made the same observations by **dynamic sonography**, while asking the women being examined to voluntarily contract their pelvis: They were thus able to observe movements of the clitoris with a reduction of the inter-crural angle, an increase in the angulation of the clitoral elbow and a forward and downward thrusting movement of the glans. With this process, they were also able to study vaginal penetration by asking the patient under examination to insert a finger into her vagina: this experiment made it possible to show that the pelvic contraction, caused by vaginal penetration, brings the clitoral crura of the anterior vaginal wall closer together.[2] Thereafter, the authors used dynamic sonography to directly study the anatomy of the bulbo-clitoral organ during intercourse, such as initiated since 1999 by the dynamic genital MRI developers (W. W. Schultz et al.).

[2] As a result, the authors have deducted the assumption that during the phase of sexual excitation, the crura of the clitoris, which have become turgescent and congestive and have increased in volume, become stimulation factors for the anterior vaginal wall (one of the assumptions about the "G spot").

The Bulbo-clitoral Organ in the Sexual Act

15

15.1 General

Although our study is mainly anatomical, it appeared useful to include some physiological data specifically related to the modifications and the role of the bulbo-clitoral organ during the different phases of the sexual act.

How does this organ behave during intercourse? Here is a question, which has remained without an accurate answer for a long time, as if it were taboo or too mysterious to be mentioned. Fortunately, what was, up until now, only assumptions or the product of people's imaginations has now been confronted with the reality of images obtained with various methods of exploration and especially techniques providing dynamic representations. The recent progress made in studies on intercourse in relation to computed tomography, NMR (publications of A. Burdens et al.), echography (work of Odile Buisson and Pierre Foldes), macrophotography and videos have finally made it possible to clarify this question.

15.1.1 Description of the Different Phases

It should be recalled that, since the work of W.H. Masters and V.C. Johnson, it is known that the sexual cycle of a woman, when it is complete (and therefore ideal!), has four phases: an excitation phase of variable duration, a plateau phase (which lasts for 3 min maximum), an orgasm phase (which only lasts a few seconds[1]) and a resolution phase of 15 min.

During the **phase of sexual excitation**, triggered by the imagination and erotic memories, a favourable context, the stimulation of various erogenous areas in a woman (including the clitoris itself) and maintained through the secretion of dopamine,[2] the bulbo-clitoral organ undergoes many significant transformations:

- Due to the blood supply to the erectile bodies, the latter will increase in volume with the intumescence of the entire bulbo-clitoral organ, at the level of the spongy bodies (which explains the rising of the labia minora) as well as at that of the cavernous bodies, until the erection of the clitoris. Such as already observed, in the chapter devoted to the physiology of the cavernous and spongy tissues, this erection, although real (rigidity and extension of the body of the clitoris), will remain with a limited amplitude, resulting in a slight rising of the descending portion of the clitoris (increase in the angular opening of the clitoral elbow, in a sagittal plane) and a potential extrusion of the glans outside the hood.
- At the same time, vasodilatation will diffuse to all the vascular networks communicating with the erectile bodies. We will therefore observe significant effects on all of the external genitalia, including, in particular, the swelling of the labia minora, but also on the urethra and the vagina (congestion of the urethral and vaginal vessels). This vaginal hyperhemia will result in an increase of the capillary pressure and an important transudation (high concentration in Na^+ and Cl^- ions) in the vaginal lumen. This transudation is, furthermore, favoured by an increase in capillary permeability, which, itself, is under the influence of neuromediators such as the VIP. This transudate becomes so significant that the capacity of reabsorption through the vaginal epithelium is saturated (R.J. Levin), which produces and maintains the vaginal lubrication essential for the copulation. Many secretions also contribute to this phenomenon: secretions of the glands of the uterine collar, of the vestibular glands (grater vestibular glands, in particular) and of Skene's paraurethral glands.[3] The entire lubrication process is under hormonal influence (essential role of the oestrogens[4]).

[1] The duration of the female orgasm (average 20 s) is greater than that of man (average 6 s)!

[2] This neurotransmitter is regarded as the true "hormone of pleasure" (J.D. Vincent).

[3] The excreting canals of the two paraurethral glands or glands of Skene are located on the distal end of the urethra, on either side of the external ostium of the urethra. In certain women, known as "fountain women", during the sexual excitation phase, the secretion of these glands can be particularly abundant.

[4] Vaginal dryness, a frequent occurrence after the menopause, is related to a hypo-oestrogenia. This phenomenon can also be observed in

V. Di Marino, H. Lepidi, *Anatomic Study of the Clitoris and the Bulbo-Clitoral Organ*,
DOI 10.1007/978-3-319-04894-9_15,

The penetration and the back-and-forth movements of the male penis in the vestibulo-vaginal tract will then generate, in women, various phenomena leading to the full completion of the sexual act. These phenomena will enable her to reach the **plateau phase**. We will successively consider the following:

The penetration and the vagino-cavernous reflex

The penetration and the movements of the deep bulbo-clitoral structures

The stimulation of the external clitoris

The stimulation of the urethra

The contraction of the levator muscles

- **The vagino-cavernous reflex** is triggered by vaginal distension caused by the penetration. This distension generates the reflex contraction of the ischiocavernous and bulbospongious muscles. The EMG studies showed that this reflex occurs during each penetration movement and that the amplitude of the muscular contraction potentials is proportional to the significance of the vaginal distension (work completed by A. Shafik with small inflated air small balloons with progressive volumes). This reflex plays an essential part for the two partners, as it contributes to maintaining the erection of the clitoris (by compressing the veins of the clitoris via the strap of the bulbospongious muscles and by compressing the erectile bodies via the action of the ischiocavernous and bulbospongious muscles) as well as the erection of the penis (compression of the dorsal vein of the penis and of the penile erectile bodies by the contracted bulbospongious muscles, erotic stimulation of the penis through the contraction sensation of the partner's peri-vaginal muscles[5]) and the quantity of sperm (according to A. Shafik, the vagino-cavernous reflex increases the volume of ejaculate while expulsing, towards the vagina, the residual sperm remaining in the urethral lumen of the urethral tract).
- **The penetration** will displace the deep structures of the bulbo-clitoral organ to a certain degree. The penis in erection (thus dilated and rigid) penetrates the vagina (thickened through vasodilatation) according to a boomerang-type trajectory[6] regardless of the position of the partners.

  During this penetration, it will push the anterior vaginal wall backwards and the urethra forwards, by applying, with a certain force, these structures on the commissure of the bulbs and on the intermediate area, which will be crushed against the pubic symphysis (Fig. 15.1). Similarly, the penis drives back the lateral walls of the vagina, thus compressing not only the bulbs but also the clitoral roots, which are all at maximum turgescence.

With an echography (work performed by O. Buisson et al.), it is possible to observe, during each penetration, an actual "displacement" of the bulbo-clitoridal organ. The arc formed by the commissure of the bulbs increases in size, while the bulbs, which are saturated with blood to their maximum capacity, become wider. The roots of the clitoris, whose volume doubles during the back-and-forth movements, are drawn apart by the dilated penile cylinder. Their junction angle is transformed into an arc. The intermediate network is also pushed back against the symphysis. Each back-and-forth movement acts as a pump and propels blood to the clitoral body through the vessels of this network.[7] Overall, during the penetration-withdrawal movements, the internal part of the female erectile bodies is the site of maximum blood saturation and intense compressive phenomena, which tend to make the blood flow back to the discharge pathways (Kobelt's network and peri-vaginal venous plexus). It obviously generates an intense stimulation of the anterior vaginal wall and of the urethra, which are caught in a vice-like grip between the female erectile bodies and the penis at maximum turgescence.

In men, the combination of vascular saturation, which induces the erection, and of the compressive phenomena generated by the turgescence of the female erectile bodies similarly maintains the stimulation and thus the erection.

There is thus a true feedback, "a sexual synergy of the two partners with reciprocally reinforced phenomena" (P. Lavoisier et al.). The penis in erection, penetrating the vagina, modifies the intra-vaginal pressure and generates the tumescence of the erectile bodies and the erection of the clitoris. At the same time, the congestion of the peri-vaginal vessels compresses the vagina and thus the penile cylinder increases the intra-cavernous pressures of the penis and maintains its erection.

The back-and-forth movements will enable the following:

- The bulbo-clitoral organ will be able to expand during each partial withdrawal.
- The vagina will be able to vary the pressures applied to its wall by increasing them with each withdrawal movement of the penis, with a positive action for the tumescence of the penis when it advances again inside the vaginal cylinder.

non-menopaused women, in relation to "dry syndromes" or during certain vulvo-vaginal mycoses. There are also other types of dryness of iatrogenic origin, following the absorption of certain drugs.

[5]It should be recalled that the medial part of the bulbospongious muscle in a woman forms what previous scientists referred to as the "constrictor cunni".

[6]The first NMR of sexual intercourse, performed in 1999 by W.W. Schultz et al., showed that the penis in erection is, on the sagittal sections, neither straight, such as represented by Leonard de Vinci, nor shaped like an S, such as represented by R.L. Dickinson in 1933, but shaped like a boomerang as it forms a 120° angle with each of its roots.

[7]Georg Ludwig Kobelt (1851), in his remarkable book, referred to the "expulsive contractions of the bulbo-spongiosus muscles" which make the blood of the bulbs flow back to the intermediate network and behave like a genuine "hydraulic apparatus (sexual heart)"!

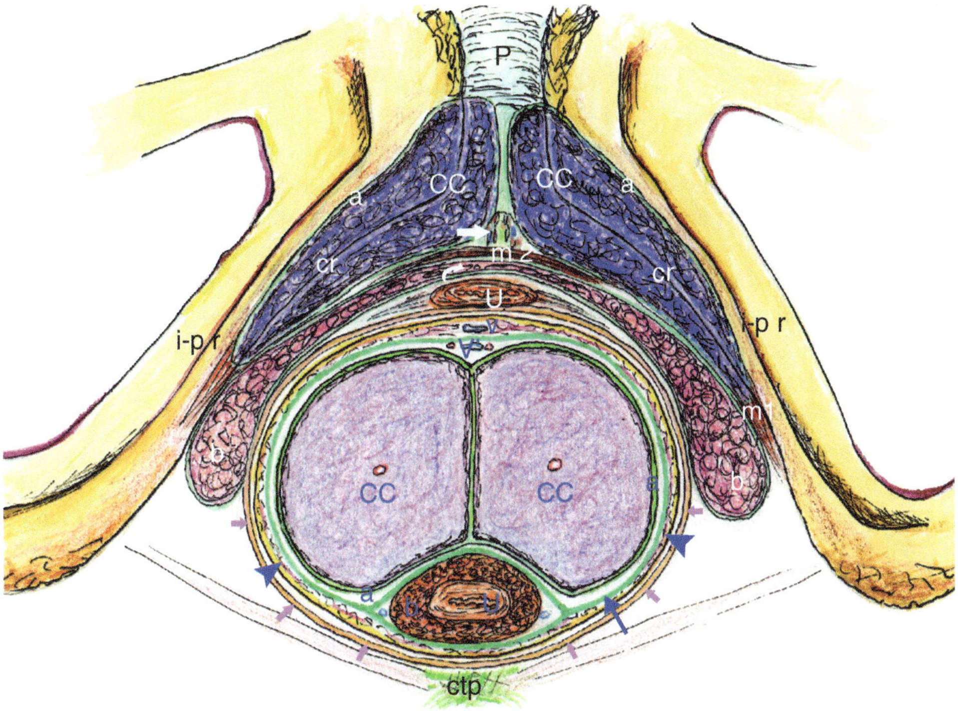

Fig. 15.1 Sketch from a coronal section of the erectile bodies during coïtus, at the time of the penis pressure (note: *white captions* for bulbo-clitoral organ; *blue* captions for penis). White captions: *a* albuginea (getting very thin because of the turgescence of corpora cavernosa), *b* bulb, *CC* corpus cavernosum, *cr* crus clitoridis, *m1* ischiocavernosus muscle, *m2* bulbo-spongiosus muscle, *U* female urethra, *right white arrow* location of the pars intermedia, *curved white arrow* commissure of the bulbs (commissura bulborum). Blue captions: *a* albuginea, *b* corpus spongiosum urethrae (spongiosus bulb), *CC* corpus cavernosum, *U* male urethra (penile urethra), *V* subcutaneous dorsal vein, *V′* deep dorsal vein, *right blue arrow* it shows the Buck's fascia (fascia penis), *Blue arrowhead* they show the erected penis. Other captions: *P* pubic symphysis, *i-p r* ischio-pubic ramus, *ctp* central tendon of perineum, *little pink arrows* they show the posterior part of the vaginal wall (the vaginal cavity is completely distended by the penetration of the erected penis)

- The penis will not undergo anoxia and will alternate phases of "external compression" (infra-systolic phases) and of "internal compression" (supra-systolic phases).[8]
- **The stimulation of the external part**, i.e. the visible part of the clitoris, is achieved by the back-and-forth movements of the penis. This stimulation is double: direct and ancillary and indirect and yet essential:
  1. **Direct stimulation**, i.e. contact between the glans clitoridis and the male penis, should not theoretically occur. The clitoral erection takes place as an extension and will move the clitoris away from the penis and place it on a higher level, parallel to the plane of penetration.[9]

     However, contacts exist between the end of the clitoris and the penis, but they mostly are brushing-type contacts and particularly concern the clitoral hood rather than the glans itself (which tends "to retract at each assault"). These contacts occur with the back of the penis, in anterior penetrations, in the traditional missionary position, and with the lower surface of the penis, during posterior penetrations, known as "doggie style". They only occur during the penetrating back-and-forth movements: the penetration draws aside the labia minora and pulls on the frenula of the clitoris and the lateral parts of the hood, thus causing a backward inflection of the descending portion of the clitoris. This inflection will be accentuated at the climax of each penetration thrust, when the pressure exerted by the anterior male perineum against the vulva is at its maximum. It is the moment when the hood (and sometimes also the glans) rubs against the surface of the penis. During each withdrawal movement, during the back-and-forth action, the pressure ceases and the clitoris goes back to its erected position, parallel and above the penis. The glans even seems to retract then because the hood, still drawn backwards, covers it.
  2. **Indirect stimulation** is much more important as the alternation of the advances and withdrawals of the back-and-forth movement will not only draw aside the

---

[8]The NMR images (A. Faix et al., W.W Schultz et al.) showed that during intercourse the penile glans reaches the anterior vaginal fornix or the posterior vaginal fornix and that this depends on the penetration position and on the topography of the uterus (ante- or retroverted).

[9]We can very logically wonder whether the finality of clitoral erection, at a simple mechanical level, is not to prevent the clitoris from hindering or being an obstacle to penile penetration.

labia minora and rhythmically pull on the frenula of the clitoris (with direct action on the corpuscles of the glans) but will also actually titillate the glans through the concomitant friction of the hood pulled backwards by the labio-prepuce system and of the glans pushed forwards by the erection of the body of the clitoris.[10] Moreover, there exists two symmetrical lateral ligamentory formations described by A. Ingelman-Sundberg,[11] which connect the lateral edge of the vagina to the neck of the clitoris and which transmit the movements of the anterior vaginal wall during the back-and-forth movement to the clitoris. The tensioning of these ligaments facilitates the action of bringing the clitoris and penis close together. This means that the descending portion of the clitoral body is doubly mobilised compared to the hood.

Overall, the penetration stimulates the clitoris as much as the vagina (with very little innervation and, in any case, without comparison to the extraordinary nerve system of the clitoris). This is why the distinction made between "vaginal" women and "clitoral" women, which had its hour of glory during the era of S. Freud, is now generally criticised.

- Many authors also consider, and rightly so, that vaginal stimulation and clitoral stimulation are closely related to a certain degree of **stimulation of the urethra**. This is not surprising when we consider what anatomy teaches us:
  - Communication and vascular balancing between the spongy bulbs and the peripheral plexuses of the urethral tract
  - Rich innervation of the urethral wall, in particular at the level of the posterior part of the urethral cylinder, in contact with the anterior wall of the vagina
  - Joining of the urethral and anterior vaginal walls, superimposed from back to front

  Thus, certain authors (G. Vanneuville et al.) located the famous G spot[12] at the level of the urethral sphincter. Other authors (J.L. Sevelly in particular) consider that the clitoris is not the single excitable organ during sexual intercourse and that the external ostium of the female urethra and the epithelium, which surrounds it, play an equivalently significant role. They believe that this area has a high erotic sensitivity,[13] comparable to that of the end of the male glans, around the external ostium of the urethra. Sevelly called it the "female glans" (in order to not mix it up with the glans clitoridis). Still according to this author and with the support of evidence, this "female glans" is particularly excited during back-and-forth movements, as the penis always encounters it during the transition between the partial withdrawal phase and the complete insertion phase and vice versa.

  In any case, all the authors agree to include, in the same erogenous groups, all anatomically connected organs, likely to be stimulated and excited during a sexual act. This leads to terminologies, which resemble and complete each other:

  O' Connell: clitoral complex (clitoral complex), O. Buisson: clito-urethro-vaginal complex (CUV), L.C. Lenck: urethro-clitorido-vulvar entity and, for us, the bulbo-clitoral organ and the urethro-vaginal unit.
- **The roles of the pubo-rectal bundles of the levator ani muscles** are better known, since the current study of these muscles by dynamic NMR, in genital cases of prolapse.[14] Their voluntary and powerful contraction modifies the diameter of the urogenital hiatus. They can thus, during the penetration, reinforce the action of the ischio- and bulbocavernous muscles. Expert women, who know their body well, can thus add an additional degree of eroticisation for their partner by repeatedly contracting these pubo-rectal bundles,[15] which will compress the penis in full back-and-forth action.

At the end of the **plateau phase**, at the maximum of the **excitation phase**,[16] the **orgasm phase** will occur, when the conditions are favourable, with many simultaneous manifestations, including first and foremost, a reduction of the lumen of the lower third of the vagina, which is saturated with blood; turgescence and a modification of the colour of the labia minora; vaginal lubrication of maximum intensity; spasmodic contractions of the uterus, vagina, perineal muscles and of all the pelvic muscles[17]; and lastly, a temporary retraction of the

[10] It is this dynamic (hood towards the rear and clitoral body towards the front), which partly explains the apparent temporary retractions of the glans under the hood, observable during the back-and-forth movements. These withdrawals also occur during orgasm when the muscular contractions of the ischio- and bulbocavernous muscles amplify the phenomenon.

[11] A. Ingelman-Sundberg was able to systematically highlight these ligaments during operations relating to the anterior perineum.

[12] The G spot (Grafenberg spot) is an area (which has still not been determined from an anatomical viewpoint) located on the anterior wall of the vagina, approximately a few centimetres from its external ostium and whose characteristic is to be extremely erogenous.

[13] It should be noted that our research showed a significant amount of nerves at the periphery of the external ostium of the female urethra, but on the other hand, a total absence of corpuscles of pleasure (Fig. 15.2).

[14] The usual function of the pubo-rectal bundles of the levator ani muscle is a lifting function: their contraction brings the median pelvic organs of the pubic symphysis closer together.

[15] The exceptional penis captivus (involuntary muscular spasm, retaining the penis in erection inside the vagina) is due to the peri-vaginal muscles (constrictor cunni) and not to the pubo-rectal bundles of the levator ani muscles.

[16] It is at the level of the limbic lobe that the pleasure sensation ends but functional MRI has shown that many areas of the brain are involved: amygdaloid body, hippocampus, accumbens nucleus, lateral hypothalamus, temporal cortex, frontal cortex with, when approaching orgasm, the production of oxytocin, which explains the small mammary discharges of this phase and the occurrence of uterine contractions.

[17] These muscle contractions occur in two phases: the first, unique and intense, the tonic phase, the second, made of short burst contractions, the clonic phase (suggestive of clonic seizure).

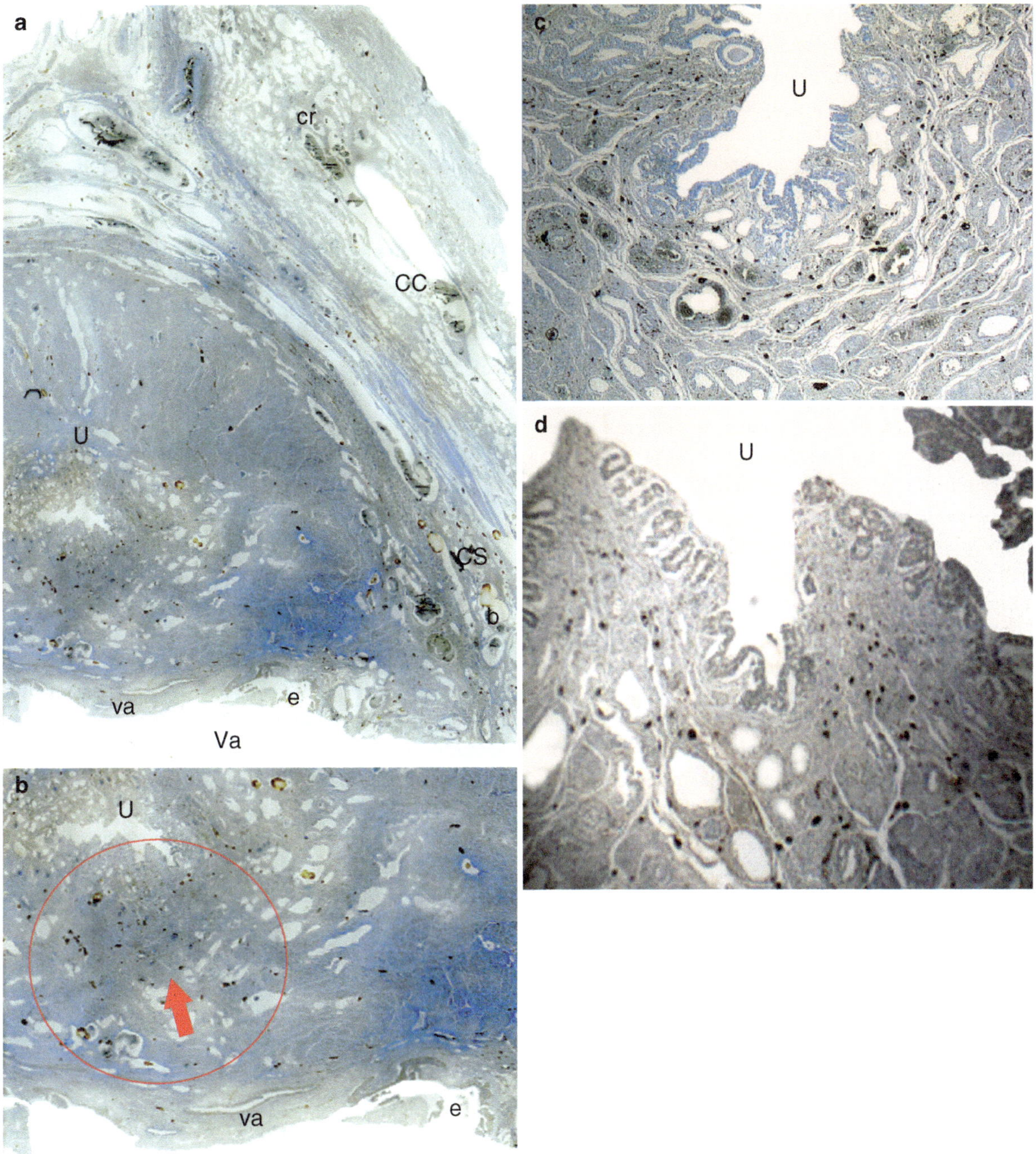

Fig. 15.2 A urethral hypothesis on G point. The *red circle* and *red arrow* show the most innervated area. PS100 staining showing the innervations of the erectile bodies and of the "urethro-vaginal unit". (**a**) Photomicrograph of a transverse section of the "urethro-vaginal unit", through the crura clitoridis (the section is perpendicular to the urethral axis). (**b**) Magnification of the "urethro-vaginal unit" to show the rich innervation of the posterior half of the urethra above the anterior vaginal wall (*red arrow*). (**c**, **d**) Two microscopic aspects of the *red-circled* area. The multiple *dark brown* spots are nerves! *b* bulb, *CC* corpora cavernosa, *cr* crus clitoridis, *CS* corpus spongiosum, *e* left lateral edge of vagina, *U* urethra, *Va* vagina, *va* anterior vaginal wall

clitoris in erection under the prepuce. Many vegetative reactions of the entire body (hyperventilation, arterial hypertension, tachycardia,[18] redness of the face and torso, mydriasis) as well as corporal modifications (increase in the volume of the breasts with erection of the nipples) accompany these pelvic phenomena. It seems that the sensation of intense pleasure, characteristic of this highlight, is related to the rapid discharge of blood, which congested all the venous plexuses and all the pelvic tissues, under the effect of the waves of reflex

[18] A respiration rate of 30/min, a systolic blood pressure of 20 and a pulse of 120 are not exceptional.

contractions of the perineal striated muscles, of the levator ani muscles, of the urethral and anal sphincters and of the smooth muscle fibres of the uterus.[19] The orgasm is a complex phenomenon that physiologists have greatly studied and which can be clarified by the use of modern exploration techniques.[20]

We should also note the significance of the neurosecretions[21] occurring at this stage, especially endorphin secretions, which explain the absence of suffering and even the actual analgesia, from which women benefit during an orgasmic phase, in spite of the multiple muscular contractions and the various pressures, which are more or less violent.

The male ejaculation (which usually corresponds to the phase of male orgasm) often does not correspond to the female orgasm phase. If he can, the man must continue, as much as possible, with the rhythmic movements to prolong the sexual excitation of the woman. However, this is only possible for a short moment. When his erection ceases, the penis tends to exit the vaginal tract. But there again, the woman's stimulation can be continued through the man's pelvic movements, thus applying pressure on the pubis and frictions of the still turgescent penis against the clitoris, which is still in erection. In positions known as "straddle positions", the woman has a more active role and she will have the initiative of these movements. By undulating her pelvis, she will know how to stimulate her clitoris against the root of the penis and the male pubis. This is a good way for her to reach orgasm.

The **resolution phase**, which occurs after the short but very intense phase of pleasure and exaltation, characteristic of an orgasm, is a phase of well-being and physical relaxation due to the pelvic-perineal vascular and muscular systems returning to their normal status. The clitoris, which was retracted under the prepuce, will return to its normal position in a few moments. However, the glans remains hypersensitive during several minutes. As for the detumescence of the body of the clitoris, it only occurs after a longer period, often exceeding 15 min. Similarly to the genitalia, the torso and the face recover their usual aspect within 10–20 min.

If no orgasm has occurred (which does not mean that there was no pleasure!), the resolution phase is shorter and lasts only around 10 min. It should be noted that a woman can also have several successive orgasms (specificity of mankind!).

15.1.2 Sexual Intercourse and Sexual Act

The significance of the bulbo-clitoral organ (whether its "internal part" or its "external part") during the various phases of the sexual act is undeniable. However, for mankind, who is provided with an extraordinary nervous system and, especially, an exceptional brain (the most developed brain in the animal world), the sexual act cannot be compared to or mixed up with a perfect succession of mechanical phenomena, nor even with the mating of animals, with an exclusively reproductive purpose.[22] The human sexual act is specific because it does not obey a simple impulse and it can be controlled by will or reason. Moreover, for humans, this act is part of an intimate context of communication: sexual intercourse. When the understanding between two partners is harmonious, this relationship will lead to an impassioned sexual act, in which each partner will not only attempt to satisfy his/her own pleasure but also that of the other. It is the brains of the two partners which will choose the most adapted words, the softest caresses and the most pleasant positions so that the union of their bodies is perfect and provides mutual and total satisfaction. At best, when the relationship benefits from a strong psycho-emotional component and that it is the expression of deep love with a wish of procreation, the conditions of the sexual act become ideal: no inhibiting cerebral factor is an obstacle, the hypothalamic endocrine secretions occur freely and dopamine, the hormone of the pleasure, is distributed abundantly. The sexual act becomes the most beautiful exchange, combining tenderness and passion, as the aim of each lover is to reach, together, the height of pleasure and exaltation, i.e. orgasm. The latter will be an "ecstatic" moment[23] where the intensity of pleasure and happiness will generally be combined with a brief moment of psychic disconnection with a decreased vigilance. This phase is so complex, especially in women, that the majority of them describe orgasm like an "indefinable" state....due to the fact that when it occurs, most of the cerebral areas are activated and several neurotransmitters are involved.

The resolution phase, which will follow these moments of great intensity, will also be controlled by the endocrine system and will be readily extended, for the two lovers, by a restorative sleep.

[19]The rhythmic contractions of the uterine musculature are controlled by the vegetative system (utero-vaginal plexus).

[20]Cerebral NMR (B. Komisaruk 2011) has shown that, during orgasm, most of the cerebral zones are activated.

[21]The most important neurosecretions during the orgasmic phase are the dopamine, the oxytocin and the endogenous opioids.

[22]However, certain specialists of the animal world consider that female cats and apes can have orgasms.

[23]Certain authors think that the exaltation and the impression of detachment of the body, felt by hundreds women during orgasm, share common features with the ecstasy phenomena observed in great mystics.

16 Mutilations of the Clitoris

Although the purpose of our study is the anatomical analysis of the bulbo-clitoral organ, we must also mention the abominable mutilations which are undergone by certain young girls, at the level of the exteriorised parts of this organ and sometimes also of the vulva. Such mutilations are still practised today.[1] These mutilations are referred to by the general term of SGM (sexual genital mutilations).

These acts have appalling consequences and are true amputations as they involve removing and injuring unique and essential anatomical parts and thus depriving the women, who have undergone such operations, of an elementary function of sexuality: female pleasure.

16.1 The Different Types of Mutilation

According to the classification of R. Cook, established in 1995 by the WHO, four types of SGM can be observed:

Type 1: Excision of the clitoral prepuce with or without partial excision of the clitoris.[2]

Type 2: Excision of the prepuce and the clitoris with partial or total excision of the labia minora.

Type 3: Excision of the clitoris with partial or total excision of the labia minora and almost complete closure of the vulvar orifice.[3]

Type 4: Any other process, combining the excision of the clitoris, complex mutilations such as scarifications, application of corrosive substances (in the vagina in particular), stretching of the labia, etc.

The most frequent cases concern type 2 (80 % of the cases).

[1]Even today, the excision continues to wreak havoc; an example: 70 % of the Mauritanian women are victims (Unicef Report 2013).

[2]Type 1 is the sunna operation, sometimes called.... "symbolic excision".

[3]Type 3, still called the "pharaonic excision", is the most serious type of excision: It only leaves a narrow orifice for urine and menstrual flow.

16.2 Brief Historical Reminder

These mutilations have a very ancient origin (5,000 years B.C., on the borders of the Red Sea). Excisions seem to have mainly existed in Egypt due to the fact that certain texts, and especially papyrus from the Ptolemaic era, kept at the British Museum, mention "young girls who had not been cut". However, formal evidence is lacking (the external genitalia of female mommies, which were found in the North of Germany, between 1915 and 1920, showed scars which may be infibulations, but their conservation condition makes observations difficult to interpret).[4] What is certain is that excisions existed before the emergence of the Catholic religion and especially before that of Islam. Therefore, this practice is not a religious rite, but more of an ancestral, clan-type, societal and customary tradition, whose origin is related to an aberrant and totally erroneous concept (a woman can only become a real woman if her masculine attributes, represented by the clitoris, are removed) or to horrible historical bases (infibulation is considered as its origin, a way of preventing women from being raped by invaders).

Another pretext, as horrifying and at the origin of this terrible operation, was to prevent young women from having the slightest desire and from masturbating, so that they are true virgins when they get married and remain faithful... Many old texts refer to the practise of clitoridectomy: Herodotus, a well-known historian (484–420 B.C.), was the first to mention, in his great book, "The investigation", the existence of excisions in 700 B.C., which were performed by the Colchidians, the Egyptians, the Ethiopians, and the Phoenicians.

Such as recalled by Mr Wehenkel in his thesis, "infibulation, a very widespread custom in the Middle East, then spread to Arabia during the pre-Islamic period".

The great physicians of Antiquity: Paul from Egina, Albucasis and Avicenna refer to excision in their texts.

[4]It is entirely different for masculine circumcisions in Egypt, for which papyrus and engravings provide irrefutable proof.

V. Di Marino, H. Lepidi, *Anatomic Study of the Clitoris and the Bulbo-Clitoral Organ*,
DOI 10.1007/978-3-319-04894-9_16, © Springer International Publishing Switzerland 2014

Aetius, as for him, described a "method for cutting the nymph[5]" (in current terms, the clitoridectomy technique), a therapy which he considers as… usable in some cases of large clitorises (reported by De Graaf[6]). In the sixteenth century, Jean Riolan, in his "Discours sur les hermaphrodites" ("Speech on hermaphrodites"), and then Ambroise Paré, in his book entitled "Des monstres et prodiges" ("Monsters and prodigies"), also recommended clitoridectomy for treating certain cases of clitoral hypertrophy. It appears that many European physicians, as well as many parents, proposed, at that time, the amputation of the clitoris, as soon as an anomaly was observed in young girls… Thus, in spite of the major scientific discoveries made during the sixteenth century in relation to the anatomy of the clitoris and its role in woman's sexuality, this organ was going to be accused of all evils, to the point that during the following centuries and until the nineteenth century, clitoridectomy was going to spread in all Western Europe, as it was supposed to cure the majority of women's diseases. This technique spread to such an extent that an English physician I.B. Brown (referred to by E. Sheean) wrote, in 1866, a book entitled "On the curability of some forms of insanity, epilepsy, catalepsy and hysteria in females" and acquired celebrity while preaching that clitoridectomy is therapeutic (see Chap. 1, p. 11). Fortunately certain young gynaecologists were going to oppose the serious and absurd theories supported by Brown, until he was discredited. Worse still, in France, during the same period (second half of the nineteenth century), well-known physicians, such as P. Garnier, or the famous neurologist, P. P. Broca (who discovered one of the language areas) dared to recommend clitoridectomy to fight against onanism, i.e. masturbation which was regarded as a psychiatric disease.[7] Since the twentieth century and the scientific study of sexuality, these cruel operations, with unthinkable indications, have completely disappeared. Only a few exceptional cases of clitoroplasty[8] persist as they are necessary in repair surgery for malformations or transgender surgery. On the other hand, standard excision operations remain in many countries (especially in sub-Saharan Africa) and even in Europe, in emigrated populations, while, at the same time, this practice is against the law in all European countries, including France, and is punished severely. According to the figures of the WHO (World Health Organization): 100–140 million young girls and women in the world have undergone this type of mutilation. Three million young girls per year, in the world, are likely to undergo these types of operations. Many cases are still reported in 28 countries in Africa and several countries in Asia[9] and the Middle East, despite the legislations aiming to prohibit this practice, which have been adopted by 13 of these countries. It is necessary to also underline the combined efforts made by major organisations (ONU, WHO, UNICEF) to put an end to these "extremely serious personal injuries", even if progress is still slow. In France, the legislator has been very efficient and considers excisions as violence causing a permanent mutilation, an offence which is punishable by 10 years in prison and a fine of 150,000 euros.[10] This offence should be dissuasive. Alas, many excisions are still performed on this territory (these mutilations, due to their illegal character, are carried out in the most hazardous conditions with a frightening lack of hygiene). In some cases, they are performed with the assistance of healthcare providers, while the law states that all healthcare providers (including physicians) must report all cases of excision, even if they have not been performed yet. However, in this country, due to the high density of the emigrated African population (at least half of this population is in favour of excisions), 12–20,000 young girls are exposed to a risk of mutilation (the mutilation is carried out when the parents go back to their country for their holidays). It is, alas, the same in all countries with a high rate of African emigration.

16.3 Consequences Related to Excisions

As it can easily be imagined, the consequences are related not only to the amputations: removal of an essential part of the female body but also to all the iatrogenic pathologies induced by the mutilation itself.[11]

[5]Aetius calls the clitoris "the nymph" and the labia minora "the small wings". For the majority of authors from the same period, the term nymphs (plural) is only used for the labia minora.

[6]"These large clitorises create a lot of trouble for women, because while they are walking or doing their housework, they rub against their clothes, which tickles them so much that they get excited and throw themselves at men in a shameless manner…"

[7]The speeches made by psychiatrists, and especially child psychiatrists, are currently and very fortunately of a very different nature: Masturbation is physiological in children: discovering their body is normal for young girls, as well as for little boys, and allows them "to develop" correctly.

[8]At present, the exceptional clitoroplasties, indicated for treating certain cases of clitoral hypertrophies, in a context of hermaphroditism or female pseudo-hermaphroditism, are performed by specialised surgeons, who know how to dissect and respect the vasculo-nerve pedicles and preserve the distal part of the clitoris, which bears the corpuscles of pleasure.

[9]The Indonesian Ministry of Health is preoccupied by the increasing trend of female sexual mutilations performed by physicians (Source: WHO).

[10]If the victim is a minor, the excision is considered as a crime and is therefore punished by a penalty of 15 years of rigorous imprisonment (20 years if the culprit is a legitimate ascendant), with or without a banned entry into France for 5 years (French Penal Code).

[11]According to researchers from the INED, "one excised woman out of ten is hindered on a daily basis when urinating, walking or wearing certain clothes".

The psychological consequences are **post-traumatic stress**: the terrible vision of the excision operation, the memory of the violent and insupportable pain felt during and after the operation and the often significant bleeding which follows generally profoundly affect the imagination of a mutilated young girl and will haunt her memory and dreams, which have become nightmares, for a long time and perhaps forever. Thus are explained the difficulties related to sleeping, sleeping disorders, anxiety attacks, loss of appetite (or sometimes bulimia), feelings of permanent fear, anxiety and deep sadness felt by mutilated women (who feel that they are no longer normal women).

Sadly, the **impact on sexuality** is only too obvious due to the fact that, during a mutilation, the organ of female pleasure is removed or damaged. Despite the difficulties encountered when investigating this matter (women do not want to reveal the problems related to their sex life due to an understandable modesty), a few studies have been conducted, especially in Egypt. They show that the extent to which the sex life of mutilated women is affected is proportional to the importance of the mutilation. The most common problems are frigidity, with the impossibility or difficulties in feeling sexual pleasure, and dyspareunia (dyspareunia with intromission[12] and deep dyspareunia).

The **consequences for the health** of an excised woman are multiple:

They can be immediate: collapse, haemorrhages and even death.

Most consequences are secondary and affect the health of excised women during their entire life:

- Urinary complications (repetitive infections, retentions, pyelonephritis which can cause renal failure).
- Genital complications (perineal cuts, vulvar keloids, neuromas, fistulas, chronic pelvic algia).

Obstetrical consequences are frequent and may alone represent a major problem for public health as they compromise childbirth by vaginal delivery and generate a significant neonatal mortality. During a pregnancy, obstetrical monitoring is complicated due to pains, to difficulties or even to the impossibility of performing vaginal examinations. Cases of dystocia are common. The surveys conducted by the WHO (World Health Organization) have shown that childbirths by excised women "have a higher risk of being complicated: caesareans, haemorrhaging of the post-partum, episiotomy, spontaneous tears and extended hospital stays". As for the children born to excised women, "they run a greater risk of dying during the birth or undergoing immediate resuscitation following their birth".

16.4 Clitoral Reconstruction

Such as already mentioned, the sensory corpuscles destroyed during an amputation cannot be recovered through performing an operation. Nevertheless, repair surgery of the clitoris can be used, in most cases, to recover a satisfactory morphological aspect and appease, at least partially, the disturbances affecting female personality, generated by the genital mutilation. A few rare teams of French surgeons (P. Foldes in particular) have the merit of studying clitoris reconstruction techniques through transposition (thesis written by Mr Mansour-Hugues).

The operation is simple and well codified: It consists in resecting integuments and scar tissue, freeing what remains of the clitoral body and releasing the suspensory ligament from its pubic attachments. The "residual" clitoris can then be lowered and the "neo-glans" can be ideally placed, slightly below the lower edge of the pubic symphysis, while protruding "by at least 5 mm in front of the vulvar plane". The complications related to this operation are rare and not serious:

Although the aesthetical result is considered as satisfactory by 2/3 of the patients, the functional results are random (no orgasm), most patients (73 %) report an improvement in their sex life.[13] As for their partner, he is generally satisfied with the operation!

16.5 A New Risk of Mutilation

We do not want to end this chapter without mentioning a major risk which, at present, threatens women who have undergone vulvar surgery, in which aesthetics and mutilation are mixed up.

Sadly, this is the consequence of the mad race towards the utopia of an ideal body (so-called ideal body) that our partners are reminded of, on a daily basis, by model icons. Plastic surgery has invaded the human body, especially the female body, and the external genitalia are also concerned. This is how surgery, referred to as sexual surgery, was born and is being increasingly developed in order to fill the pockets of unscrupulous physicians thanks to the excessive naivety of many women. Thus, the clitoral hood is to be resected so that the glans is more accessible, a labium minus is to be resected as it is too long, the edges of the vulva are to be corrected so

[12]Dyspareunia with intromission is a major problem in the case of a type 3 excision and we can imagine the atrocity of the first sexual relationships of infibulated women.

Deep dyspareunia is often related to vestibular or vaginal stenosis when corrosive substances have been inserted inside the vestibule or the vaginal tract.

[13]It is probable that women, who have undergone clitoral reconstruction, have other satisfaction motives of a psychological nature. Through the reconstruction, they can forget about the trauma resulting from a horrible mutilation and rediscover their female identity.

that they are perfectly symmetrical with respect to one another, etc., and we are forgetting that these operations are mutilations, which are freely agreed, perfectly accepted and very expensive, whereas they cause the loss of corpuscles (see Chap. 9, p. 90 and 92) and nerves, which are essential for a happy sex life. More serious still, we are reverting back to the absurd ideas of certain Greek physicians, who worked during the era of the emperor Trajan and who considered that the vulva was a deform shameful part and that it had to be resected. And some women, in all this absurdness, will even ask a surgeon to more or less completely resect their labia minora (total or subtotal nymphectomy, nymphoplasty) and to make the labia majora flatter. This is the "increasing trend of vulvar plastic surgery", which is destructive and which, 1 day, might even affect the clitoris, still relatively preserved[14] by surgeons working in the above-mentioned speciality.

[14]We have used the term "relatively", as the poor glans has already been exposed to a certain trauma during piercings which obviously destroy nervous ramifications and corpuscles and thus sadly damage the lovely jewel that nature has given to women (in addition to the infections and vascular risks which may affect this fragile organ and cause discomfort to women with a piercing)!

17 The Clitoris in Art

17.1 General

Such as said by Matisse[1]*: "Art is the reflection of the soul; all the means employed for expressing it, are good: poetry, painting, sculpture, music etc....".*

Like all the parts of the female body, the vulva has always inspired artists from the most various horizons. But what about the end of the clitoris, the only visible part of the bulbo-clitoral organ? How has it been perceived, understood, described or represented by the artists in the various fields of art[2]? We will attempt to provide a brief insight while knowing that some of our readers will happily pursue this overview.

17.1.1 The Clitoris and Literature

17.1.1.1 The Clitoris and Anatomical Literature

It is obviously the first observers who provided the most spontaneous and naïve, but sometimes also the most judicious, considerations concerning this organ. For instance, we can quote Aristotle: "Women feel pleasure when they are touched in the same place, where men also feel pleasure" (see Chap. 1).

We can also quote the good advice provided by Ovid in his famous booklet: The art to loving, book II: "When you find a place where a woman likes to be caressed, your sense of decency should not prevent you from caressing it!".

A few centuries later (sixteenth century), it is the enthusiastic and exalted thoughts of the anatomists, who discovered the clitoris, which marked literature. We will therefore recall the lines (written in Latin) that Realdo Colombo devoted, in his book "De Re Anatomica" (Fig. 1.3), to the description of what he believed to have discovered and whose translation is provided in the following section: "This protuberance emerges from the uterus, close to the opening, which is called 'mouth of the matrix' … is the site female delight … As soon as women are sexually aroused, and not only when this part is vigorously rubbed by a penis, but also as soon as it is touched with a finger, sperms flows therefrom more rapidly than air, due to pleasure, even if they don't want it….. Therefore, considering that nobody has observed this protuberance and its function, I can name it as I discovered it, and it will be called **Amor Veneris**".

This enthusiasm was not shared by all and it is with a certain astonishment that we can read the following text written by Ambroise Paré, an exceptional anatomist and surgeon (he invented arterial ligations and is the founder of modern surgery), in the second book describing his work (written in old French): "Les récents anatomistes, comme Colombus & Fallopius, ont fait mention d'une particule qui est tout en haut des parties honteuses mesme sus le conduit de l'urine…. Fallopius luy accomode le nom Grec cleitoris, duquel est dérivé le verbe infâme cleitorizein. Et parce que la dicte partie est fort obcene, ie renvoye le lecteur à Colombus & Fallopius" ("Recent anatomists, such as Colombus & Fallopius, mentioned a particle, which is located right at the top of the shameful parts mesme sus the urinary duct….Fallopius has given it the Greek name cleitoris, from which has been derived the infamous verb cleitorizein. And because the said part is significantly obscene, i.e. the reader is referred to Colombus & Fallopius"). It is to be believed that the author had certain problems in relation to the female sexual organ.

During the seventeenth century, the well-known anatomist, Jean Riolan, physician to Marie de Medicis and Louis XIII, expressed again, in chosen terms, all the magic of the clitoris, in his book "De l'Anthropographie" ("About Anthropography") written in French: "On tient que c'est en cette partie où l'amour fait la résidence, laquelle chatouillée bien à propos est capable d'en allumer le feu dans les cœurs les plus refroidis" ("It is considered that it is in this part that

[1] Henri Matisse (1869–1954), a famous French painter and sculptor. He was the leader of the Fauvism period.

[2] The twentieth century has added four arts to the five major arts of the nineteenth century (literature/poetry, architecture, painting/drawing, music, sculpture): performing arts (theatre, dance, mime, circus), (cinema), (radio, TV, photo/comic strips). The last art (comic strips) has only recently been included with its eight older sisters.

V. Di Marino, H. Lepidi, *Anatomic Study of the Clitoris and the Bulbo-Clitoral Organ*,
DOI 10.1007/978-3-319-04894-9_17,

love takes up residence, which when properly tickled, is capable of lighting fire in the coldest hearts"). However, his contemporary, the wise anatomist, Regnier De Graaf, pointed out in his "Histoire anatomique des parties genitales" ("Anatomical history of the genitals…") (also in old French): "L'usage du clitoris est de tenir en état la partie honteuse, de l'empêcher de tomber et de réveiller l'appétit amoureux par le sentiment exquis et délicat de son gland qui donne un plaisir si tendre et si charmant aux femmes qu'elles croient avoir bon marché de l'acheter au prix d'une grossesse de neuf mois, d'un accouchement douloureux et souvent mortel et des soins de l'éducation des enfants". ("The clitoris is to maintain the shameful parts in condition, prevent them from falling and awaken a loving appetite through the exquisite and delicate feeling of its glans, which provides a pleasure so tender and so charming for women, that they believe that it is cheap at the price of a 9 month pregnancy, a painful and often fatal childbirth and all the care involved in child education").

It was necessary to wait, as we saw in the History chapter (Chap. 1), for the nineteenth century when GL. Kobelt, a German anatomist, finally provided the most accurate and careful description of the external female genitalia in the second part "De l'appareil du sens génital chez la femelle ("From the apparatus in the genital sense in women") in his remarkable book concerning male and female sexual organs, written in German.

Beyond the woman's anatomy, Kobelt studied the compared anatomy of the clitoris, observed on animal females and, furthermore, analysed the physiology of human intercourse with interesting considerations on the "hydraulics of the erectile bodies of women…", the "expulsive contractions of the compressor of the bulb"… the action of the "sexual core", concepts that have been confirmed by modern studies, based on sophisticated exploration techniques.

17.1.1.2 The Clitoris and General Literature

Extra-medical literature took interest in the clitoris as soon as it was better known thanks to the work of the anatomists. The clitoris therefore appears in libertine texts as from the seventeenth century, for the pleasure of many male readers and a few rare female readers, who managed to access them!

Among the many texts referring to the clitoris, we want to recall, during the eighteenth century in the memoirs of Casanova ("History of my life"), the embarrassment and apprehension of this great seducer when he falls in love with a so-called castrato, with an exceptional voice, the so-called Bellino, who will prove (to the great relief of Casanova, who is viscerally heterosexual) to be in fact, a false-castrato and a true woman, named Theresa, who has a beautiful voice… a woman's voice, but also a hypertrophic clitoris (undoubtedly hermaphrodite).

The Marquis de Sade, his contemporary, makes significant use of the terminology "clitoris" in his books, especially those describing the busy lives of his heroine, Justine, and then of her sister, Juliette. He positions the clitoris at the top of the "peristyles of virtue". However, the book includes a certain number of scandalous misfortunes described in elegant terms, which show a remarkable use of the French language.

Between the nineteenth century and twentieth century, erotic literature continued to exist (Les 11000 verges) (e.g. "The 11000 penises" written by G. Apollinaire) but became less popular due to the appearance and development of pornography. The clitoris is increasingly referred to and described, but with a lesser global literary quality, due to the fact that most authors (moreover, generally women) mostly think about the sales of their books in station halls rather than about winning a literary prize.

But what happened during the twenty-first century? Which books mention the clitoris or make it their main topic?

- *Firstly, all of the literary essays*, aim to promote this organ, which has been forgotten for too long, in order to "rehabilitate it", to explain its role in the blossoming of a woman, its function in a personal sexual life, its importance in the success of life as a couple, to recall that it is a constant source of female pleasure and to wonder how society accepts this reality.

  We can quote, in these fields, some interesting French books[3]: "La revanche du Clitoris" ("The revenge of the clitoris") (M. Mazurette and D. Mascaret) with an evocative title "Un petit bout de Bonheur" ("A small part of happiness" (R. Pujol), with a malicious subtitle written by the author "Small handbook on clitology", which recalls that female pleasure does not have an age limit, "La caresse de Vénus" "Venus' caress" (G. Leleu), subtitled "Le clitoris, mode d'emploi" ("User manual for the clitoris")… according to the actual accuracy of the author (published in 1997).

  We can quote, among the essays written in English,[4] "The clitoral truth" by the American author, R. Chalker, a small booklet with an imaged title "Tip of the iceberg: a book

---

3

- La revanche du clitoris. M. Mazurette et D. Mascaret. Ed. La Musardine, 2008
- Un petit bout de bonheur. Petit manuel de clitologie. R. Pujol. Ed. J.C. Gawsewitch, 2007
- La caresse de Vénus. Le clitoris, mode d'emploi. G. Leleu. Ed. Leduc.s, 2009.

4

- The clitoral truth : The secret of your fingertips. R. Chalker. Publisher : Seven Stories Press, 2000
- The clitourist; A Guide to one of the Hottest Spots on Earth. K. Salmansohn; illustrated by Tr. Krauss, Ed. Universe, 2002
- Tip of the iceberg : a book about clitoris. L. Szumowski. Publisher Zoo-Mouse-Key Press, 2012
- An intimate Geography. N. Angier. Publisher: Houghton Mifflin Harcourt, 1999.

about the clitoris" by L. Szumowski, "The clitourist: A Guide to One of the Hottest Spots on Earth" and the very interesting chapter that N. Angier, a scientific journalist from the New York Times, devoted to the clitoris in her book: "Intimate Geography".

Furthermore, a few good quality essays concern female masturbation. The most recently published, "Nouvel éloge de la masturbation" ("New praise for masturbation"),[5] written by the psychiatrist and sexologist Ph. Brenot, not only retraces the history of this natural act (wrongly considered, over the centuries, as taboo) but also recalls the benefits of autoerotism, for the two sexes.

- Novels and short stories
  - Erotic: such as "Clitomotrice" (S. Jabès) or "la tentation du clitoris" ("The temptation of the clitoris") (R. Jauffret) or even "El clitoris of Camille" ("Camille's clitoris"), a book by D. Medrano, written in Spanish[6] or
  - Historic: (see the Sect. 17.1.6)

Other books only mention the clitoris here and there, maybe to attract the interest of the readers for their books, by stimulating their erotic imagination. For instance, we can mention a known and acclaimed French writer, Michel Houellebecq.[7] In "Les particules élémentaires" ("The elementary particles"), the heroine, Christiane, following a lovemaking session, clarifies a few points for her partner, Bruno, about the sensitive corpuscles of the clitoris. In another of his novels, "Plateforme" ("Platform"), the same author shows his hero, Michel, who is observing the clitoris of his partner Valerie and realising that until then, he had never looked at it. And many other examples are provided in many contemporary novels, to the extent that a new type of erotic literary style is emerging, with numerous situations involving the clitoris and its caresses.

17.1.1.3 The Clitoris and Poetry

It is really in this art that the clitoris thrives. The clitoris, itself, is poetry as it "expresses softness, pleasure, joy of living and pleasure". Moreover, the morphology of the glans overhanging the neighbouring labia minora, easily calls for a comparison with various wonders of nature, which poets likes to celebrate: flowers (Fig. 1.3), berries, jewels, gemstones, and even the facial appendices of certain exotic birds.[8] However, poets introduced the term "clitoris" into their books only gradually so that it is especially as from the eighteenth century that we observed the most beautiful parts of anthology with the introduction of this remarkable feminine attribute. It has been given various names: "rosebud, bead of love, crimson bilberry, shining ruby, living coral, pink hermit under an animated hood etc.". It should also be noted that although the clitoris is the subject of one or two verses of a poem (e.g. in "l'épouseur de famille" ("the Family spouse") from the collection of libertine poems by Th. Gautier, in "Poésies libres" ("Free poems") by Guy de Maupassant or in "poèmes à Lou" ("Poems for Lou") by G. Apollinaire or even in "parallèlement, les amies" ("In parallel, friends") by P. Verlaine), it is rarely the title and the subject of the entire poem. This shows how precious poems entitled "the clitoris" are, especially as well-known poets have written them. We can especially quote "le clitoris" ("the clitoris") by P. Louÿs, which is one of the most erotic poems written by this author, in which he refers to this organ as the "pistil chair" ("flesh stigma"), "rubis mystérieux" ("mysterious ruby") or "bouton de rose" ("rose bud"). We can also refer to three more famous clitorises: "Le clitoris" ("The clitoris") by G. Apollinaire, that of F. Arrabal and that of H. Cantel, from which we want to quote four verses (the first two and the last), which are particularly beautiful, for the reader's benefit:

Le clitoris en fleur que jalousent les roses The budding clitoris which makes the roses jealous

Aspire sous la robe, à l'invincible amant Aspires under the dress, to the invincible lover

....

Et le désir en flamme ouvre amoureusement And the enflamed desire lovingly opens

Le clitoris en fleur qui jalouse les roses The budding clitoris which is jealous of the roses

Our twenty-first century also includes numerous unknown poets, who have written poems about the clitoris. We discovered one poem, written by an anonymous author, which we found randomly during our research (Fig. 17.1).

17.1.2 The Clitoris in Painting and Drawing

17.1.2.1 The Clitoris in Painting

It is not necessary to specify that the naked female body is, since the beginnings of painting, and in particular of oil painting, one of the favourite subjects of successive artists. However, even if representations of the "hypogastric region", of the harmonious curves of the mons Venus, of the pubic triangle and of the hips are admirable and numerous, paintings

[5] Nouvel éloge de la masturbation. Ph. Brenot. Ed. L'Esprit du temps, 2013.

[6]
- Clitomotrice. S. Jabes. Ed. J.C. Lattès, 2005
- La tentation du clitoris (The temptation of the clitoris). R. Jauffret. Ed. Publie.net, 2010
- El clitoris of Camille (Camille's clitoris). D. Medrano, Leading Seix Baral, 2006.

[7]
- Les particules élémentaires (The elementary particles). Mr. Houellebecq. Ed. Flammarion, 1998
- Plateforme (Platform): Mr. Houellebecq. Ed. Flammarion, 2001.

[8] Such the astonishing Volturius amplus "landica".

Fig. 17.1 *Le Phalaenopsis*
Tes cuisses lègères se sont doucement écartées,
offrant à mes yeux enivrés, ton petit nid de volupté.
Ineffable bonheur
de voir si belle fleur,
divin phalaenopsis
qu'est ton clitoris

showing a woman lying down with her thighs apart and her perineum exposed are rare.

In addition, the vulva is more or less hidden inside abundant hair. Most people have already seen at least one picture of the very beautiful work of Gustave Courbet: "L'origine du monde" ("The origin of the world") and most people know about the vicissitudes of this painting, whose successive owners have preferred to dissimulate from a non-specialised public.

In the same style, but more recent, we can mention a painting by Ch. Camoin: "La saltimbanque au repos" ("The resting travelling female acrobat"), or a few naked female portraits by Lucian Freud (including that of Kate Moss and another person, directly inspired by Courbet), which are very realistic, with spread thighs and no spared details. It is, however, the painting by Courbet, which remains the most explicit (superb labia majora bordering an admirably painted vulvar cleft, from which directly emerge the magnificent labia minora coloured with a magnificent incarnate). However, the vulvar cleft is not open and the clitoris remains invisible.

A few painters have represented more close-up views of the vulva without showing the clitoral anatomy. This is the case of the erotic watercolours by Rodin, paintings by Picasso (in which the vulva is generally represented by a lengthened ellipse bordered by coarsely positioned hairs) or certain recent paintings, in which the vulva is simply represented as a coloured biconvex trace, painted on a canvas, or even as a longitudinal notch in the canvas, which has been coloured beforehand (this appeals to the imagination of the viewer, who must rebuild in his brain… his own virtual painting!).

Lastly, we only know of one famous artist, who has painted, on two occasions, a vulva with a clitoris. It is E. Schiele, who successively painted, in 1910, "Mâdchen in schwarzem kleid put geispreizten beinen", and in 1911, "Die traumbeschaute", gouache paintings showing women, who are spreading their thighs and revealing their vulva and clitoris (a good anatomical representation).

Overall, painters have preferred representations of masturbation, of the clitoral caress and of autoerotism rather than direct paintings of the clitoris. As such, their paintings are part of the representations of "la femme dans tous ses états" ("the woman in all her states"). The most representative painting of this type is that by the famous Venetian painter, Tiziano Vecelli, referred to as Titian: Venus d' Urbino (The Venus of Urbino), which nowadays is hanging in the splendid Uffizi Gallery. This Venus is awake (as opposed to the Venus de Giorgone, undoubtedly painted by the same Titian and who is represented asleep). The painting was ordered from the painter by the Duke of Urbino, Guidobaldo della Rovere, who wanted to buy it for his young wife aged only 14 years old and whom he had married 4 years earlier. This painting represents, for a few rare art critics, a young woman shyly hiding her sex with her left hand. However, for most other specialists, her action can be mistaken, she is masturbating[9] and is looking at the viewer to excite or invite him. This second assumption is the most plausible due to the fact that, such as underlined by Rona Goffen,[10] a historian, during the sixteenth century, masturbation is recommended for women by physicians and the Church to ensure that their arousal is maximum during intercourse (this state of pleasure is the absolute condition to ensure that intercourse has a fertilising outcome). And it is undoubtedly what Guidobaldo wished for, as he had just barely started having a sexual relationship with his young wife!

[9]On this topic, read the excellent and diverting study of this painting in a book written by D. Level: "On n'y voit rien: Descriptions (Nothing there to see: Descriptions)

Denoël Ed. 2005, Chap. La femme dans le coffre (The woman in the trunk).

[10]R. Goffen :" Titian's Venus of Urbino "Publisher R. Goffen, Cambridge 1997".

Fig. 17.2 "A soft moment". Drawing of a Venus masturbating. Charcoal, sanguine and white chalk drawing; hardback grey paper 21 × 29.7 cm

17.1.2.2 The Clitoris in Drawing

Female masturbation is also represented in drawings, which provides the draughtsman with the opportunity of creating a doubly erotic representation, as a drawing of the naked female body, in a lascivious attitude, is combined with a drawing of clitoral caresses and an impression of great happiness on the face of the woman performing this exercise. Such drawings are rare but are always very beautiful. It is the case of the "nu au fauteuil" ("The naked woman in the armchair") by Andre Derain, which represents a young woman, who has collapsed in a chair, with her lower limbs spread apart, and who is masturbating, or even the impressive drawing by E. Schiele, with an non-ambiguous title "Masturbierendes Madchen" or also the recent drawing "A soft moment" (Fig. 17.2).

However, drawing has mainly been used to support the descriptions of the external genital apparatus in anatomy books, and we are in awe of the excellence, quality and accuracy of the illustrations that can be found. We can especially refer to the drawings by De Lairesse, which were used for the illustrations of at least three books written by well-known anatomists and from which we reproduced the chart representing clitorises and labia minora (Fig. 1.4). More closer to us, the illustrations from the chapter entitled "A woman's external genitals" in the book by J.M. Bourgery and N.H. Jacob[11] leave us speechless.

17.1.2.3 The Clitoris in Comic Strips, Mangas and Other Styles

Comic strips, such as mangas, are another means employed by artists to represent reality. Sex is very present as the comic strips, which address this topic, sale well and are widely accessible as free reading in large-surface bookshops and interest many teenagers, who use them as a way of furthering their sexual education and, especially, their anatomical knowledge of the opposite sex.

As anatomists, we must acknowledge the fact that the designers of erotic comic strips, mangas and Chinese or

[11] J.M. Bourgery and N.H. Jacob: "Traité complet de l'anatomie de l'homme" ("Complete treatise on man's anatomy"); eighth volume, Chap. "Woman's sexual organs", Ed. L. Guerin, 1866–1867.

Indian erotic drawings know how to perfectly represent the details of the vulva and that the clitoris itself is sometimes illustrated accurately.

17.1.3 The Clitoris and Sculpture

The same comments made in relation to painting can also be made for sculpture. Since ancient times, genitals have been well represented on sculptures, statues and low reliefs, as our sculptor ancestors perfected their knowledge of the naked body. However, while male genitals are perfectly visible, female genitals are only represented by a closed vulvar cleft, bordered by two labia majora, which are joined and hardly distinct, so that the clitoris is not visible (which corresponds to reality) and that the labia minora are never highlighted by the artist.

This "way of not seeing or hiding the woman's genitals" is starting to change and progressively female genitals are being revealed.

Thus, very faithful representations of the labia minora and the clitoris have been found in contemporary exhibitions. We can recall, for example, the first exhibition, "Mapping the studio", of the Fr. Pinault Foundation, in the exceptional site of the Dogana in Venice, in 2009, where we could admire the extremely controversial work of Paul McCarthy, an American artist "who has no limits", which is both brilliant and creative, but also often obscene or scatological: miniature models of female human bodies, with a very good flesh colour imitation (with a deformed humanoid head), in a sitting or gynaecological position, thighs spread apart, thus exposing their vulvas, with an impressive reality and an exceptional anatomical accuracy. It is true that this artist uses plastic mouldings on the various parts of the body to obtain such impressive facsimile effects.[12]

Another example, The "Great Wall of Vagina", completed by an English artist, Jamie McCartney, a series of low reliefs, which significantly shocked the English public during his exhibition at the Festival of Brighton in 2011. The work of this sculptor includes 10 panels of 40 plaster mouldings representing vulvas (i.e. 400 overall) from women of 18–76 years old (each panel measures 9 m long). As an anatomist, three remarks are essential:

The name of the sculpture shows that the artist mixes up the vulva and the vagina and, therefore, has introduced an erroneous terminology in the vocabulary of his visitors. A simple glance on the panels obviously shows the incredible diversity of vulvar morphologies (well-known of gynaecologists and anatomists). Lastly, an attentive examination shows that very few clitorises are visible when the labia are drawn apart, which is the case of the vulvas reproduced on the panels.

The sculptures, which we have referred to as examples, are acknowledged as works of art, even if they are based on moulding techniques rather than on sculpture techniques. We cannot end this paragraph without recalling the numerous sculptures representing the genital apparatus, which can be found in museums and anatomy laboratories with a glorious past, and the plaster mouldings painted with anatomical care and waxes, especially the mouldings from the museum of Specola in Florence,[13] where all the anatomical components (from the clitoris and the labia minora) are represented accurately and which deservedly earn the admiration of students, of their masters and of the public, who are lucky enough to see them. In this museum, we can especially observe the anatomical preparations of vulvar resections, of excellent accuracy, an osseous pelvis with a dissected clitoris, in position, and vulvas on entire bodies, with thighs spread apart, which seem so true that they could be mixed up with the living if they were not abnormally open.

17.1.4 The Clitoris and Photography: The Clitoris in Radio and Television

Radio and television are generally classified as photographic art. We however prefer to isolate them from this type of art because their relationships with the "clitoris" topic are different.

17.1.4.1 The Clitoris and Photography

Although there are numerous photographs of vulvas on the Internet, in a context of sexual education or pornography, specific pictures of the clitoris are infinitely rarer. If we consider artistic photographs of the vulva, i.e. photographs made by an expert or well-known photographer, in a purely artistic aim, to be included in an art book or to be shown in an exhibition, for instance, the number is much smaller. However, if we are searching for artistic photographs of the clitoris, the number is even smaller. It should be said that taking such photographs is certainly difficult, that it requires macrophotography techniques, that the lighting of the area to be reproduced is difficult to achieve and that, at the very least, a certain intimacy is necessary between the photographer and his model.

[12] We would have readily acquired a copy of one of these pelvises for the anatomical education of our future doctors if their cost was not prohibitive, such as most of the work of non-standard contemporary artists.

[13] The Museum of Specola in Florence was founded in 1775 by Pierre-Léopold of Habsbourg-Lorraine, Grand Duke of Tuscany, from a collection belonging to the Medicis family, whose successive generations were great lovers of art and science. The collection was continuously increased with the addition, to the Museum, of various departments (botanical and the like) and of an institute of wax moulding and thanks to the remarkable work of the Masters of wax moulding, including Gaetano Giulio Zumbo, from the famous school of anatomy in Bologna.

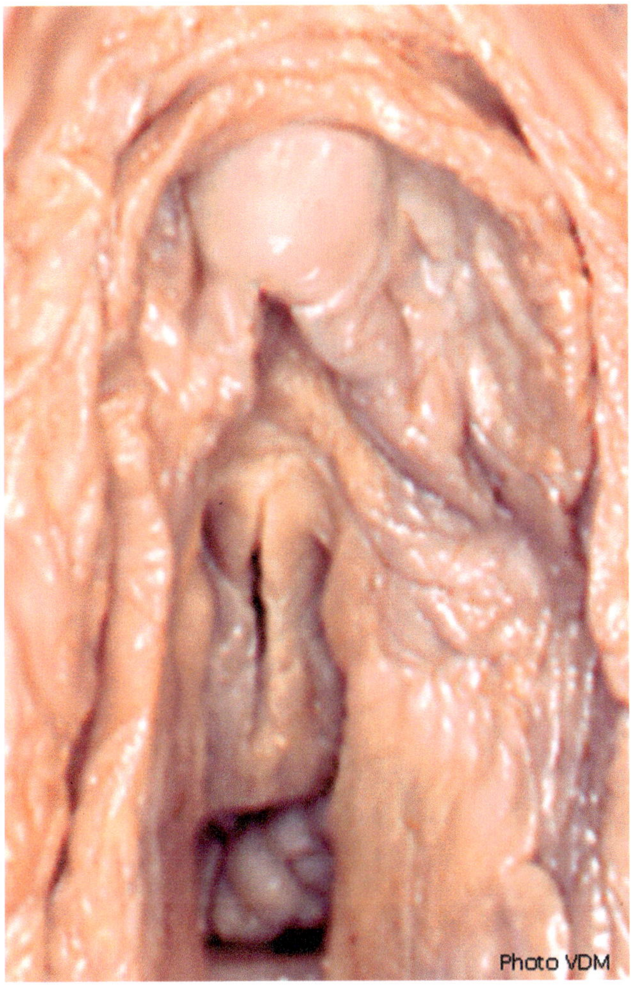

Fig. 17.3 Photography of the glans clitoridis, at the anterior end of the vulva, overhanging the external urethral orifice and the introïtus. Box Minolta XG9, macro lens 1200 AF, 50 mm, Minolta, diaphragm 22, speed 60, ring flash Minolta, Film fujichrome, 50 asa

Moreover, some good quality photographs are the work of female photographers, who have taken photographs of their own intimacy (in a view from above). The rarity of such photographs makes them all the more interesting. Whether they are taken in black and white or in colour (Fig. 17.3), with natural or artificial lighting, outside or beneath a thin layer of water, from the top or laterally, the shots that we have seen are exceptional and show the incomparable beauty of the glans covered by its hood and its relationship with the labia minora . We must mention, in the surprising field of contemporary art, the three books of the photographer, Frannie Adams, "Pussy Portraits", in which this artist shows "portraits from the top and the bottom" (i.e. photos of frontal views of the face and vulva) of young women having accepted this….double identity photographs. It should be noted that each photograph representing a face is placed next to the vulvar photograph enlarged to the same format as the facial photograph. This approach provides sufficiently large vulvar photographs in which the clitoris can be seen whenever it is not covered by its hood.

We must also mention an interesting essay (which sadly is unavailable), which takes into account the multiplanar structure of the clitoris by means of 3D photos: "The clitoris—essays on female genitals" (with four View-Master real in 3D) by Th. P. Lowry and Th. Sn. Lowry.[14]

17.1.4.2 The Clitoris in Radio and Television

In spite of their immense possibilities for disseminating knowledge, artistic as well as scientific, the radio and television have been more timid than photography in their approach of the clitoris. The radio was the first to dare broach this subject, in the form of answers provided by doctors to questions asked by listeners, during programmes primarily dedicated to stay-at-home wives, during hours when their children are at school and their husbands are at work. French television, which for a certain time was embarrassed and… timid, finally programmed, in 2004, a documentary called "Le clitoris, ce cher inconnu"[15] ("The clitoris, this dear unknown"), which was broadcasted during the Thema evening "Women's sex". This good programme, which is well documented from a scientific viewpoint, with a schematic artistic support of excellent performance, was a great success. Sadly, as far as we know, no similar programme has been made since then, which is regrettable. Teenagers searching for knowledge on sex, too often fall back on what they discover on pornographic websites and this is very bad for the development of their sexuality and their personality. Such as repeated by Professor Israel Nisand, a gynaecologist-obstetrician of the Faculty of Strasbourg, during his interventions in the educational field, pornography, by reducing the sexual act to a technique "is a treatise on virility, which promotes penetration and creates a prison, in which women are subordinated to the male desire, which will inevitably have consequences on the capacity of young people to develop a stable loving relationship in the future…"

17.1.5 The Clitoris and Cinema

It is obviously in pornography that the external female apparatus has been filmed the most. However, the clitoral views are generally ignored in favour of penetration.

Nevertheless it is also thanks to the cinema that sexual education films have been created and that scientific studies could be conducted on the erection and especially on the

[14]"The clitoris—essays on female genitals—with real 4 view masters in 3D". Th. P. Lowry and T. Sn. Lowry, Published Warren H. Green, 1976.

[15]"The clitoris, this dear unknown", a documentary by Mr. Dominici, V. Moszinski and St Firmin, 2003, co-produced by Cats and Dogs Films, Sylicone and Arte France.

clitoris, on the movements of the penis and the clitoris itself and finally on intercourse.

It is also the cinema which made it possible to broadcast the history (fictionalised) of the "discovery" of the clitoris by R. Colombo (see Chap. 1), by bringing to the screen[16] the book written by F. Andahazi: "L'anatomiste" ("The anatomist").

It should also be recalled that a few mainstream films dared show or evoke female masturbation, a topic which is still taboo,[17] even nowadays. Very recently, a film by a French filmmaker, Fr. Ozon, "Jeune et jolie" ("Young and pretty"), showed a very erotic scene where you can see the brother of the heroine, who is entering, by a slightly open door, into his sister's room and is shocked to discover her back view while she is becoming aroused by means of a cushion.

We can also mention, for example, a scene in the film "Le silence" ("The silence") by the regretted I. Bergman, a scene in "Elles" ("They") by M. Szumowska, a scene in "une femme coréenne" ("a Korean woman") by I. Sang-soo or a fugacious scene in Black Swann by D. Aronofski. Lastly, we will refer to the "Antichrist" by Lars von Trier, a film including a scene involving an extremely testing situation, during which the heroine, who is desperate, is sinking into madness and wants to end everything which represents femininity in her body (which she considers as the cause of the drama in her life), mutilates her clitoris…

17.1.6 The Clitoris in History

Such as observed in the "History" chapter (Chap. 1) of this book, the history of the clitoris is based on the history of anatomy, as well as on the history of medicine. It is therefore not surprising that the clitoris has inspired the thoughts and stories of the historians.

Among the historical books referring to the clitoris, we will especially recall the recent book written by Jean-Claude Piquard, "La fabuleuse histoire du clitoris" ("The fabulous history of the clitoris"),[18] whose title already enables to foresee the long and beautiful history of the visible part of the bulbo-clitoridal organ through the centuries, a hardly believable, epic tale. Two novels must also be mentioned as they provide fictionalised versions of the notable phases of the history of the clitoris:

The first, "l'Anatomiste" ("The Anatomist")[19] (by F. Andahazi), which tells the exceptional story of the life of Mr Colombo, who discovered the clitoris. Andahazi (who is a psychiatrist) won a prize for this book as best novel, which was awarded by the Fortabat Foundation in Argentina. However, the book was withdrawn from the market for pornography (and underwent censure, such as Colombo, the hero of the book, who was exposed, four centuries and half earlier, to the anger of the Inquisition for the description he made of the clitoris in his book "De Re Anatomica", published in 1558). However, since this eventful beginning, the novel has been translated into 15 languages and is considered, in the United States, as a bestseller.

The second, "les 200 clitoris de Marie Bonaparte" ("The 200 clitorises of Marie Bonaparte")[20] concerns more recent facts: Marie Bonaparte, the granddaughter of a nephew of the emperor, wife of Prince George of Greece, a student and disciple of Freud, was convinced that her frigidity was due to an excessive distance (CM) between the glans clitoridis and the urinary meatus (external orifice of the urethra). Convinced of the veracity of this theory, she recruited 200 women in order to measure the said distance and to check if their ability to feel pleasure is connected to a smaller CM length. She published the results of the study under the pseudonym of A.E. Narjani and was even operated on (sadly, she did not achieve the expected results) in order to shorten her CM distance. It is this amazing history, which is embellished by the romantic and odd character of Marie and the disconcerting friendship, which she had with Freud, that the author, A. Lemel, describes in this enthralling novel.

17.1.7 The Clitoris and Music

Although music is nice to listen to during lovemaking, as far as we know, there is no music sheet entitled "Ode to the clitoris". On the other hand, songs, which are closely related to poetry, have sometimes contributed to celebrating the clitoris and the pleasure it provides. This is the case of a song by Léo Ferré "Le faux poète" ("The false poet") written in 1986, in which the author quotes the clitoris in one of the verses. However, in most cases, it is student's songs and songs sung in guardrooms that refer to the clitoris, and most "carabins" can recall singing, during heavy drinking sessions, the famous song "Le plaisir des Dieux"

[16]"The anatomist", a film produced in 2011 by G. Tagliavini, an Argentinian screenwriter.

[17]This is certainly due to the fact that masturbation is an intimate process and is not suited for the cinema with many viewers, except in the case of pornography films.

[18]"La fabuleuse histoire du clitoris" ("The fabulous history of the clitoris") Essay, J-C Piquard, Ed. Blanche, 2012.

[19]"L'anatomiste" ("The anatomist"). F. Andahazi, Edit. R. Laffont (1998)

Note: published in Agentina in 1997.

[20]"Les 200 clitoris de Marie Bonaparte" ("The 200 clitorises of Marie Bonaparte") A. Lemel. Ed. Fayard/Mille et une nuits, 2010.

("The pleasure of the Gods")[21] (dedicated to Bacchus). We can also recall that there is a parody of the song by Pierre Perret "Vous saurez tout sur le zizi" ("You will know everything about the zizi"), entitled "You will know everything about the clito".

17.1.8 The Clitoris and Theatre

Who would have believed that, during the very avant-gardist twentieth century, somebody could write and especially act in a play with such an unusual and more or less shocking title (for the era!) as "Les monologues du vagin" ("Monologues of the vagina")? It is, however, what Eve Ensler dared with the creation of a play in Broadway, in 1996,[22] with the immense success which followed: translation into about 50 languages and staging of the show in 130 countries. In France, it was necessary to wait for the year 2000 for the play to be hosted in Paris and starring Fanny Cottençon. Since then, considering the required performance, a version with three actresses and the figures speak for themselves:

3,500 representations overall (Paris + province), 70 well-known French actresses, who have successively played the part and theatres, which are always full 13 years later.

The bet was won and it is undoubtedly what pushed the author and director to attempt the adventure with, this time, the clitoris at the epicentre. This is how the show, "Le bouton de rose" ("The rosebud"), also called, with a lot of humour "Le clito rit" ("The clito is laughing") was imagined, acted and sung by the actress, Sophie Accaoui, and staged by Laurent Levy. The first show took place in Paris, on February 14, 2011, right on Saint Valentine's day. Since then, the show has toured France so that the clitoris has been to the recent Festival Off of Avignon 2013. All of the audience appreciated the actress's performance as she told the story with enthusiasm and sung a cappella. As for the critics, they were unanimous and acknowledged that this show was very funny and made people laugh a lot, but that it was also educational.

So we can actually wonder, when will a show called "The labia majora and minora" be designed and produced?

17.1.9 The Clitoris and the Decorative Arts

All forms of art have been connected to the clitoris; decorative arts are no exception. They have even made a significant contribution if we dare include clitoral piercings in the category of ornaments for the female body.[23]

Fashion itself is also concerned as in a fashion show; a great designer simulated the vulva by means of an evening dress with folds and layers of red or black fabric provided at the model's pelvis level.

Fashion accessories follow the same trend and many young women, especially in the United States, wear pendants on small chains or cords around their necks, which are plaster and miniature reproductions of their own vulva. It is possible that one day, these pendants without much commercial value will give way to real jewels (medals or pendants), which will reproduce the same genital themes.

We would also add, that in certain countries (Japan in particular) the "onahole" (holes with onanism), which are facsimile female genitals made of silicone or other material, achieve, among other qualities an astonishing visual perfection as they are exact replications of reality, are actual works of art and are obviously provided with a visible clitoris. Apparently, they are the most sold item in sex shops!

---

[21] Hospital and Latin anthology. Collection of Songs from guard-rooms… (collected by Courtepaille). Ed. Bichat, 1911.

[22] Recall also that the other side of the Channel, the English comedian, Roy Chubby Brown, known for his sarcastic humour, created in 1995 a show aptly named "Clitoris allsorts", but the author, very controversial because of his sometimes outrageous mockery against women, did not have the expected success.

[23] We do not refer to tattoos of the said organ, whose extremely reduced visible surface and exacerbated sensitivity appear to be serious reasons for contraindicating this practice.

Conclusions

We believe that, by the end of this book, we have responded to the legitimate wishes of all women and men who have deplored the lack of an updated scientific study of the clitoris and its connections.

We hope that we have shared our sense of wonder with everyone reading this book, a sense that we felt all throughout our research, during dissections and microscopic examinations. The extraordinary sensory corpuscles, the incredible extent of the nerve supply, the importance of the hidden part of the bulbo-clitoral organ, the ingenious vascular network equipped with gates (or locks) and overflow systems that enable the clitoris to fill or empty as desired according to the needs, the small size of the glans clitoridis and the erectile function of the clitoral tissue that work together so as not to hinder penetration during intercourse while ensuring the stimulation of sensory corpuscles via the movement of the frenula and the extension reserves of the suspensory ligament and the nerves are only some of the fascinating aspects of the bulbo-clitoral organ.

We hope that the images presented in our book will convince women of the admirable nature of their intimate anatomy. Each and every woman possesses a priceless sensorial treasure: the bulbo-clitoral organ.

We also hope that men can use this book to discover or learn more about the feminine genital anatomy, to better appreciate their companions' capacity for pleasure and ultimately learn to really share this pleasure with them, even if female and male orgasms rarely occur at the same time.

The creation of this book has also made us think about this dichotomy of the female genitals which obviously includes two parts, each with a very specific purpose: a birth canal (ovaries, fallopian tubes, uterus, vagina) entirely devoted to reproduction and a bulbo-clitoral organ for the sole purpose of pleasure. This observation is of course a matter of serious thought about philosophical questions!

We also thought a lot about certain authors who have lumped everything together and who only think of the clitoris as a miniature penis. We object to this over-simplistic idea. The bulbo-clitoral organ is not a small model of the penis, and parents should not give credence to this misconception, especially when their girls ask them about it (e.g., after taking a bath with their little brother). The bulbo-clitoral organ is a specialised organ and only has a few similarities with the male penis, due to a common early embryology. The bulbo-clitoral organ has a single function and is therefore specialised: its only purpose is to provide sexual pleasure. The penis is multifunctional and therefore unspecialised (remember, it is used to void urine, release semen, as well as for penetration and sexual pleasure). This did not escape the great anatomist L. Kobelt who, when asked at the end of his book on the sensory genital apparatus of men and women about "the amount of pleasure or orgasm that each of the sexes experience during the act of copulation", replied without hesitation: "the large size of the female bulbs, compared to the volume of the glans of clitoris, their immediate action on this organ, the energetic compression they are subjected to from the penis and especially the large number of nerves concentrated in a small area (multus in minimo), all of that in addition to the high overall sensitivity of women, are all reasons for us to admit that the proportionate share of the female individual is greater".

Such an exceptional organ is unfortunately too often the object of abominable mutilations, female excisions, which we insisted on discussing in our anatomical study because this abuse results in anatomical harm with a definitive amputation of one of the most basic functions of a living individual: the right to sexual pleasure.

In return, we devoted a chapter to the clitoris in art in order to show that the clitoris, an integral part of the woman and her beauty, has every right to have its own place within the circles of artistic creation.

December 2013 The authors

V. Di Marino, H. Lepidi, *Anatomic Study of the Clitoris and the Bulbo-Clitoral Organ*,
DOI 10.1007/978-3-319-04894-9, © Springer International Publishing Switzerland 2014

Clitoris Bibliography

The reader may also consult: "Clitoris; A Bibliography, Medical Dictionary, and Annoted Research Guide to Internet References", ICON Health Publications, 2004.

Achtari C, McKenzie BJ, Hiscock R, Rosamilia A, Schierlitz L, Briggs CA, Dwyer PL (2006) Anatomical study of the obturator foramen and dorsal nerve of the clitoris and their relationship to minimally invasive slings. Int Urogynecol J Pelvic Floor Dysfunct 17:330–334

Alsaid B, Moszkowicz D, Peschaud F, Bessede T, Zaitouna M, Karam I, Droupy S, Benoit G (2011) Autonomic-somatic communications in the human pelvis: computer-assisted anatomic dissection in male and female fetuses. J Anat 219(5):565–573

Anson BJ, Mac Vay CB (1971) Surgical anatomy, vol 2. Saunders, Philadelphia

Aristote (1883) Histoire des animaux. 3 vol. Traduit en français et accompagné de notes perpétuelles par J. Barthelemy-Saint Hilaire. Hachette éd., Paris

Aristote (1887) Traité de la génération des animaux, vol 3. Traduit en français et accompagné de notes perpétuelles par J. Barthelemy-Saint Hilaire. Hachette éd., Paris

Aubin A (2001–2002) Approche anatomique du clitoris. Mémoire de Maîtrise en Sciences biologiques et médicales, Nantes

Baggish MS, Steele AC, Karram M (1999) The relationships of the vestibular bulb and corpora cavernosa to the female urethra: a micro-anatomic study. Part 2. J Gynecol Surg 15((4):171–178

Bartholin Th (1651) Anatomia, ex Caspari Bartholini parentis Institutionibus, omniumque recentorium et propiis observationibus tertiùm ad sanguinis circulationem reformata. Cum iconibus novis accuratissimi. Liber I, Cap. XXXIV, 186–189. apud Franciscum Hackium, Lugd. Batav

Baskin LS (2004) Anatomical studies of the female genitalia: surgical reconstructive implications. J Pediatr Endocrinol Metab 17:581–587

Baskin LS, Erol A, Wu LY, Kurzrock E, Cunha GR (1999) Anatomical studies of the human clitoris. J Urol 162:1015–1020

Bell J (1811) The anatomy of the human body, vol 3. 3ème éd. Longman and Rees Ed., London: Londres

Ben Pasky PH D (1986) Embryologie humaine. Traduit de l'américain par Horn Gary F. Ellipses

Benett AH (1994) Diagnosis and management of erectile dysfunction. WB Saunders Company, Philadelphia

Berengario da Carpi J (1523) Isagogae breves per lucidae ac uberrimae in anatomiam humani corporis a communi medicorum academia usitatam…Benedict Hector édit., Bologne

Berkow SG, Amboy P (1953) The corpus spongiosum of the urethra: its possible role in urinary control and stress incontinence in women. Am J Obstet Gynecol 65(2):346–351

Bertolini R (1966) Über die bindegewebige Befestigung der Clitoris des Menschen. Anat Anz Bd 118(2):117–121

Bonnel F (2004) Epopée historique des premières dissections françaises à Montpellier. Morphologie 88:281–292

Borderes R (1905) Des tumeurs malignes primitives du clitoris. Monographie. G. Firmin Montane et Sicardi édit, Montpellier

Bourgery JM, Bernard C et Jacob NH (1866–1867) Traité complet de l'anatomie de l'homme (en 10 tomes). Tome cinquième: Appareil de nutrition, splanchnologie. L. Guérin, Paris

Buisson O, Foldes P, Jannini E, Mimoun S (2010) Coïtus as revealed by ultrasound in one volunteer couple. J Sex Med 7:2750–2754

Burnett AL, Calvin DC, Silver RI, Peppas DS, Docimo SG (1997) Immunohistochemical description of nitric oxide synthase isoforms in human clitoris. J Urol 158(1):75–78

Calabrisi P (1996) The nerve supply of the erectile cavernous tissue of the genitalia in the human embryo and fetus. Anat Rec 125: 713–723

Cavalotti C, Malinovsky L, D'andrea V (1995) Ultrastructure of lamellar complexes in glomerular Krause spherical corpuscles. Int J Anat Embryol 100(Suppl 1):573–578

Chaperon S (2007) From anaphrodisia to frigidity: landmarks in history. Sexologies 16:189–194

Chaperon S (2012) "Le trône des plaisirs et des voluptés": anatomie politique du clitoris, de l'Antiquité à la fin du XIX e siècle. Cahiers d'histoire. Revue d'histoire critique 118:41–60

Cohen M (1978) The mysterious origin of the word "clitoris". In: Lowry TP (ed) The classic clitoris historic contributions to scientific sexuality. Nelson Hall, Chicago

Cold CJ, Taylor JR (1999) The prepuce. BJU Int 83(Suppl 1):34–44

Colombo R (1559) De re anatomica libri XV. Nicolo Bevilacque edit, Venise

Cotgrave R (1611) A dictionarie of the French and English tongues compiled by Randle Cotgrave. Adam Islip Anno, London

Cuneo et Marcille M (1901) Note sur les lymphatiques du clitoris. Bulletins et mémoires de la Société Anatomique de Paris. LXXVI ème année, 6ème série, Tome III. G. Stenheil édit., Paris

De Graaf Regnier (1649) Histoire anatomique des parties génitales de l'homme et de la femme qui servent à la génération… Traduit en français par Mr. N.P.D.M. Bâle, chez J.G. König, Paris

DeLancey JOL (1986) Correlative study of paraurethral anatomy. Obstet Gynecol 68(1):91–97

Dionis P (1718) Traité général des accouchements qui instruit de tout ce qu'il faut faire pour être habile accoucheur. C.-M. d'Houry, édit., Paris

Dogiel AS (1893) Die Nervendigungen in der Haut der äusseren Genitalorgane des menschen. Arch Mikro Anat 41:585

Dorschner W, Biesold M, Schmidt F, Stolzenburg JU (1999) The dispute about the external sphincter and the urogenital diaphragm. J Urol 162(6):1942–1945

V. Di Marino, H. Lepidi, *Anatomic Study of the Clitoris and the Bulbo-Clitoral Organ*,
DOI 10.1007/978-3-319-04894-9,

Estienne CH (1546) La dissection des parties du corps humain. Simon de Collines édit., Paris

Faix A, Lapray JF, Courtieu C, Maubon A, Lanfrey K (2000) Magnetic resonance imaging of sexual intercourse: initial experience. J Sex Marital Ther 27(5):475–482

Faix A, Lapray JF, Callede O, Maubon A, Lanfrey K (2002) Magnetic resonance imaging of sexual intercourse: second experience in missionary position and initial experience in posterior position. J Sex Marital Ther 28(Suppl 1):63–76

Fallope G (1561) Observationes anatomicae. Venise. Cologne and Paris 1562. Keruer I. Ed. Paris 1562

Fallope G (1600) Gabrielis Falloppi Mutinensis Physici ac chirurgi praeclarissimi...opera omnia. apud haredes Andrea Wecheli, Francofurti

Favre L (1875) Dictionnaire historique de l'ancien langage français ou glossaire de la langue franxoise depuis son origine jusqu'au siècle de Louis XIV. Publié par les soins de L. Favre

Fay EW (1907) Greek and Latin word studies. The classical quarterly , 1: 13–30; Cambridge University Press: Cambridge, England

Finger W (1866) Ueber die endigungen der Wolustnernen. Ztschr Rat Med 3(28):222

Foldes P, et Buisson O (2007) Clitoris et point G: liaison fatale. Gynécologie Obstétrique et Fertilité 35:3–5

Foldes P, et Buisson O (2008) Sonography of the clitoris. J Sex Med 5(2):413–417

Foldes P, et Buisson O (2009) The clitoral complex: a dynamic sonographic study. J Sex Med 6:1223–1231

Forster A (1903) Beiträge zur Anatomie der äusseren männli Geschlechtsorgane des Menschen. Zeit f Morph u Anthrop 6:435–502

Galenus Cl (1490) Opera (2 vol.). 1ère 2dition imprimée, en latin. Filippo Pinzi édit., Venise

Ginger VA, Cold CJ, Yang CC (2011) Surgical anatomy of the dorsal nerve of the clitoris. Neurourol Urodyn 30(3):412–416

Goldstein AMB, Padma-Nathan H (1990) The microarchitecture of the intracavernosal smooth muscle and the cavernosal fibrous skeleton. J Urol 144:1144–1146

Goldstein AMB, Morrow JW, Meehan JP, Buckley PA, Rogers FA (1984) Special microanatomical features surrounding the intracorpora cavernosa nerves and their probable function during erection. J Urol 132(1):44–46

Goldstein AMB, Morrow JW, Meehan JP, Buckley PA, Rogers FA (1985) The fibrous skeleton of the corpora cavernosa and its probable function in the mechanism of erection. Br J Urol 57: 574–578

Gragasin FS, Michelakis ED, Hogan A, Moudgil R, Hashimoto K, Wu X, Bonnet S, Al Haromy A, Archer SL (2004) The neurovascular mechanism of clitoral erection: nitric oxide and cGMP-stimulated activation of BKCa channels. FASEB J 18:1382–1391

Hippocrate (1839–1861) Oeuvres complètes (10 Vol.). E. Littré trad. (Traduction nouvelle avec le texte grec en regard) Vol. 7: De la génération; Vol.8: Des chairs. J-B Bailliere édit., Paris

Hruby S, Ebmer J, Dellon AL, Aszmann OC (2005) Anatomy of pudendal nerve at urogenital diaphragm – new critical site for nerve entrapment. Urology 66(5):949–952

Huard P, Imbault-Huart MJ (1980) André Vésale; iconographie anatomique (Fabrica, Epitome, Tabulae sex). Roger Dacosta, Paris

Ingelman-Sundberg A (1997) The anterior vaginal wall as an organ for the transmission of active forces to the urethra and the clitoris. Int Urogynecol J 8:50–51

Jaeger H (1944) Recherches histologiques sur les terminaisons nerveuses de la peau normale des organes génitaux externes humains. Dermatologica 90:49–74

Jayle Felix Léon (1918) La gynécologie. Tome I: L'anatomie morphologique de la femme. Masson, Paris

Kamina P (1984) Anatomie gynécologique et obstétricale, 4ème éd. Maloine édit., Paris

Kamina P (2009) Anatomie clinique. Tome 4. Organes urinaires et génitaux. Pelvis. Coupes du tronc, 2ème edition. Maloine, Paris

Kantner M (1954) Studien über den sensiblen Apparat in der Glans clitoridis. (1. Die clitoris der Greisin). Ztschr Mikr Anat Forsch 60:388–398

Katsuyoshi Y (1951) On the sensory nerve terminations in clitoris in human adult. Tohoku J Exp Med 54:163–174

Kobelt GL(1851) De l'appareil du sens génital des deux sexes dans l'espèce humaine et dans quelques mammifères au point de vue anatomique et physiologique. Traduit de l'allemand par Kaula H. Labé, Paris

Komisaruk BR, Whipple B (2005) Functional MRI of the brain during orgasm in women. Annu Rev Sex Res 16:62–86

Krantz KE (1958) Innervation of the human vulva and vagina. A microscopic study. Acta Obstet Gynecol Scand 12(4):382–396

Krause W (1866a) Uber die Nervenendigung in den Clitoris. Göttinger Nachristen, Avril

Krause W (1866b) Uber die Nervenendigung in den Geschlechtsorganen. Zeitschrift f rate Med 28:86–88

Larsen William J (2004) Embryologie humaine. Traduit de l'anglais par Dhem A, 3ème édition. De Boeck, Bruxelles

Lavoisier P, Aloui R, Schmidt MH, Watrelot A (1995) Clitoral blood flow increases following vaginal pressure stimulation. Arch Sex Behav 24(1):37–45

Lenck LC, Vanneuville G, Monnet JP, Harmand Y (1992) Le sphincter urétral (point G). Correlations anatomo-cliniques. Revue Française de Gynécologie et D'obstétrique 87(2):65–69

Mansour-Hugues M (2007) Transposition du clitoris: Résultats anatomiques et fonctionnels sur 42 patientes. Thèse Méd., V-René Descartes, Paris

Maravilla KR, Yang CC (2008) Magnetic resonance imaging and the female sexual response: overview of techniques, results, and future directions. J Sex Med 5(7):1559–1571

Maravilla KR, Heiman JR, Garland PA, Cao Y, Carter WO, Peterson BT, Weisskoff RM (2003) Dynamic MR imaging of the sexual arousal response in women. J Sex Marital Ther 29(Suppl 1):71–76

Marjolin JN (1815) Manuel d'anatomie (2 vol in 8). Mequignon-Marvis, Paris

Martin-Alguacil N, Pfaff DW, Shelley DN, Schober JM (2008a) Clitoral sexual arousal: an immunocytochemical and innervation study of the clitoris. BJU Int 101(11):1407–1413

Martin-Alguacil N, Schber JM, Sengelaub DR, Pfaff DW, Shelley DN (2008b) Clitoral sexual arousal: neuronal tracing study from the clitoris through the spinal tracts. J Urol 180(4):1241–1248

Mirilas P, Skandalakis JE (2004) Urogenital diaphragm: an erroneous concept casting its shadow over the sphincter urethrae and deep perineal space. J Am Coll Surg 198(2):279–290

Mondino de' Liuzzi (1532) Anathomia mundini. 1ère éd. 1478 (manuscrit datant de 1316) - Cy est Lanathomie de maistre Mundin boullonoys, e naguere traduicte de langue latine en vulgaire ou francoyse…. Lotrian A. et Janot D. édit., Paris

Moszkowicz D, Alsaid B, Bessede T, Zaitouna M, Penna C, Benoit G, Peschaud F (2011a) Neural supply to the clitoris: Immunohistochemical study with three-dimensional reconstruction of cavernous nerve, spongious nerve and dorsal clitoris nerve in human fetus. J Sex Med 8(4):1112–1122

Moszkowicz D, Alsaid B, Benoit G, Peschaud F (2011b) Etude anatomique et immuno-histochimique de l'innervation pelvi-périnéale de la femme avec reconstruction 3D. Peut-on encore diminuer les séquelles sexuelles et urinaires de la dissection rectale pour cancer? e-mémoires de l'Académie Nationale de Chirurgie 10(2):046–051

Narjani AE (1924) Considérations sur les causes anatomiques de la frigidité chez la femme. Brux Med 27:768–778

Oberfield SE, Mondok A, Shahrivar F, Klein JF, Levine LS (1989) Clitoral size in full-term infants. Am J Perinatol 6:453–454

O'Connell HE, DeLancey JOL (2005) Clitoral anatomy in nulliparous, healthy, premenopausal volunteers using unenhanced magnetic resonance imaging. J Urol 173:2060–2062063
O'Connell HE, Hutson JM, Anderson CR, Plenter RJ (1998) Anatomical relationship between urethra and clitoris. J Urol 159:1892–1897
O'Connell HE, Anderson CR, Plenter RJ, Hutson JM (2004) The clitoris: a unified structure. Histology of the clitoral glans, body, crura and bulbs. Urodinamica 14:127–132
O'Connell HE, Sanjeevan KV, Hutson JM (2005) Anatomy of the clitoris. J Urol 174(4, Part 1 of 2):1189–1195
O'Connell HE, Eizenberg N, Rahman M, Cleeve J (2008) The anatomy of the distal vagina: towards unity. J Sex Med 5:1883–1891
Ohmori D (1924)- Ueber die Entwicklung der Innervation der Genitalapparate als peripheren Aufnahmeapparat der genitalen Reflexe. Zeits. F. Anat. u. Entwicklungsgeschichte, Bd.LXX, 347
OMS (2011) Rapports du secrétariat la 61ème assemblée mondiale de la santé (20 Mars 2008) et de la 128ème session (7 Janvier 2011)
Oribase (1851–1876) Œuvres complètes avec texte grec et traduction française établie par les Dr. Ch. Daremberg et Bussemaker. Tome 3, Des noms des parties de l'homme. Imprimerie Nationale, Paris
Paré A (1971) Des monstres et des prodiges. Jean Céard édit, Genève
Park K (1997) The rediscovery of the clitoris. French medicine and the tribade, 1570–1620. In: Hillman D, Mazzio C (eds) The body in parts. Fantasies of corporality in early modern Europe. Routledge, New York, pp 171–193
Paturet G (1958) Traité d'Anatomie humaine. Masson édit, Paris
Pernkopf E (1943) Topographische Anatomie des Menschen. Urban und Schwarzenberg, Wien
Petkovic S (1953) Phlébographie pelvienne chez la femme par le clitoris (Pelvi-venographie). Gynaecologia 13(1):43–51
Phikipp M, De Boer C, Pilpel D, Karplus M, Sofer S (1996) Clitoral and penile sizes of full term newborns in two different ethnic groups. J Pediatr Endocrinol Metab 9(2):175–179
Poirier P, Charpy A (1901) Traité d'anatomie humaine, vol 5. Masson édit, Paris, p 580
Portal A (1770–1773) Histoire de l'anatomie et de la chirurgie, contenant l'origine et les progrès des sciences. Tome sixième. Première partie. Didot le jeune Ed, Paris. 654–655
Puppo V (2006) Re: clitoral anatomy in nulliparous...... (J. Urol. 2005, 173 (6), 2060–2063). J Urol 175(2):790–791
Rees MA, O'Connell HE, Plenter RJ, Huston JM (2000) The suspensory ligament of the clitoris: connective tissue supports of the erectile tissues of the female urogenital region. Clin Anat 13:397–403
Riolan (le jeune) J (1614) Discours sur les hermaphrodites. Pierre Ramier édit, Paris
Riolan J (1628–1629) Les œuvres anatomiques deM. Jean Riolan, revues et augmentées d'une cinquième partie de cette édition...le tout rangé, divisé, noté et mis en français par M. Pierre Constant... Tome I, De l'anthropographie, livre second, 435–436. Denys Moreau édit, Paris
Robert R, Labat JJ, Riant T, Louppe JM, Hamel O (2009) The pudendal nerve: clinical and therapeutic morphogenesis, anatomy, and physiopathology. Neurochirurgie 55(4–5):463–469
Rufus d'Ephèse (1879) Oeuvres collationnées sur les manuscrits traduits pour la 1ère fois en français , avec une introduction. Publication commencée par le Dr. Ch.Daremberg et terminée par Ch. Emile Ruelle. BNF, Imprimerie Nationale, Paris
Sane K, Pescovitz OH (1992) The clitoral index: a determination of clitoral size in normal girls and in girls with abnormal sexual development. J Pediatr 120(2 Pt 1):264–266
Schultz WW, van Andel P, Sabelis I, Mooyaart E (1999) Magnetic resonance imaging of male and female genitals during coïtus and female arousal. BMJ 319:1596–1600
Sedy J, Nanka O, Spackova J, Jarolim L (2008) Clinical implications of a close vicinity of nervus dorsalis penis/clitoridis and os pubis. J Sex Med 5(7):1572–1581
Sevelly JL (1987) Eve's secrets. A new perspective on human sexuality. Bloomsbury, London
Sfameni P (1904) Sulle terminazioni nervose nei genitali femminili esterni e sul loro significato morfologico e funzionale. Arch Fisiol I(4):346–384
Sfameni P (1905) Sur les terminaisons nerveuses dans les organes génitaux femelles externes et sur leur signification morphologique et fonctionnelle. Arch Ital Biol, 39:75–80
Shafik A (1993) Vaginocavernous reflex. Clinical significance and role in sexual act. Gynecol Obstet Invest 35(2):114–117
Sheehan E (1981) Victorian clitoridectomy: Isaac Baker Brown and his harmless operative procedure. Med Anthrpol News 12(4):9–15
Sinclair AH, Berta P, Palmer MS, Hawkins JR, Griffiths BL, Smith MJ, Foster JW, Frischauf AM, Lowell-Badge R, Goodfellow PN (1990) A gene from the human sex-determining region encodes a protein with homology to a conserved DNA-binding motif. Nature 346(6281):240–244
Solère M (1969) Atlas d'embryologie humaine. Maloine, Paris
Soranos d'Ephèse (1988) Maladies des femmes (gynakeion pathon A), Livre I (traduction Burguière P, Gourevitch D, Malinas Y.) Les Belles Lettres édit., Paris
Stefani R, Serra GP, Loffredo F, Spiga S (1988) Sulla struttura e sull'origine del glande clitorideo. Arch Ital Anat Embriol 93(4): 277–295
Stein TA, DeLancey JOL (2008) Structure of the perineal membrane in females: gross and microscopic anatomy. Obstet Gynecol 111(3): 686–693
Stringer MD, Becker I (2010) Colombo and the clitoris. Eur J Obstetr Gynecol Reprod Biol 151(2):130–133
Suh DD, Yang CC, Cao Y, Garland PA, Maravilla KR (2003) Magnetic resonance imaging anatomy of the female genitalia in premenopausal and postmenopausal women. J Urol 170:138–144
Tagatz GE, Kopher RA, Nagel TC, Okagaki T (1979) The clitoral index: a bioassay of androgenic stimulation. Obstet Gynecol 54: 562–564
Tello JF (1932) Contributions à la connaissance des terminaisons sensitives dans les organes génitaux externes et de leur développement. Travaux du Laboratoire de recherches Biologiques de l'Université de Madrid 28:1–59
Temesvary N (1924) Die Regio clitoridis. Arch Gynakol 122:102–128
Testut L Traité d'Anatomie Humaine (1911). Tome 4ème. 6ème édition. O.Doin et Fils, édit, Paris
Toesca A, Stolfi VM, Cocchia D (1996) Immunohistochemical study of the corpora cavernosa of the human clitoris. J Anat 188:513–520
Van Turnhout AA, Hage JJ, van Diest PJ (1995) The female corpus spongiosum revisited. Acta Obstet Gynecol Scand 74:767–771
Vaze A, Goldman H, Jones JS, Rackley R, Vasavada S, Gustavson KJ (2009) Determining the course of the dorsal nerve of the clitoris. Urology, 2008; 72(5):1040–1043. Re: Sedy R. Urology 73(3):684
Verkauf BS, Von Thron J, O'Brien WE (1992) Clitoral size in normal women. Obstet Gynecol 80(1):41–44
Vesale A (1555) De humani corporis fabrica libri septem. Basilea: per J. Oporinum
Veyrunes F, Chevret P, Catalan J, Castiglia R, Watson J, Dobigny G, Robinson TJ, Britton-Davidian J (2010) A novel sex determination system in a close relative of the house mouse. Proc R Soc B 277(1684):1049–1056
Vincent JD (2007) Voyage extraordinaire au centre du cerveau. Ed. Odile Jacob, Paris
Vincent JD (2010) Le sexe expliqué à ma fille. Ed. du Seuil, Paris
Wallen K, Lloyd EA (2011) Female sexual arousal: genital anatomy and orgasm in intercourse. Horm Behav 59(5):780–792

Webster JC (1891) The nerve-endings in the labia minora and clitoris, with special reference to the pathology of pruritus vulvae. Zeitschrift f Gyn 62:35–44

Wehenkel M (2007) Dépistage et prévention des mutilations sexuelles féminines et de leurs complications en médecine générale. Thèse Med., Faculté St Antoine (Univ Paris 6), n°2007PA06G057

Wespes E, Schulman CC (1986) Rôle hémodynamique de l'albuginée dans l'érection. Acta Urol Belg 54:114–122

Williams PL, WarWick R, Dyson M et al (1989) Gray's anatomy. Churchill Livingstone, Edinburgh

Winkelmann RK (1957) The mucocutaneous end-organ. AMA Arch Derm 76(2):225–235

Winkelmann RK (1959) The erogenous zones: Their nerve supply and significance. Mayo Clin Proc 34(2):39–47

Yamada K (1951) Studies on the Innervation of Clitoris in 10th Month Human Embryo. Tohoku J Exp Med 54(2):151–157

Yilmaz U, Kromm BG, Yang CC (2004) Evaluation of autonomic innervation of the clitoris and bulb. J Urol 172(5 Pt 1):1930–1934; discussion 1934

Yucel S, De Souza A Jr, Baskin LS (2004) Neuroanatomy of the human female lower urogenital tract. J Urol 172(1):191–195

GPSR Compliance

The European Union's (EU) General Product Safety Regulation (GPSR) is a set of rules that requires consumer products to be safe and our obligations to ensure this.

If you have any concerns about our products, you can contact us on ProductSafety@springernature.com

In case Publisher is established outside the EU, the EU authorized representative is:

Springer Nature Customer Service Center GmbH
Europaplatz 3
69115 Heidelberg, Germany

Batch number: 10731510

Printed by Printforce, the Netherlands